Edward Hull

The Geology of the Burnley Coal-Field

and of the Country around Clitheroe, Blackburn, Preston, Chorley, Haslingden, and Todmorden. Quarter sheets 88 N. W., 89 N. E., 89 N. W., and 92 S. W., of the 1-inch geological maps.

Edward Hull

The Geology of the Burnley Coal-Field
and of the Country around Clitheroe, Blackburn, Preston, Chorley, Haslingden, and Todmorden. Quarter sheets 88 N. W., 89 N. E., 89 N. W., and 92 S. W., of the 1-inch geological maps.

ISBN/EAN: 9783337212926

Printed in Europe, USA, Canada, Australia, Japan

Cover: Foto ©berggeist007 / pixelio.de

More available books at **www.hansebooks.com**

MEMOIRS OF THE GEOLOGICAL SURVEY.

ENGLAND AND WALES.

THE

GEOLOGY

OF THE

BURNLEY COALFIELD

AND OF THE

COUNTRY AROUND CLITHEROE, BLACKBURN, PRESTON, CHORLEY, HASLINGDEN, AND TODMORDEN.

(QUARTER SHEETS 88 N.W., 89 N.E., 89 N.W., AND 92 S.W. OF THE 1-INCH GEOLOGICAL MAPS.)

BY

EDWARD HULL, M.A., F.R.S., J. R. DAKYNS, M.A.,
R. H. TIDDEMAN, M.A.,
J. C. WARD, W. GUNN, AND C. E. DE RANCE.

TABLE OF FOSSILS, BY R. ETHERIDGE, F.R.S.L. & E.

PUBLISHED BY ORDER OF THE LORDS COMMISSIONERS OF HER MAJESTY'S TREASURY.

LONDON:
PRINTED FOR HER MAJESTY'S STATIONERY OFFICE,
AND SOLD BY
LONGMAN & Co., PATERNOSTER ROW,
AND
EDWARD STANFORD, 6, CHARING CROSS, S.W.

1875.
[*Price Twelve Shillings.*]

THE following description of the Burnley Coalfield and the adjacent formations is the 14th memoir published by the Geological Survey on the Coalfields of Great Britain. Of these one is descriptive of the Geological Maps of the Cheshire Coalfield by Mr. Hull and Mr. Green, and four are descriptive of the Lancashire Coalfield by Mr. Hull. This memoir on the Burnley District, therefore, completes the account of the Lancashire Coalfield and the adjoining formations, the coal-measure maps of which are published on a scale of six inches and of one inch to a mile with illustrative sections.

I do not doubt that the book will be welcome to all those interested both in the mining industries, and in the more purely scientific geology of the country described.

ANDREW C. RAMSAY,
Director-General.

H.M. Geological Survey Office,
Jermyn Street, London.

The country of which the Geology is described in the present Memoir was surveyed by Mr. Edward Hull and Messrs. Green, Tiddeman, Dakyns, Ward, Gunn, De Rance, and Strangways, under the direction of Professor Ramsay.

This area, comprised in quarter-sheets 89 N.W. and N.E., 88 N.W., and 92 S.W., includes the highly developed carboniferous series of North Lancashire, the small but rich coalfields of Burnley and Blackburn, and the Coppull district of that part of the Lancashire coal-measures which is known as the Chorley Coalfield.

The Permian and Triassic rocks, on the west side of the great fault which cuts off the carboniferous rocks, are also described; as well as the drift and post-glacial deposits of the plain about the estuary of the Ribble and which form the seaboard of that part of Lancashire.

The greatest portion of the Memoir was written by Mr. Hull, by whom also most of the ground was surveyed; a considerable part being, however, from the pen of Mr. De Rance. Mr. Tiddeman deserves especial mention as not only having contributed to the Memoir, but as having acted as general editor of the whole; in the performance of which duty he has spared neither pains nor time.

A very valuable stratigraphical list of the fossils occurring in the district has been drawn up by Mr. Etheridge, Palæontologist to the Survey, as an appendix; while a list of the works and papers bearing upon the geology of the district of which the Memoir treats has been added by Messrs. Whitaker and Tiddeman as a second appendix, fully illustrating the bibliography of the subject.

The index has been drawn up by Mr. H. B. Woodward, the references to localities having been supplied by Mr. Tiddeman.

       HENRY W. BRISTOW,
         Director for England and Wales.

H.M. Geological Survey Office,
 28, Jermyn Street, London.
  1st December 1874.

ANALYTICAL TABLE OF CONTENTS.

CHAPTER I.
PHYSICAL GEOGRAPHY.

	PAGE
AREA I.—THE TRIASSIC PLAIN	5
AREA II.—THE COALFIELDS OF BURNLEY, BLACKBURN, AND CHORLEY, AND THE ROSSENDALE ANTICLINAL	5
Principal Watersheds	6
Watersheds crossing Valleys	
Sabden Valley	
Dean Valley, Great Harwood	
Roddlesworth and Belmont Valley	6–8
Whitworth Valley	
Calder Head	
Todmorden	
Principal Elevations, Table of	8
AREA III.—COUNTRY EAST OF THE ANTICLINAL FAULT:—	
Boundaries	
Physical Features	8–10
Drainage	
AREA IV.—PENDLE RANGE AND RIBBLE VALLEY.	
ANTICLINALS:—	
1. The Clitheroe and Skipton Anticlinal	
2. The Slaidburn Anticlinal	10–11
3. The Sykes Anticlinal	
Drainage	11–12
The Great Watershed	
Physical Features of the Ribble Valley	
,, ,, Pendle Range	12–13
,, ,, Country East of Gisburn	

CHAPTER II.
SEDIMENTARY ROCKS BELOW THE COAL MEASURES ALONG THE RIBBLE VALLEY.

	PAGE
THE CARBONIFEROUS LIMESTONE:—	
Thickness	
Range	
Withgill	13–16
Clitheroe, Salt Hill, Coplow	
Chatburn (Analysis of Lime)	
Gisburn, and Bolton-in-Bowland	
THE SHALES-WITH-LIMESTONE	16–17
THE PENDLESIDE LIMESTONE	17–19
THE LOWER YOREDALE GRIT	20–21

CONTENTS.

	PAGE
THE BOWLAND SHALES:—	
Characteristics	
A warning to coal seekers	21-22
Thickness	
Weets	
THE UPPER YOREDALE GRIT:—	
Nomenclature	
Characteristics	
Range:—Mellor, Ribchester Station, Whalley, Sabden	22-28
"Burst" on Pendle	
Form of Pendle on the Map	
Range East of Pendle	
SHALES ABOVE THE UPPER YOREDALE GRIT	28
LOWER CARBONIFEROUS ROCKS EAST OF GISBURN	28-32
THE MILLSTONE GRIT SERIES:—	
"FOURTH," OR KINDER-SCOUT GRIT	33-6
SABDEN VALLEY SHALES	36-7
"THIRD GRIT" SERIES	37-42
"SECOND GRIT," OR HASLINGDEN FLAGS	42-3
"FIRST GRIT," OR ROUGH ROCK	43-6
THE COUNTRY ON THE EASTERN SIDE OF 92 S.W:—	
The Carlton Synclinal	
The Lothersdale Anticlinal	46-52
The Reedshaw Moss Synclinal and Watersheddles Anticlinal	
Borings near Laneshaw Bridge	

CHAPTER III.

THE BURNLEY COALFIELD.

Introductory	53
Authorities	
LOWER COAL MEASURES:—	
Cliviger Section	
Flagstones	54-6
"Old Lawrence Rock," Catlow Quarries	
"Woodhead Hill Rock"	
COAL SEAMS:—	
Gannister Coal, or Mountain Mine	56
The Forty Yards, or Upper Mountain, Mine	
Darwen Coalfield	56-7
Turton and Quarlton	58
Littleborough and Wardle	58-60
Bacup Coal District	60-2
Accrington and Blackburn Districts	63-4
Harwood, Huntroyde, Marsden, and Worsthorn Districts	65-70
Estimate of Future Supply from the Lower Coal Measures	70
MIDDLE COAL MEASURES OF THE BURNLEY BASIN:—	
Sections at the Fulledge, Cliviger, and Gawthorpe Collieries	71-3
The Arley Mine	73-7
The Dandy Bed, or Cally Coal of Cliviger	77-8
The China Bed, or Cliviger 2-feet Coal	78
The Slaty Coal	78
The Great, or Bing, Mine	78-80
Fulledge Thin Bed	80
Cannel Seam	80
Low Bottom or Blindstone Coal	80-1
The Lower Yard Coal	81-2
The Old Yard	82
The Burnley Four-foot	82
The Shell Coal or Vicarage Mine	82
The Kershaw Coal	82

CONTENTS.

	PAGE
RESOURCES OF THE BURNLEY BASIN	82-3
FAULTS OF THE BURNLEY BASIN :—	
Little Harwood Fault	
Harper Clough Fault	
Oakenshaw Fault	
Altham Clough Fault	83-4
Hapton Hall Fault	
Padiham Green Fault	
Habergham Fault	
Cliviger Fault	84
Fulledge Fault	84-5
Thieveley Fault	85
Deerplay Hill Fault	86
Faults at Rowley Colliery	86
Hollins Fault	86
Worsthorn Fault	87
Marsden and Colne Districts	
Padiham and Gawthorpe	87
The Ightenhill Fault	
The Great Anticlinal Fault	88-9
Blackstone Edge	

CHAPTER IV.
THE CHORLEY COAL FIELD.
(Coppul District.)

Extent	90
LOWER COAL MEASURES, OR GANNISTER BEDS	91
Section at Charnock Richard	91-2
MIDDLE COAL MEASURES :—	
The Arley Mine	92-4
The Smith Coal	
The Bone Coal	94-5
The Yard Coal	
Cannel and King Coals	
(Sections in Lower Coal Measures, Blackburn and Chorley Railway)	95
Blainscough Hall Section	95-6
Faults and Borings	96-8

CHAPTER V.
COUNTRY EAST OF THE ANTICLINAL FAULT.

LITHOLOGICAL DESCRIPTION :—	
Rocks and Thicknesses	98
Blackstone Edge	99
STRATIGRAPHICAL DESCRIPTION	100
Anticlinal Fault	100-1
DETAILED DESCRIPTION :—	
Yoredale Shales	101-2
,, ,, and Grit	102-6
,, ,, ,, in Widdop	106
,, ,, ,, in Hey Slacks Clough	106-7
Kinder-Scout Grit	107-12
Third Grit Series	113-19
AGRICULTURAL FEATURES	119-20

CONTENTS.

CHAPTER VI.
THE PERMIAN SYSTEM.

	Page
Roach Bridge Section	120-1
Waddow Hall Section	121-2
Yarrow Section	122

CHAPTER VII.
THE TRIASSIC ROCKS.

Bashall Brook Section	122
Waddow and Thornyholme Fault	122
Pebble Beds around Preston	123-6
Upper Mottled Sandstone	126-7
Keuper Marls	127-8

CHAPTER VIII.
GLACIAL AND POST-GLACIAL DRIFT DEPOSITS — 128-33

Glaciated Rock Surfaces	133
Whalley, Chatburn	133-4
Horwich Moor	134
Ribble Valley	135-7
COUNTRY NORTH AND EAST OF COLNE	137-9
RIVER TERRACES AND ALLUVIUM	139
GLACIAL DRIFTS OF THE DISTRICT AROUND PRESTON AND CHORLEY	140-54
Lower Boulder Clay	140-1
Middle Drift	141-51
Upper Boulder Clay	151-4

CHAPTER IX.
POST-GLACIAL DEPOSITS OF THE WESTERN PLAIN.

High Level Alluvium	155-6
Shirdley Hill Sand	156-8
Lower Cyclas and Scrobicularia Clays	158-9
Ancient Low-level Fluviatile Alluvium	159
Peat	159-61
Martin Mere	161-2
Recent Fluviatile Alluvium	162-3
Recent Tidal and Estuarine Alluvium	163-8
Sequence of Post-Glacial Deposits	168

CHAPTER X.
RELATIVE AGES OF THE PENDLE AND PENINE CHAINS — 168-9
FAULTS AND LIE OF THE ROCKS IN THE WESTERN PLAIN — 169-71

CONTENTS.

CHAPTER XI.
IGNEOUS ROCKS.

	PAGE
Trap Dykes - -	171-2

CHAPTER XII.
MINERALS, METAL MINES, &c.

Minerals - - -	172-3
Lead Mines of the Ribble Valley	173-5
Fire-clays, &c. - -	175

APPENDIX I.

LIST OF FOSSILS - -	176

APPENDIX II.

A LIST OF WORKS AND PAPERS RELATING TO THE GEOLOGY OF LANCASHIRE	188

NOTICE.

THE country described in this Memoir cannot be considered as a separate tract distinguished from its surroundings, either by physical features or geological structure, nor as containing throughout it any common characteristics which will serve to connect it together as a whole. At the same time it is naturally divisible into four well-marked areas.

1. The western consists of a low tract * about the estuary of the Ribble, a portion of the plain which forms the Lancashire sea-board. It is composed of soft Triassic and Permian rocks, for the most part concealed by Drift, and is bounded on the E. by a large N.N.E. fault, which brings up the Carboniferous rocks.

2. Running obliquely from this in a N.E. direction, the Pendle range of hills stands out in marked contrast to the lower tract of country on the S.E. formed by the softer Coal Measures of Chorley, Blackburn, and Burnley. These consist of an irregular basin, the highest beds of which are in the neighbourhood of Burnley. From beneath them rise towards the S. of the district the successive members of the Gannister and Millstone Grit series, the latter forming several elevated tracts separated by valleys. This area is bounded on the E. by a line of fracture called "the Anticlinal fault" from its nearly coinciding with the axis of the Penine chain.

3. On the other side of that fault the rocks, consisting in the main of the Millstone Grit series, dip gently to the E. beneath the Yorkshire Coalfield, and constitute our third division.

4. The Pendle range of hills overlooks to the N. the valley of the Ribble, which is excavated in the Lower Carboniferous rocks along the axis of the Clitheroe anticlinal, and with it forms our fourth division.

J. R. D.
R. H. T.

---

* Contained in Quarter-sheet 89 N.W.

GEOLOGY

OF THE

BURNLEY COALFIELD, AND OF THE COUNTRY AROUND CLITHEROE, BLACKBURN, PRESTON, CHORLEY, HASLINGDEN, AND TODMORDEN.

CHAPTER I.
PHYSICAL GEOGRAPHY.

AREA I.
THE TRIASSIC PLAIN.

The district comprised under this head consists of a plain, gradually sloping upwards from the flats bordering the estuary of the Ribble to the hills of the Pendle range.

The continuity of the plain is broken by the valley of the Ribble and by the narrow valleys of the various small brooks running into that river, as well as by the shallower valleys of its larger tributaries, the Darwen and the Douglas or Asland.*

A small tract of country to the north-west is drained by brooks flowing into the river Wyre.

West of the Douglas the country is very low, often, indeed, beneath the sea-level, and is a continuation of the great peat-moss plain which has been described in the Survey Explanations of Maps 90 S.E. and N.E., as extending between Liverpool and Southport. The undulating plain of Glacial Drift above has an average elevation of 120 feet above the sea, towards which it slopes at a very low angle, terminating abruptly at Blackpool in a line of cliffs about 70 feet in height.— C. E. R.

AREA II.
THE COALFIELDS OF BURNLEY, BLACKBURN, AND CHORLEY, AND THE ROSSENDALE ANTICLINAL.

Physical Features.

This district extends from the Triassic Plain on the west to Sowerby Bridge on the east, and from Bury and Rochdale on the south to Colne † on the north. It is for the most part made up of hill and dale, one of these latter attaining the proportions of a broad valley, stretching in a north-easterly direction along the base of the Pendle range of hills, from Fenniscowles to the vale of the Calder at Padiham and Burnley. In this vale the important town of Blackburn is situated; and it may here be premised that the outward trough-shaped aspect of this tract is only

---

* In the Report of the Royal Rivers Pollution Commission, the area drained by the Ribble is given at 412,480 acres, with a population of 387,839, and the Douglas at 109,760, with a population of 101,337. The total area being 816 square miles. —C. E. R.

† The Roman *Colunio.*

an index to its geological structure, as it is almost continuous with the Burnley and Blackburn coal-basin.

To the north-west of the Burnley basin rises the long range of the Pendle Hills. The basin is bounded to the eastward by the ranges which culminate in Boulsworth Hill and Black Hambledon, and to the south by a series of broken hills and slopes, deeply indented by valleys, but in the main forming a line of watershed which, stretching from Anglezark Moor on the west to Dirpley Hill on the east, throws off to the northward and southward the affluents of the Ribble and the Mersey respectively.

This range consists of formations of older date than those of the Burnley basin, rising into flat-topped elevations ; and, considered as a whole, forms the axis of a low arch or anticlinal ; for the beds roll over to the south and sink down below the South Lancashire coalfield.

Thus, this district may be regarded as formed of one great stratigraphical undulation, consisting of a trough on the north, with a corresponding arch on the south ; the axes of both lying in a direction nearly parallel to each other, viz., north of east and south of west. The trough has been long known under the name of "the Burnley coal-basin," and the parallel arch I propose to call "the Rossendale anticlinal."

WATERSHEDS.—The district contains two important lines of watershed. One of these parts asunder the waters flowing down into the German Ocean on the east from those which find their way through the channels of the Ribble and Mersey into the Irish Sea on the west. Ranging southward from Boulsworth Hill by Black Hambledon, it crosses the Portsmouth valley at Calder Head (758 feet); thence taking the line of the ridge east of Bacup as far as Shore Moor, turns to the eastward and crosses the vale of Todmorden at Dean Head, about two miles north of Littleborough. The line then ascends to the summit of the moorlands which form the boundary between the counties of York and Lancaster, and then takes a southerly direction by Blackstone Edge, almost parallel with the county boundary. The other line of watershed has a general E.N.E. direction, corresponding with the hills of the Rossendale anticlinal. It separates the waters which flow into the Ribble from those which flow into the Mersey. Commencing at Anglezark Moor, it crosses the valley north of Belmont, a short distance south of Hollingshead Hall (950 feet), then ascending the highest parts of Darwen Moor it descends on the opposite side, crossing the valley at Cranberry Moss. It thence takes an easterly course to Orrell Moss, then northward to Pike Low and Haslingden Moor, and crossing the valley about a mile north of Haslingden ascends the ridge of Hambledon Hill and Hapton Park, crosses the valley on the east by Horelaw Nook* (984 feet), and passing along the crest of the ridge south of the valley of the Calder, forms a junction with the north and south watershed already described.

WATERSHEDS CROSSING VALLEYS.—Amongst the physical phenomena of the district perhaps none are more worthy of attention than the examples afforded of the parting of the streams in opposite directions, not only along the crests of ridges or elevations, but in the valleys themselves. I shall here notice some of the cases referred to.

Sabden Valley.—The watershed here crosses the valley from north to south by a farm called Moss Nook, near New-Church-in-Pendle. At this spot the valley is about a quarter of a mile wide, hollowed out in shales between two beds of Millstone Grit, and has a smooth, almost

---

* Corrupted into "Whoolaw Nook."

level, surface. On either side of the saddle two small streams take their rise, one, Sabden brook, flowing towards the south-west, the other, Dimpenley Clough, flowing towards the north-east and falling into Pendle Water. Both streams ultimately unite in the river Calder. The elevation of this watershed is 780 feet above the sea.

Dean Valley, Great Harwood.—A similar case occurs at the source of Dean Brook, Great Harwood. The valley is here about a quarter of a mile in breadth, and formed in the same beds of shale as that of Sabden. The saddle is here very flat and smooth, and has an elevation of 625 feet.

Roddlesworth and Belmont Valley.—The watershed here lies at the summit of a broad valley, and is only slightly removed from the horizontal for some distance on either side. It divides the Roddlesworth brook, which flows northward into the Ribble, from a stream which flows in an opposite direction into the Irwell and Mersey; it has already been alluded to as being part of the main watershed of the country. The elevation is 950 feet.

Whitworth Valley.—Near Under Shore farm, at the head of Whitworth Valley, the watershed, dividing the sources of the river Spidden from one of the affluents of the river Irwell, crosses a well-formed valley, with steep banks descending into a nearly level channel. The elevation is about 980 feet.

Calder Head.—This is a remarkable case, from the fact that at this spot two streams of the same name take their rise and flow towards opposite coasts of England. It seems not improbable that this confusion of two streams under the same name has arisen from their sources being in the same valley ; and the name being in each case transmitted downwards along the course of the stream, it came to pass that the people living on the banks were, without knowing it, applying the same name to different streams.

At the summit of the valley is situated one of the pits of Cliviger colliery, the shaft of which passes through 60 yards of gravel, probably drift, as boulder clay is shown in the cutting of the railway close by. The valley here lies in the line of a large fault, and on the southern side the cliffs of Millstone Grit rise in the form of a steep escarpment. The descent on both sides of the watershed is pretty rapid, and the elevation is 758 feet above the sea.

Todmorden.*—The last case which I shall describe is that of the watershed at Dean Head. Here it crosses one of the deepest and most perfectly formed valleys in Lancashire, bounded by steep banks, which rise on either side to an elevation of more than 300 feet above the valley. The change of inclination in the ground on crossing the saddle is almost imperceptible to the eye, and the bottom of the valley presents

---

* I venture here to offer a suggestion regarding the origin of this remarkable name. Taking the last syllable first,—" den :" this is the Saxon, and means (according to Mr. T. T. Wilkinson), a deep dale, very properly applied in this case. Now for " Tod " and " mor." I do not believe they have anything whatever to do with " death," but are capable of two derivations ; either " Tor," the British for a cliff, or pinnacle of rock, of which there are striking examples in the case of the Bride Stones, the Hawk Stones, and Chisley Stones, all overhanging the valley, or " Tod, " the Saxon name for a fox. " Mor " I take undoubtedly to be British, signifying great or large. Hence we have either the " dale or valley of the great rock or Tor," or " the dale of the great fox." The latter I take to be the more probable, though it requires the conjunction of a Celtic adjective with a Saxon noun. The termination of the Saxon " den " to the " Tod-mor " need not surprise us.—E. H.

a smooth surface, which, though not at present containing any stream, has clearly been levelled by the action of water. The elevation of the summit of the valley is 600 feet above the sea level.

ELEVATIONS.—The following is a list of the principal elevations in this area, or its margin, in the order of their respective heights,* commencing with Pendle Hill, which properly stands at the head of the list :—

Name.	Formation.	Altitude.
Pendle Hill	Yoredale Grit	1,831
Lad Law, Boulsworth Hill	Millstone Grit	1,700
Black Hambledon	,,	1,572·8
Blackstone Edge	,,	1,551·5
Whittle Hill	Lower Coal Measures	1,534
Winter Hill	Millstone Grit	1,495
Bartle Cwm Colliery	Lower Coal Measures	1,458
Coupe Lowe	Millstone Grit	1,436·8
Deerplay, or Dirplay Hill	Lower Coal Measures	1,427·4
Tooter Hill	,, ,,	1,419
Knowl Hill	,, ,,	1,378
Ramsden Hill	,, ,,	1,346·2
Hambledon Hill	Middle Coal Measures	1,341·2
Darwen Moor	Lower Coal Measures	1,318
Lowe Hill (Portsmouth)	Millstone Grit	1,219·7
Hartley's Pasture (Cliviger)	Middle Coal Measures	1,180
Deerplay Moor (watershed)	Lower Coal Measures	1,166
Holcombe Hill (monument)	Millstone Grit	1,162
Caster Cliff, Colne	Lower Coal Measures	906·5
Black Hill (Padiham Heights)	Millstone Grit	865·2
Grant's Tower	,,	825
Billinge (Blackburn)		806
Moss Nook (Newchurch-in-Pendle) (watershed)	,,	780
Revidge (Blackburn)	,,	751
Roman Camp, Mellor	Yoredale Grit	731·7
Whalley Nab	Millstone Grit	605·7
Dean Head, Todmorden Valley (watershed)	,,	600
Hoghton Tower	,,	575
Port-field (Roman camp), Whalley	,,	400

E. H.

AREA III.

COUNTRY EAST OF THE ANTICLINAL FAULT.

Physical Features and Drainage.

This division comprises the country bounded on the west by the Anticlinal Fault, on the north by the great Fault running between Boulsworth and Crow Hill, and thence, by the watershed of the rivers Aire and Calder, by Withins Height, Stairs Hill, Sun Hill, Yeoman Hill, and the Oxenhope Moors, as far as Nab, the northern end of Cold Edge; and on the east by Cold Edge and the hills bordering the Luddenden valley on the east as far as Sowerby Bridge, and thence by the hills on the right bank of the Ribcurne, and on the south by Mosleden Height and Way Stone Edge.

* These are taken from the 6-inch ordnance maps.

PHYSICAL FEATURES.

The country thus circumscribed forms a well-marked area, parted both physically and geologically from the Burnley basin on the west by the Anticlinal Fault; and, though in a less marked manner, still sufficiently well from the surrounding country by the line of hills indicated above.

The Anticlinal Fault runs in a direction N. by W. along a well-marked mountain wall, which forms generally the watershed between the rivers that flow east into the German Ocean and those that flow west into the Irish Sea. This watershed generally coincides pretty well with the anticlinal axis of the Penine chain, as it is called, save where the latter is broken through by valleys, as is conspicuously the case at Todmorden, where the Yorkshire Calder cuts across the axis, and thus interrupts for the space of about two miles the mountain wall which is generally formed by the Anticlinal Fault bringing the hard and massive Kinder Grit against the various beds of grit and shale which form the higher part of the Millstone Grit series.

This wall of Kinder reaches from Boulsworth, in the extreme north, by Greystone Hill, Black Moor, and Gorplestones, along Black Hambledon, Langfield and Walsden Moors, and Blackstone Edge, to Axletree Edge, in the extreme south. It attains the following heights in feet above the sea level:—1,700 at Lad Law, Boulsworth; 1,516 at Crown Point, Grey Stone Hill; 1,535 at Gorple Stones; 1,574 at Hoof Stones Height, on Black Hambledon; 1,434 at Bride Stones; 1,139 on Langfield Edge; 1,219 at Gadden Reservoir; 1,360 on Byron Edge; 1,553 on Blackstone Edge; and 1,356 at Stokes in the Moss. It thus attains an average extreme height of 1,438 feet.

The country falls away eastward with an inclination equal to or slightly less than the dip of the beds, which come on one above the other in a succession of terraces sloping eastward, whose steep sides, formed by the escarpments of grits and sandstones, face westward, as the general strike of the beds is N. 22 W., which is also very nearly the direction of the Anticlinal Fault. Thus as we go eastward we come on to higher and higher beds, which successively attain the heights of 1,172, 1,405, 1,311, 1,316, 1,369, 1,518, 1,253, 1,050, 1,258, 1,425, and finally, in the Rough Rock of Cold Edge, of 1,450 feet above the sea.

With the exception of a small portion about Boulsworth, in the extreme north-west, which belongs to the Ribble basin, this district is drained by the Yorkshire Calder and its affluents. These are, on the south, the river Ribourne and the St. John's Vale stream; on the north, the river Hebden and its affluent from Horsebridge Clough, the Colden Clough beck, and the stream which flows down the Luddenden valley. The river Calder divides the country conveniently into two portions, which we will describe separately. This river is formed by the junction at Todmorden of two branches, one coming from the N.W., which bears the name of Calder, and one from the south along the vale of Todmorden, which leads by Gauxholme towards Rochdale. From Todmorden the united streams flow eastward to Sowerby Bridge. The river Hebden rises on the moors of Black Hambledon, Black Moor, Widdop Moor, and Jackson Ridge; it is formed by the union at Blackden Bridge of the Gorple water, the Widdop water, and streams flowing down Greave Clough and Walshaw Dean. It flows south with an easterly set, and is joined at New Bridge by a stream flowing due south down Horsebridge Clough, and the two united join the Calder at Hebden Bridge. The Colden Clough stream drains the moors between the basins of the Hebden and Calder. Another affluent from the north flows down the Luddenden valley to join the Calder at Luddenden Foot. On the south a large beck flows down the vale of St. John in a N. by E. direction to Mytholmroyd; and at Sowerby Bridge the Calder is joined by the Ribourne from the south, which is formed by the union of two streams flowing east from the moors of Blackstone Edge.

In the north-west the stream draining Hey Slacks Clough runs into the Lancashire Calder, and the northern slope of Boulsworth likewise drains into that river.—J. R. D.

AREA IV.

PENDLE RANGE AND RIBBLE VALLEY.

Our fourth division forms a part of one great compound anticlinal, which separates the coalfield of Burnley, on the S.E., from that of Ingleton and Black Burton, on the N.W. by a great upheaval, or rather series of parallel upheavals of lower and earlier rocks. Measured across the strike from coalfield to coalfield, it occupies a breadth of at least 22 miles.

Only a portion of this comes into the present district, but as it affects the whole country it will be necessary to describe it briefly.* It consists of three principal anticlinals, with others of less importance intercalated which are not so continuous. The general direction of their axes is E. 35 N., but they seem to run more easterly as they approach the Penine Chain. The three greater decrease in size and importance from S. to N.

1. *The Clitheroe and Skipton Anticlinal* is seen above Roach Bridge, in the river Darwen, where its further westerly extension is concealed by the Permian and Triassic rocks. In the other direction it runs by Mellor to Bonny Inn, where it enters the valley of the Ribble and greatly enlarges in breadth. Thence its axis runs through the low ground where, as a popular couplet tells us—

"The Ribble, the Hodder, the Calder, the *Rain*,
All flow into Mitton Demesne."

Then pretty much up the Ribble under the red Permian rocks to Waddow. As the Limestone shows, its course is tolerably clear by Chatburn to Downham, beyond which we have great complications, faults, and minor folds, which greatly obscure it, but appear to shift the chief axis further north. It lies S. of the road from Sawley to Gisburn, and on by Skipton towards Bolton Abbey. Its length is more than 35 miles.

Two conspicuous smaller anticlinals, with their associated troughs, occur South of that of Clitheroe at the two extremities of the Pendle Chain. One is E. of Bamber Bridge, and probably brings on in the synclinal N. of it at Maudsley Fold, a small area of coal-measures. The other is the anticlinal of Lothersdale, which has on its north side a trough of Upper Yoredale Grit, forming Elslack Moor. Neither of them has a long range.

A third, enclosed between two faults, and traversed by another, occurs about Barnoldswick and Thornton. Its main direction is N.E.

2. *The Slaidburn Anticlinal* lies along the valley of the Loud, north of Longridge Fell and its W.S.Wly. continuation, embracing Chipping and Whitewell. At the latter place its continuity is interrupted by faults throwing down the grit of Birkett, and Hodder Bank, Fells. It resumes its course at Knowlmere Manor, runs by Newton and Slaid-

---

* This great anticlinal was briefly described by Mr. Hull in a paper read to the British Association in 1867, partly from information given by me, but my knowledge of it was then necessarily imperfect.—R. H. T.

burn, and passing N. of Champion and S. of Tosside Chapel, is most likely continuous with that which forms under Malham Cove, the cradle of the Aire.

The trough between these two great anticlinals embraces the basin of Millstone Grit which lies between Longridge Fell on the N., and Mellor on the S., the grit fells of Waddington, Easington, and Champion, and, across the Ribble, "The Weets" E. of Malham.

3. *The Sykes Anticlinal* occurs at the head of the valley of the River Brock, about Admarsh. It runs E.N.Ely. through the Fells of Bowland, not making a continuous valley, as do the two greater anticlinals, but bringing up the lower rocks in the valleys which cross it, as at Sykes and Brennand, and in Croasdale above the House of Croasdale, and continues across the vale of the Hodder from Lamb Hill to Fair Hill. Thence easterly it is somewhat indistinct beneath the grits, but probably is the same as that which thrusts up the Limestone on the S. side of Stockdale near the Ryeloaf, in the neighbourhood of Settle.

Between this and the Ingleton Coalfield there are many minor undulations, but none so continuous or so marked in their effect as those just briefly described.

It will be seen that the anticlinals generally form valleys, whilst the synclinals stand out in relief between them.

The part of this great compound anticlinal with which we are now more immediately concerned is contained in the 1-inch quarter-sheet 92 S.W., of which it forms about four fifths, in the N.W. corner of 89 N.E., and in the northern half of the east border of 89 N.W.

Drainage.—It is drained for the most part by the Ribble, with its southern tributaries the Darwen and Calder, and the Hodder which runs into it from the N.W. The chief of the lesser affluents of the Ribble in 92 S.W. from the N. are Skirden Beck and Holden Brook, which join it at Bolton Park, Grindleton, Bradford, and Waddington brooks, which water the villages whose names they bear, and Bashall Brook, which is of some length and has a tortuous course. From the S., Stock Beck enters it near Gisburn, and Swanside and Ings Beck united near Sawley. The latter forms the county boundary from Rimington Moor to the Ribble. Worston Brook runs through Clitheroe, is joined by Standen Brook at Primrose Mills, and enters the Ribble at Henthorn Mill.—R. H. T.

The Great Watershed between the rivers running into the Irish and North seas crosses the district from north to south. It runs from Marton Scar and Cranoe Hill over the low Craven country by West Marton, Gill, and Rain Hall. At the canal, north of Gill Church, it is less than 450 feet above the sea, and this is probably the lowest point on the watershed between Derbyshire and the Cheviot Hills.* It again crosses the canal east of Barnoldswick, at a height of about 490 feet, and runs up on White Moor to a height of 1,200 feet, descending rapidly again to about 530 feet near Foulridge. Henceforth it traverses high ground, its winding course being generally near the county boundary. Reedshaw Moss 875 feet, Watersheddles, S. of Combe Hill, about 1,120 feet, and Widdop Cross 1,286 feet, may be mentioned as passes across the watershed. All the country west of the watershed is drained into the Ribble, and that which lies east of it into the Aire. Stock Beck, which falls into the Ribble near Gisburn, drains the low

---

* The part of the watershed crossed by the Newcastle and Carlisle Railway is about as low as this.

ground west of the watershed. This stream is formed by the junction of several small streams rising in the high ground of Weets, swelled by others rising near Gledstone. The high ground west of the watershed is drained by feeders of the Lancashire Calder (an affluent of the Ribble), the longest branch of which, the river Laneshaw, rises on Combe Hill, and forms for some distance the county boundary. It receives in its western course the Wycoller and Trawden brooks, which drain the northern slope of Boulsworth, and two miles below Colne it is joined by the Pendle Water, a considerable stream, which with its branches drains the southern slope of the high ground between Pendle Hill and Weets.

The little streams near Crow Hill are the sources of the river Worth, which joins the Aire at Keighley. The Ickornshaw, Gill, and Lothersdale becks unite to form Glusburn Beck, which runs into the Aire below Kildwick. Several small becks near Carlton run directly into the Aire. Broughton Beck, formed by the junction of County Beck and Lancashire Gill in the low moss north of Foulridge (which is probably the site of an old lake), receives, near Kelbrook, Earby, and Elslack, feeders coming down from the high ground of Roger, Thornton, and Elslack moors, and at Elslack Bridge a feeder, which drains the low ground east of the watershed.—W. G.

Physical Features.

The valley itself of the Ribble, lying between the Pendle Range and the Grit Fells, which separate it from the basin of the Hodder, is covered for the most part with a tolerably thick coating of glacial drift and other surface deposits, which have a general slope from the Fell-sides to the river, but rise in certain parts into rounded hummocks, as *e.g.* the long gravel ridge which runs from Waddington to Bashall Hall, and resumes its S.W. course, after an interval of about a mile and a half, beyond the Hodder, in the neighbourhood of Stoney-hurst. Through this surface envelope the limestone rises in many places in low hills, rounded for the most part, but in others with terraces following the outcrop of successive beds of limestone. Of the first I may mention Withgill, Worsaw, Gerna, and Wybersey, and of the latter the beautiful wooded glades about Downham, Gledstone, and Waddow. Above Sawley for a considerable distance the drift has the predominance, and only along the river course itself, which is here a deep ravine, and in a few brook channels, are the rocks exposed. This low country lies chiefly beneath the 500 feet level, and is entirely laid out in pasture.

On the S.E. side of the Ribble, at higher elevations, alternations of hard and soft beds of limestone shale and grit, no longer masked by drift, form marked features along the hillsides. This is especially notable under Pendle, where they run very regularly, and the same sort of thing may be seen for a long distance to the S.W. About Whalley the ridges of gritstone have a very characteristic rounded form, and cannot fail to suggest, to the eye experienced in ice-phenomena, a former extensive glaciation of this country by ice coming from the N. On the other side of the valley the features are not so marked, the rocks being contorted and irregular, and therefore not so well able to influence the course of the furrows made by the "ice-plough," the general result being merely a low dome. This is also very much the state of things on the side of the Hodder valley S.E. of Newton.

The Pendle Range from Mellor to Roughlee consists of a double ridge of gritstones with valleys between them. One ridge is made up of the "Rough Rock," and subdivisions of the "Third Grit," the other of the Kinder-Scout and Yoredale Grits. The hollow between though continuous is drained by little brooks running to the N.E. and S.W. alternately, and is excavated, as might be expected, in a very thick bed of soft shale. N. of Colne there is also a small valley between the "Rough Rock" and the "Third Grit." From Hoghton Tower to Whittle-le-Woods the grits form but one ridge, the Kinder-Scout and Yoredale Grits not being brought up to the surface by the anticlinal. It ends abruptly in the Lancashire plain at the latter place.—R. H. T.

The portion of the district lying east of Gisburn and north of Barnoldswick which is a part of the celebrated Craven pasture ground contrasts in a marked manner with the large portion lying to the south of it. It is an undulating plain of an average height of from 500 to 600 feet above the sea, furrowed with many small winding streams, and studded with an immense number of low rounded hills, which have sometimes a marked linear arrangement, and not seldom coalesce to form definite ridges. This low ground sends from Barnoldswick a long tongue southward to Foulridge. Barnoldswick, Foulridge, Kelbrook, Earby, Elslack, and Carlton are all on its outskirts. It reaches at the ordnance station east of Gill Church a height of 638 feet, the highest point in that neighbourhood. North of Gledstone, however, the ground is still higher, and rises to over 700 feet in Cranoe Hill and Marton Scar. The lowest ground, the alluvium of the Aire, is a little over 300 feet. Nearly the whole of this low ground is made up of the lower Carboniferous rocks, viz., Carboniferous limestone and shales. The higher ground to the south of this, which is composed of higher Carboniferous grits and shales —from the Yoredale Grit up to the Lower Coal Measures—forms a series of ridges with intervening valleys, running in the main parallel to the general N.E. and S.W. strike of the beds. The high ground of Yoredale Grit which rises in Weets to 1,300 feet above the sea ends off in a marked manner near Barnoldswick, but sets in again to the eastward in the high ground of Thornton and Elslack moors, which rises to 1,175 feet at Bleara Lowe, and to 1,274 at the Beacon on Pinnaw. This great eastward shift, so to speak, of the high ground corresponds to a great fault which runs in a W.N.W. direction by Barnoldswick. South of the valley of Lothersdale, in which the limestones and shales are brought up by an anticlinal, the Yoredale Grit rises to the height of 1,175 feet at Kelbrook Wood and Hawshaw Moor. The ridge of Kinder Scout Grit rises to 1,018 feet on Blacko Hill, Noyna 980, Piked Edge 1,165, Sweet Brow, near Cowling, 1,125. The highest point on the 3rd grit ridge is about 1,100 feet at Colne Moor. Knoll Hill (3rd grit) is about 825. The chief heights attained by the Rough Rock are, Monk Edge 972, Scars Top, near Cowloughton Dam, 1,153, Combe Hill 1,357,[*] Little Wolf Stones 1,454, Crow Hill 1,501 ; the last being the greatest height it reaches in this district.—W. G.

CHAPTER II.
SEDIMENTARY ROCKS BELOW THE COAL-MEASURES.

The Carboniferous Limestone.

The lowest of the rocks which occupy the district described in this memoir is the Carboniferous Limestone, and probably it underlies the other Carboniferous Rocks over the whole of the area. It consists of a great thickness of beds of limestone with beds of shale, the limestone predominating. Its base is nowhere seen on the S.W. side of the Craven Faults until we get so far north as Kirkby Lonsdale, but on

---

[*] Wrongly given as 1,557 on the 1-inch map.

the upcast side we see it reposing on Silurian Rocks in the valley of the Ribble north of Settle, near Malham Tarn, in brooks above Ingleton, and in other places, a description of which will be given in another memoir.

Its greatest apparent thickness in this district is obtained from a section taken from the axis of the Clitheroe anticlinal arch at a point south-west of Gisburn, towards Park House in a north-north-westerly direction, and is at a low estimate 3,250 feet. These are only the lowest beds visible, and so we must add an unknown quantity to take us to the actual base. Another section without a base, from the axis across the southern side of the arch through Chatburn to Worsaw, gives a thickness of 2,600 feet, and here dips are very abundant, and there seems little likelihood of faults repeating the beds.

The line which we have taken as its upper limit may only at best be considered an artificial boundary. It represents the horizon above which shales predominate over limestone, and below which the limestones preponderate along the Clitheroe anticlinal, and the line is in this valley doubtless a true geological horizon. It would be rash, however, to assert its identity with the line which we have taken as the boundary in the parallel valley of Slaidburn. It is very likely that the top of the thick limestones there is higher in the series, i.e. that some of the calcareous shales near Clitheroe above the limestone have become more calcareous and less argillaceous further north, the currents which brought the mud having been more prevalent to the south, a supposition rendered more likely by the fact of the total thickness of the limestone at Ingleborough being only 600 feet. and containing no shales.*

In the west of this district we get the limestone at Withgill, where it rises in a small boss, with a quaqua-versal dip from beneath the overlying shales. It appears to be bounded by a fault, however, on the north side. The limestone is greyish white and full of fossils, which may be easily extracted. One of the commonest is *Conocardium*. On the west side of the old quarry may be seen a mineral vein containing fine cubical crystals of fluor spar, which is not a common mineral in this area.

We next come to the main mass of the limestone at Clitheroe, its west-south-westerly extension being cut off by a large fault which crosses the Ribble below Waddow Hall, and throws down the Permian Sandstone resting on Yoredale shales and limestones, against some beds much lower in the series. The Carboniferous limestone of the Clitheroe anticlinal is concealed or very obscure all along its northern border, and the many folds into which it has been thrown, an instance of which, with a fault, is given in Fig. 1 (p. 15), have rendered its boundaries difficult to map with accuracy. On the southern side of the arch, however, we have a succession of very good rock exposures, which afford facilities for its study. It contains two very distinct members. The lower consists of very black and pure bituminous limestone, and sometimes contains beds of black calcareous shale. It is almost always very distinctly and evenly bedded, and forms in its range a very straight and very well-marked ridge, which commences at Horrocksford Quarries and continues in an east-north-easterly direction by Ridding

---

* Indeed it is highly probable that the main mass of the limestone in the Pendle country, together with the Shales-with-Limestones which lie above it, and the Pendleside limestone, are all represented in point of time by the comparatively pure limestone of the "Great Scar" of Penyghent and Ingleborough. There are several reasons for supposing this to be the case. I merely throw this out as a suggestion, and must reserve for a future memoir on the Craven country a more definite opinion.

Hey and Bold Venture Limeworks, and then along the north side of Downham Hall Demesne and Twiston Lane to the old lead mines of

Fig. 1.

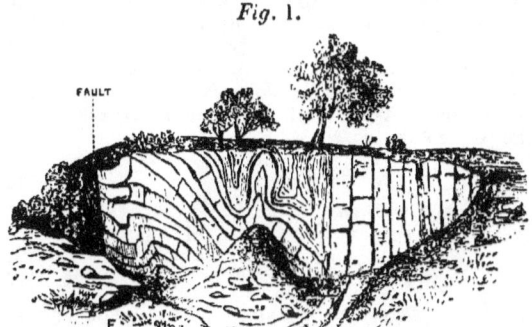

Quarry near the Road between Sawley Bridge and Grindleton.

Skelhorn or Skeleron. Beyond these it is terminated by a fault bringing down the Yoredale shales. The most numerous set of joints in it ranges from N. 30 W. to N. 10 W.

Immediately above the Black Limestone is a band of shales containing fossils, of which *Fenestellæ* are the most abundant. It was well seen at the front of a quarry on the north side of Twiston Lane, a short distance from Downham; crinoid heads could also be obtained there in good preservation. The shales at Knunk Knowles by the road cutting going down to Brungerley Bridge, near Clitheroe, are probably the same.

These shales underlie the upper light-coloured limestone of Coplow, Salt Hill, Worsaw Hill, Twiston Quarry, &c. It is very massive and never shows such distinct bedding as the Black Limestone. It is singular also that although resting on the latter, which has so regular a strike, its own surface presents a most tortuous boundary. This, no doubt, is to a certain extent due to the crumpling of the beds by lateral pressure; but I cannot help thinking that the apparent sudden changes in thickness are not altogether to be referred to this, nor to the accidents of original deposition, but in some measure to the unequal erosion of the beds by rainwater percolating the mass underground, the original crumpling of the beds determining the greatest flow of the water and the areas of subsidence. A good example of this subsidence of beds may be seen on the top of Farleton Fell in Westmoreland.

The Upper Limestone presents a curious appearance at Salt Hill. The highest, 40 feet, consist almost entirely of the remains of *Crinoidea*. It is, however, too friable to be utilized as an ornamental marble in the way that similar deposits in Derbyshire and at Dent have been, but it occupies the humbler position of garden gravel. This thick deposit of encrinites seems to be only local, for, although we get the same or a similar bed at Whitewell to the north-west, nothing resembling it occurs about Worsaw on the same horizon. The commonest species is *Actinocrinus triaconta-dactylus*.

The lowest beds of the light-coloured limestone are of a light bluish grey colour, massive, hard, and somewhat nodular. They may be well studied at Coplow Quarries near Clitheroe, where they contain a bed of light sea-green clay with many fossils. The limestone there is remarkable for the abundance of a fossil, *Amplexus coralloides*, and also for the occurrence of *Palæchinus multipora*, as I am informed by my friend, Mr. James Eccles, F.G.S.

The limestones of the Forest of Pendle, both black and white, are of great importance as supplying lime to a wide extent of country, and for this purpose are extensively worked, the railways direct from the quarries affording great facilities of transport. "The Limes of Clitheroe" are all excellent in quality compared with such as are to be found in "other quarters."* I have been favoured by Mr. Dixon Robinson with a copy of an analysis, made by Mr. John Just, of the limes burnt in the Bold Venture Limeworks at Chatburn. The average of six specimens gives as the per-centage of—

Carbonate of lime	98·72
Silica - -	·55
Protoxide of iron	·34
Residue - -	·39
	100·00

This indicates a very high degree of purity. The lime from the Black Limestone is preferred in all cases where whiteness is an object, the bituminous matter which gives the colour being all expelled in the burning.

To the east of the fault above mentioned the limestone and overlying shales are in great confusion, the beds being much faulted and contorted, and a spread of drift making them more difficult to trace; the range given to them on the map, however, is probably the nearest possible approximation to the truth.

The great mass of limestone on which Gisburn stands shows itself well in Wyhersey Hill in Bolton Park. Thence the top boundary runs along in an east-north-easterly direction on the right bank of the river for the most part under drift, but exhibiting sections in several tributary brooks. It crosses the river at the north end of Kirk Wife Wood near Gisburn Park, and then appears to be thrown forwards by a north-westerly fault to the north side of the river again by Paythorne, and remains on this side to a short distance north of Nappa, whence it curves round with a northerly dip to join the south side of the anticlinal.

The south side of the limestone at this portion of the anticlinal appears to be a fault boundary, for we only find the beds beginning to turn over along a line a little north of Great Dudland and Eel Beck, and this is close to where the main mass of the shales comes on. The boundary moreover is in the direction of the great Barnoldswick fault. The legitimate inference is that it is continued along this line, a supposition which is strengthened by the great contortions shown in the beds to the south of this line.

The Shales-with-Limestones.—The next beds above in order occupy the greater portion of the low country and rise up gradually on all sides of the valleys to where the hills begin to give a more decidedly steep feature. They consist of alternations of shales various in character with beds of limestone, cementstone, mudstone, and sometimes thin ironstone. They frequently give rise to springs containing sulphuretted hydrogen gas. One with a bath-house attached occurs close to Clitheroe, near the Shaw Bridge Road. There is another in Standen Brook, not far from the footbridge; others occur south of Worsaw, in Twiston Brook, south-east of Ravensholme; and in Holden Brook, under the work-house. In the bed of the Hodder, north of Longridge Fell, they are

---

* Vide "A Lecture on the Value and Properties of Lime for Agricultural Purposes," delivered to the Bury Agricultural Society, by John Just, Esquire. 1850.

SHALES—WITH-LIMESTONES.

very numerous, giving unmistakeable evidence of their presence by their smell when the water is low.

The shales at the base of this division consist of a thick series of layers of hard and sandy and argillaceous mud, often micaceous, with some thin limestones, and frequently good fossils may be obtained both in the shales and limestones. These may be well seen and searched in several places, but I may mention the Pendle branch of Worston Brook by the lane east of Worston, and the brook running from Lower Gills into Ings Beck, the county boundary near Skeleron Mines. On the north side of the anticlinal, they may be well seen in the Ribble beneath the circular earthwork of Castle Haugh near Gisburn. In the Slaidburn valley they are exposed in Hamerton Brook, not very far from its junction with the Hodder.

Higher up in the series these shales everywhere become more calcareous, limestones abounding, though not often of very good quality. Some contain a great admixture of clay and graduate off into mudstones; others have a large proportion of iron and some magnesia, and form a kind of cementstone; but whether they would be serviceable for cement it is impossible to say without analysis being made, or their being submitted to experiment. In the upper part of these Shales-with-Limestones the beds are in very regular alternations, the limestones being one to three feet thick, and separated from each other by about six feet of clayey shale, in brooks forming a succession of of pretty little waterfalls, as in one above Angram Green near Worston. The Shales-with-Limestones between Worsaw and the base of the Pendleside Limestone give a thickness of 2,500 feet. As a group they should be reckoned to extend higher, i.e., to the base of the Lower Yoredale Grit, or if that is absent, to the horizon at which it occurs when present, viz., the base of the Black Bowland Shales. This would give under Pendle an additional thickness of 725 feet, making a total thickness of 3,225 feet.

The Pendleside Limestone.

In all parts of this series there is a tendency towards thick and moderately pure limestone coming on, but for this the most marked horizon is near the top and below the Lower Yoredale Grit. It is this which forms what may be called the buttresses of the north-westerly slope of Pendle Hill, and from its great developement there I have ventured to call it the *Pendleside Limestone*. It may be well seen in any of the cloughs which there descend into the comparative plain, but perhaps the best section of it is in that which runs down past Little Mearly Hall.

A portion of that section in descending order is as follows :—

> Black and dark-grey crystalline limestone, containing black chert bands, with beds of black shale and occasional cement stones, the limestone weathering into anvil forms
> Compact light-brown crystalline limestones quarried for lime with a thin bed of calcareous shale   -   -   30 feet.
> Gap   .   -   -   -   -   30 feet.
> A scar of compact light-brown crystalline limestone, with here and there bands of white chert; once quarried for lime; in general appearance much like the scar limestone; fracture irregular; contains fossils which weather out on exposed surfaces; near the top of it a thin bed of fine sea-green clay
> Cementstone and black shale with cherty black limestone   -
> Yellowish-grey cementstones with bands of colour shown on a fresh surface interbedded with soft light grey muddy shales containing fossils.

Here it shows a thickness, with interbedded shales, of some 350 feet. may be traced in its south-westerly range for some distance by Knowl

Top, and across the brook above Pendleton Hall, and some distance above Pendleton. It probably crosses the lane between Wiswell and Pendleton below the Audley Reservoir, and quickly dying away to smaller dimensions is lost under the drift to the north-west of that road. To the north-east of Little Mearly Hall Clough its maintains its character and thickness for some distance, presenting in the clough above Hook Cliff the following section:—

Dark-hard crystalline limestone with regular beds of chert.
Light-brown cream-coloured limestones with a little chert.
Dark hard finely-crystalline limestone with regular bands of chert.
Impure thick bedded limestone.
Clayey shales with cement-stones containing *goniatites*.

From this point it appears to dwindle towards the north-east. It is crossed by a fault (downthrow to the south-west) west of Ravensholme, where it has been much quarried and seems very bituminous in character, and by another with a similar downthrow at Brownlow. It is much thinner in Pendle Hill Brook, where it gives rise to a sulphur spring. From this point eastwards it cannot be so easily traced, and although limestones frequently come on somewhere along this horizon they cannot with certainty be identified with it.

West of Wiswell nothing at all answering to it is to be found until we come to a set of beds of somewhat similar character, and at about the same horizon along the northern escarpment of Longridge Fell, there too forming the lower buttress to the higher scarp of the Upper Yoredale Grit, as in Pendle Hill, but less marked.

Around the sides of the moorland which comprises the Fells of Waddington, Easington, Grindleton, and Bradford, Limestones are frequently to be seen underlying the Black Shales on which rest the Upper Yoredale Grit. The Lower Yoredale Grit being absent over a good part of this district these limestones lie on very much about the same horizon as the Pendleside Limestone; but owing to the great contortions to which they have been subjected, an instance of which is here given (Fig. 2), and the thick drift which conceals them in many places,

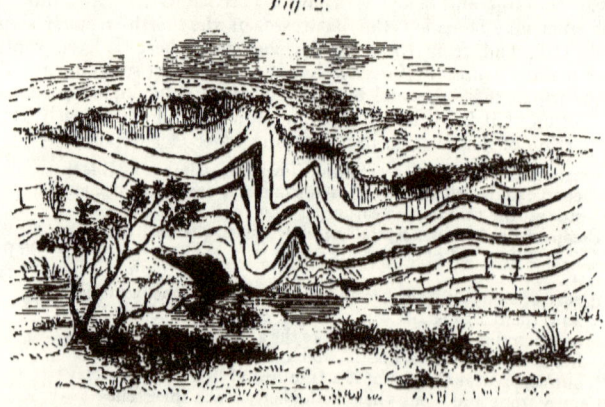

Fig. 2.

Contortion in West Clough, West Bradford, in Shales and Limestones.

it has not been found practicable to map them, and they have been massed with the Shales-and-Limestones.

Thus beds of massive brown limestone occur at New-a' Nook west-south-west of the Mooreock Inn on Waddington Fell, and have been quarried of late years for mending the road which runs from Waddington to Newton. They may be traced some distance to the west by the line of pot-holes. They may be seen in West Bradford Brook a quarter of a mile below Hanson's, and come on in force below Steelands, forming an anticlinal in Grindleton Brook, and are probably the same limestones which run through that village. They are again seen in Holden Clough above the house, where they are dipping to the north-west and are again brought up "to the day" above the Victoria Lead-mine by a fault which crosses the brook at the waterfall and runs in a S.S.Ely. direction, the lead veins being nearly parallel to it. These have only been hitherto worked in very much thinner limestones which are higher in the series and are brought against the thick limestones by the fault.

On the north-west side of this moorland we come to the curious little patch of limestone which protrudes from beneath shales at Ashnot. It bears a stronger resemblance to the Great Scar Limestone than any other beds which I know in a like position. It is a good cream-coloured limestone, massive with scarcely any visible bedding, and in places is quite white and soft, almost resembling chalk. There are signs of disturbances in the neighbourhood, but hardly I think sufficient to lead one to suppose it to be a part of the main body of limestone thrust up from below. It is more probable that it is a portion of the same beds that we have been endeavouring to trace. Lead has been worked in it some years ago and a level driven up from the brook. The lead I believe was found chiefly in "nests" and "pipes," but a thin vein is visible in the face of the rock with a S.Ely. direction. To the north-east of this these Limestones do not make any feature for some distance, although other Limestones of lower horizon may be seen in the brooks; but they come on again below the Lower Yoredale Grit of Champion, and also along the north-west side of the Slaidburn anticlinal. These localities, however, lie in a district assigned to a future memoir.

In the Little Mearley Hall section and other sections adjoining a considerable thickness of beds with shale predominating lies over the Pendleside Limestone and below the Lower Yoredale Grit. They contain impure laminated ironstones and thin beds of Limestone. Also a bed of nodular calcareous ironstones occurs here, which may be seen in several of the brook-sections. It is remarkable for a somewhat uncommon result of weathering, the surfaces exposed being pitted with little hollows in the form of inverted cones in concentric groups, several being arranged about the same axis, as shown here in section and plan.

Fig. 3.

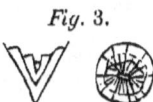

The Limestones though thin are tolerably pure, and one appears to contain many fragments of crustacea. These Limestones answer in horizon and appearance to the two beds of Limestone which form the bearing beds in the Victoria Lead-mine at Holden Clough. This set of beds may be regarded as a transition from the deep water and quiet sea represented by the organic deposits which formed Limestone to the shallower and less pure water which brought the mechanical impurities of the Lower Yoredale Grit. Indeed the existence of plant remains in several places in these beds in greater abundance than below prepares us for a change, and speaks of currents from the land, though that may have been and probably was far distant.

The Lower Yoredale Grit.

The Lower Yoredale Grit is of all recognisable beds in this district the most inconstant in its extent although very constant in its horizon; *i.e.* though frequently absent altogether, wherever it does occur, it lies at the base of the black Bowland Shales and above all the thick Limestones, forming a marked boundary between them. It consists of grits and sandstones with shales and ironstones interbedded, and very rarely has in it beds of conglomerate. In general character the beds of sandstone which predominate are very fine and hard, and sometimes might almost be called *quartzites*. In appearance they are often not unlike the Gannister Rock of the Lower Coal-Measures, though not containing the root beds of that rock; plant remains, however, are of common occurrence in them. On Pendle they consist of two principal sets of beds of sandstones and grits with shales between, but sometimes a triple arrangement occurs. Their topmost bed in Little Mearley Hall Clough is a well-marked conglomerate, but this character has not been noticed to persist either to the north-east or south-west. The fine beds of this subdivision from their closeness of grain and hard texture stand the weather remarkably well, and being of a character peculiarly well adapted for receiving and retaining marks upon their surface, the ice scratches which are abundant over this country are particularly well shown on these rocks. Indeed wherever Boulder clay rests on this rock you are pretty sure to find a glaciated surface at the line of junction, and usually ice-scratches, showing very plainly the line of direction of the ice-current at that particular point. The quarter from whence the ice came is not always so apparent on the ground; it is only when we accumulate observations of these scratches on the map that we see whence and whither the ice seems to have flowed.

I may mention as an instance of the hardness of these beds a fact kindly communicated to me by Mr. Ralph Assheton, of Downham Hall, M.P., that some person employed by an ancestor of his to bore for coal near Alders Clough, which runs down from the "big end" of Pendle, during some part of their operations could only sink the bore one inch per diem.—R. H. T.

The Lower Yoredale Grit immediately west of the Great Barnoldwick Fault seems to be composed of two bands of fine grit with a thick band of shale between.

	Feet.
Upper Grit, brown and grey, about	250
Shale	200
Lower Grit, white, quartzose	300

This is from a section near Weets.

The middle shale, however, is little seen. The whole seems to be thinning eastwards, so that near the Barnoldswick Fault the whole thickness is probably not 600 feet.

The lower fine white grit may be seen in numerous quarries between Crag and Coal Pit Lane, and also east of Coverdale Clough, where the junction with the underlying shale may be seen. A hollow marks for a considerable distance the middle shales, and the upper band of grit is marked by a distinct ridge as far as * New Field Edge, where it begins to be covered up by drift. The beds near Springs have some of them a dark grey colour. Near the great fault they are much disturbed, being

---

* There are no less than six detached houses called New Field Edge in this neighbourhood.

in one place vertical. The fault crossing the grit near Coal Pit Lane is inferred from the shifting of the two ridges of grit. It throws down on the S.W.—W. G.

The Bowland Shales.

Of all the rocks in the Forest of Pendle, the Forest of Bowland, and adjoining country, perhaps none plays so important a part in giving character to the scenery as the Black Shales which lie between the Lower and Upper Yoredale Grits. They have been named by Professor Phillips the Bowland Shales, and most appropriately, for nearly all the steep slopes of the fell-sides in Bowland are due to their disintegration.

They are usually black but sometimes grey, and in some places are coloured yellow by a thin wash of hydrated oxide of iron, but this only permeates the joints and fissures and does not extend to their substance, being an accident of infiltration or oxidation and not of original deposition. They are thinly laminated and easily split, almost always calcareous, very seldom sandy. In some localities they smell strongly of mineral oil and appear very bituminous. They are full of fossils which are much crushed or flattened and seldom in good preservation. In the thin ironstones which are common in them good specimens, however, may frequently be obtained. By far the commonest fossil is *Posidonomya Gibsoni*. Orthoceratites, Goniatites, and fish remains are very frequent in them; the last are usually very fragmentary. Some nodules somewhat resembling Coprolites * occur in Riddle Clough near Knowlmere Manor. The fish remains may be found dispersed throughout, but a notable horizon for them lies a short distance below the Upper Yoredale Grit. Vegetable remains may frequently be seen.

A Warning to Coal-seekers.—It is a most unfortunate thing for the landowners of the county in which these shales occur that they bear a strong resemblance to some of the shales of the productive Coal Measures. Their blackness, the general resemblance of their fossils, their beds of ironstone, their "Canker," or chalybeate, water, to which we may add the rude resemblance of the hard fine sandstones below them to the Gannister Rock, all serve to excite hopes; and when an ignorant miner or "practical man" comes over from a neighbouring coal-field and expresses his conviction, perhaps honestly enough, that there is coal to be had if they go deep enough, the delusion is perfect.

Boring commences and is continued or a shaft is sunk until either limestone is reached or the employer is disheartened. It would be thought that the evil would stop here, and that one or two such trials in a district would be sufficient to prevent a repetition of such failures; but it takes a good deal to make a man despair of his country, and soon some patriotic inhabitant suggests that the miners were bribed or bought off by the proprietors of the nearest colliery. This quickly gains credence, and before many years are out the inhabitants are all agog for a fresh attempt. Another frequent excuse for a failure is that "they were drowned out." Here is a characteristic story of this sort :— Near Baygate above Holden Clough some years ago a trial was made by means of a shaft. On making inquiries subsequently I was told by one man that just as they were getting *near* the coal they were drowned out by the water. Another man stated more boldly that they had reached a bed of good coal 4 feet thick when the water came in so quickly that they left behind them their picks and shovels and barrows, "and," he added,

---

* My attention was called to these by Mr. Jonathan Peel of Knowlmere Manor, to whom I am indebted for other information.

with a truly artistic touch, "they lie there yet." The statement so far as regards the tools I cannot deny; all traces of the existence of the shaft having vanished, but the 4-feet coal I can certainly affirm has no existence, for Holden Clough close by gives a very clear section of the beds which would be passed through by the shaft, and no coal bed is to be seen.

These stories are so rife wherever unsuccessful trials have been made, that in those who are frequently meeting with them they will only provoke a smile, but to the unwary they may prove a serious matter. I have therefore thought it better to call attention to them; to be forewarned is to be forearmed.

It is lamentable to see the numerous remains of shafts all along the basin of the Ribble and Hodder wherever black shale occurs. For mining "practical knowledge" of coal mines is a good thing; but unless backed by an acquaintance with the structure of the country, the succession of the rocks, and the former range of each coal bed, it is worse than useless for the discovery of new coalfields.

On the north-west face of Pendle these shales exhibit a thickness of about 700 feet. On the opposite side of the anticlinal, between Higher Heights near Holden and Beacon Hill, they are lying at about 10° and give a thickness of 600 feet.—R. H. T.

A carefully measured section shows these shales to be about 1,000 feet thick near Weets, and good sections are to be seen in most of the little streams about there. The shales are very dark towards the top and contain one or two bands of dark impure shaly limestone. One of these bands has been worked at the quarry marked on the 1-inch map about 400 yards S.E. of Cold Weather House.

North-east of Weets we get the following series in descending order:—

About 1,000 feet.
{ Upper Yoredale Grit.
Black shales.
Black shales with dark shaly limestones.
Dark shales.
Shales with compact earthy limestones.
Shales with sandstones.
Shales with limestones.
Lower Yoredale Grit.

The limestones in the lower part of the shales are more compact and several thin bands are seen. Some thin sandstones occur in the shales a considerable distance above the top of the Lower Yoredale Grit. Between New Field Edge and Edge, and again near Springs, where the junction of these beds with the underlying Lower Yoredale Grit is observed, a limestone is seen at or near the base of the shales.

The average dip near Weets is 20° to 30°, but it increases eastwards, being from 40° to 50° near Springs. The average direction is south-east.—W. G.

Upper Yoredale Grit.

The black shales which we have been describing are very seldom sandy, but on the contrary usually contain fossils, and have a tendency occasionally to run into black limestones. Above these, however, are shales of a sandy nature, and grits which contain no fossils, except of plants. When we trace any of these sandy shales for any distance, we find that they invariably show a tendency to turn into sandstones, and reciprocally the sandstones frequently turn to sandy shales. We have therefore found it often necessary, in mapping, to draw the boundary between calcareous shale and sandy shale, to preserve anything like a true geological horizon, the one being more nearly related in lithological character and the conditions of its formation to limestone, the other to grit. When we speak of the " Upper Yoredale Grit," then, we include the beds of sandy shale which are intercalated with it and are often of considerable thickness, but which so frequently change to grit and back again to shale, that any attempt to separate them by surface features or lithological character would be useless, and, without clearer and more continuous sections, give rise to serious errors in estimating thicknesses, horizons, &c.

The beds which crown the summit of Pendle and form the crest of Longridge Fell, and also constitute the greater part of the Fells around the Ribble and Hodder basins, were long accepted as the lowest beds of the Millstone Grit series, having been so called by Professor Phillips. It may seem presumptuous to disregard the nomenclature of so high an authority, but it has been found, in tracing the geology from the south, that these beds are the equivalents of others which have been known in the Survey publications as the " Yoredale Grit," and therefore, for the sake of uniformity, this name has been retained, only adding to it "Upper" to distinguish it from the Lower Yoredale Grit, which is probably (as pointed out by Mr. A. H. Green), but not undoubtedly, the equivalent of the " Yoredale Sandstones " of North Derbyshire.*

The grit at the top of Pendle is represented in Phillips' Geology of Yorkshire, both in the map and letterpress, as an outlier; this, however, it is not, as may be seen when descending from the plateau of Pendle in a southerly direction, by Ogden Clough or the tributary brooks which run into it from the hill top. The black Bowland Shales, which underlie the grit summit, may be traced clearly round from the " Big End " to a farm called Buttock, beneath the grit the whole way, and from Buttock to Barley they are dipping under it to the south at angles of from 27° to 35°. The same arrangement of these beds continues along the strike from Barley up to Weethead Height and further, and these shales never come up again between this and the coalfield. The black fossiliferous shales which are seen here and there in the Sabden Valley are not the same, for they are seen to lie above the Kinder Scout Grit in several places, and that in turn lies over the Pendle Grit, so that they belong to a much higher horizon although in lithological character they are identical.†

The Upper Yoredale Grit is somewhat variable in texture, showing every gradation from a close fine grit to a conglomerate, this latter form being only exceptional. Most commonly it is a moderately fine freestone, sometimes grey pink or greenish, but usually white or rusty yellow. It contains plates of mica and often much felspar, which is pink, white, or yellow. Brown or yellow bands of irregular form, but generally arranged more or less in relation to its bedding and joints, are usually seen in section in a broken block. This is due to iron, and large ovoid concretions of the same character, which are harder than the surrounding matrix and fall out in quarrying, are very common in it. Examples of them may be well seen in the quarries on the south side of Nick of Pendle, and in a quarry on the left of the road from Bonny Inn to Ribchester (1-inch map 91 S.E.) These concretions often 2 to 4 feet in diameter do not, so far as I am aware, commonly occur in this district in any other grit but the Upper Yoredale Grit. Conglomerates, when they occur in these beds, generally differ from those of the over-

---

* See Geology of North Derbyshire, &c. (Memoirs of the Geological Survey), p. 164.
† See a similar remark by Mr. Gunn as to the shales in the Foulridge Tunnel, p. 37.

lying Kinder Scout Grit in not having such large pebbles, and still more in the pebbles being scattered through a less coarse matrix than is the case with that grit; but this is a distinction which is probably only of local value and cannot be expected to hold good for any distance. —R. H. T.

The most westerly point along the Clitheroe anticlinal, where the Upper Yoredale Grit may be seen, is a little east of Salmesbury Hall, near Mellor, where it is rather coarse grained and massive. The dip is here from S.W. to S. at 5°. At Mellor Brook the position of the anticlinal may be very well determined along the banks of the brook itself. At the quarry by Resburn Fold the dip is south, while at the village it is north. The Yoredale Grit is again shown in the quarry at Lower Abbott House dipping northward at 20°. The symmetry of the arch is here interrupted by the presence of a fault which occupies the place of the anticlinal axis. Sections are again shown in a quarry at Harwood Fold; at Mellor below the camp; at Moor Edge; and in the cuttings of the railway north of Ribchester station. Here the beds are repeated by a fault ranging east and west, a few yards from the southern end of the tunnel. The beds are here violently contorted, and as it seems to me pushed over laterally on each other. The rock itself is, however, irregularly bedded, and some of the apparent foldings may be owing to this. At the mouth of the tunnel the dip is from the fault northward, but at the opposite entrance the dip of the underlying shales is in a southerly direction, so we must suppose that the grit returns to the southerly dip and crops out in the tunnel. The section in the cutting at the north entrance to the tunnel is interesting.

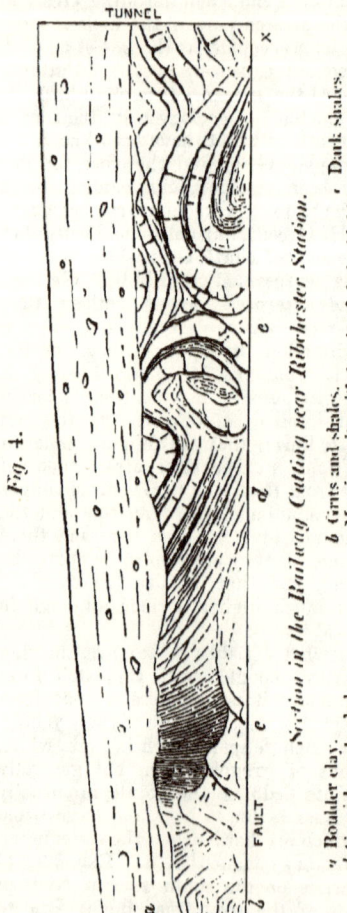

Fig. 4.

Section in the Railway Cutting near Ribchester Station.

a Boulder clay.
b Greyish sandy shales.
c Grits and shales.
d Massive grits with bands of shale, highly contorted.
e Dark shales.

The beds consist of dark shales with bands of siliceous ironstone and fine grit in the upper part, and argillaceous[s] limestone in the lower. They form a complete arch, and indicate the position of the axis of the Clitheroe anticlinal. We were unable to find any fossils at this spot.—E. H.

From this place the Upper Yoredale Grit runs along the north side of Billington Moor by Hollin Hall to Whalley. Its lower boundary is well seen, and the underlying black shales with ironstones are exposed in most of the little watercourses which run down the steep hill sides. The general dip of the black shales, Upper Yoredale Grit, and Kinder Grit is high to the S.E., varying from 40° to 65° or 70°, and increasing towards the coalfield. It is interesting to note the increasing thickness and importance of the Upper Yoredale Grit to the north-east. West of the Calder it forms, relatively to the Kinder Scout Grit, quite a secondary feature, the latter overtopping it all along the strike. In the neighbourhood of Whalley it becomes much thicker very quickly. At Clerk Hill the two grits form about equal features, but east of this the Upper Yoredale Grit asserts itself and taking possession of the ridge top retains and increases its superiority, which culminates in Pendle End at a height of 1,831 feet above the sea. Its inferior rival, the Kinder Scout Grit, from Clerk Hill to the N.E. only forms a subordinate feature, and flanks the north side of the Sabden Valley, attaining its highest elevation in this neighbourhood on Spence Moor, where the Ordnance cairn built upon it lies at a height of 1,498 feet.

A fault crosses the Pendle range near Sabden, running from near the Audley Reservoir to the E.S.E. by Parsley * Barn to Heyhouses, throwing down the Upper Yoredale and Kinder Grits to the south. A lead vein runs parallel to it at the top of the ridge and has been worked. This fault is pretty clear upon the ground from the interruption in the grit ridges, and also from opposing dips in the road E.S.E. of the barn. The same cannot be said of the very intricate piece of ground near Deerstones north of Sabden, which has not only been subjected to great displacements of the solid crust, but also suffers from frequent landships, which obscure the few sections which might give a clue to its interpretation.

"*Pendle's brasted hissel*" is a not uncommon announcement in the neighbourhood of Clitheroe, and as the phenomenon to which this term is applied appears to be regarded with a good deal of mystery,† I will state what I believe to be the cause of its occurrence. During heavy and long-continued rain it sometimes happens that a considerable body of water bursts out of the north side of the hill near the top of the escarpment, and carrying with it a *débris* of shale, blocks of grit, sods of grass, &c., rushes down the hillside and does great havoc in the enclosed lands immediately below. I saw the effects of one of these bursts soon after it had occurred in 1868, and was much astonished at the force displayed by it. The short-lived torrent had left most unmistakeable traces of its descent, having bared the hillside of grass and soil, and also cut into the shale. Lower down it had destroyed a wall and a hedge, and on arriving at a more gentle slope of ground deposited its spoils upon the meadow land. These bursts always occur at the junction of the grit and shale, and only, as stated above, in very wet weather. The grit, besides being itself porous, is full of joints and fissures, which readily receive the water which falls upon the surface and transmit it to the lowest level. It is, however, stopped, as shown in the diagram (Fig. 5, p. 26) by the shale along the line A C. It is conceivable that more rain may be falling on the grit than it can convey away along the dip of the beds towards the south. Under these circumstances the water will accumulate in the fissures until it rises to A, the highest point of the shale, when it may be compared to the water in a reservoir overflowing its banks. The top of the shale dam at A is thin and easily displaced, and so soon as the water begins to overflow it will readily

---

* A corruption of Paslieu, the name of an Abbot of Whalley who built it. See Whitaker's "History of Whalley."
† See "The Lancashire Witches," by Harrison Ainsworth.

yield and break away, and thus a considerable area of the waterlogged portion, shown by the thick lines in the engraving, will be drained in a

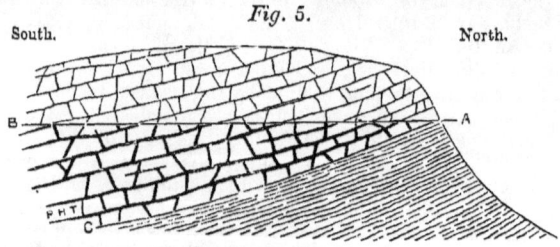

Diagram to illustrate a "Burst" on Pendle.

short time. I may mention that it is currently reported that Brast Clough, on the north-west side of Pendle, was hollowed out in this manner in a single night; but as this is a clough of considerable dimensions, and the lower part is excavated in solid limestone, my own opinion is that not even the Ribble itself, if it could be turned over the top of Pendle in flood-time, would effect so great a work in so short a time.

Form of Pendle on the Map.—The great expanse of Upper Yoredale Grit which lies north-north-west of Sabden and forms the plateau of Pendle Hill appears on the map to form an exceptional arrangement in the long straight range of the Pendle chain. Its extent to the north is probably due to the comparatively low angle which these beds assume north of Spence Moor. Although they dip beneath the Kinder Grit at considerable angles, they quickly change to a low dip at a short distance from that grit, and on the highest part of Pendle they are nearly flat (see Horizontal Sections of the Geological Survey, Sheet 85, No. 2); but still further north they must have risen at one time much higher, judging by the dip of the beds below. Denuding forces, whether subaerial or marine, seem to have reduced all the carboniferous country from here to Lancaster * to a general level so far as its hill-tops are concerned. Supposing, for the sake of argument, that this was a plain of marine denudation, out of which the valleys were afterwards excavated along the strike of the softer materials by subaerial denudation (and this theory appears to coincide with all observed facts†), we may conclude that the northerly expanse of the grit here is owing to its low angle, and consequent low elevation, having sunk it below the influence of that early pre-Permian sea. On the subsequent emergence of the whole district this capping served as a shield protecting the soft materials below it from the corrosion of rain and rivers, whilst the adjacent soft parts, from which the shield, owing to its greater elevation, had been removed by the sea, crumbled away beneath their destructive influences, and formed, as now, lower ground.

The course of the Upper Yoredale Grit from Pendle eastwards is as already described (p. 23) up to Weethead Height. Here it is thrown

---

* And probably further to the north east and south.
† I do not mean to imply that the removal of the great mass of the Carboniferous Rocks in this district previous to the Permian period was effected solely by marine agency, but only that at any rate the last stage of the operation was performed by a sea, whether inland or oceanic. See Prof. Ramsay's paper, Q. J. G. S., vol. xxvii., p. 241.

down on the north-east by a fault which passes close to Firbar House, and, throwing down the Lower Yoredale Grit also, is continued across the Ribble valley by Skeleron, Rimington, and Bolton Park. Another fault, with a downthrow to the south-west, runs between Claude and Burn Moor, and a disturbance, which is probably owing to this, may be seen in one of the cloughs below Martin Top.—R. H. T.

This grit west of Barnoldswick is generally of a yellowish brown or reddish brown colour, sometimes light grey or whitish. It is coarser than the Lower Yoredale Grit, often coarse enough to be called a *coarse grit.* But only in one place in this district can it be called a very coarse grit. This is in the north bank of the County Brook between Foulridge and Barnoldswick, and about 200 yards west of the canal. Here very coarse thick bedded grit is seen, with good sized pebbles in it, the whole looking much more like Kinder Scout Grit than ordinary Yoredale Grit. The beds seen must be near the top of the Upper Yoredale Grit.

The base of the Upper Yoredale Grit forms a good escarpment from the Gisburn road round to near Barnoldswick. Its junction with the underlying black shales may be seen at Weets, the base having several bands of shale interstratified with the sandstone. The actual junction between Weets and Barnoldswick was nowhere else seen, but grit and shale were seen in several places along the line within a few yards of each other. Near Gillians, however, the base is obscured by drift.

The upper boundary is less certain. Between White Hough Water and the road between Offa Hill and Stank Top, grit is seen in one place, with a hard feature near where the top should be. In the road above mentioned the junction is seen and the line is good to Brown Hill. Henceforth it is everywhere covered up by drift, and is to some extent conjectural. The base of the Kinder Scout Grit, which is pretty good to Blacko Hill, is some guide so far.

In numerous quarries on Stank Top Moor the grit is seen, in some places fine, in others coarse; sometimes both coarse and fine grits are seen in the same quarry. If the quarry is small, often none but rubbly grit is seen, and the beds at or near the surface have a dip which on closer inspection is seen to be not the true one. When the true dip is 30° or 40° the beds at top may dip as much as 50° to 80° or more, and often they may be seen to dip in the opposite direction to the beds below. This is always merely a surface derangement, and is probably due to ice action. The dips taken here show that there must be from 1,000 to 1,200 feet of grit.

Several sections are given in the road that leads from Lanefield by the end of Weethead plantation; but the best section by far is that on the county boundary given by Claude Clough and by Gingerbread Beck, which runs into it. The beds dip steadily and continuously to the S.E. and E.S.E. at angles varying from 20° to 35°, the average being 25° to 30°. Beginning towards the north end of Gingerbread Clough, opposite Claude, we get rather thin bedded grits, then massive grit, then shale is seen, afterwards rather coarse grit with shale in it. Then comes a gap, which may be shales; but sections in grit on either side are seen not a very long way from the beck. Lower down the clough we get grits—thick bedded coarse grits, grits, shales with thin ironstone band, grit, shale, pinkish grit, and grit again—without seeing the top. A carefully plotted section shows 1,400 feet of Yoredale Grit here without either base or top.

The grit is seen in quarries on Burn Moor and in the gully above

Higher Admergill. The dip is here less, 12° to 18°. The base of the grit is seen at the upper end of Greystone Plantation, opposite Greystone Inn, where several bands of shale are seen interbedded with the grit, as at Weets. We suppose a fault to range up Greystone Clough, the fault which is seen in a gully on the west side of the road, opposite Admergill Pasture.

On the eastern side of the section there exposed grit is seen dipping steadily about east by south, and on the western side are layers of reddish and black clay, much contorted and mixed with crushed stone. I have seen nothing in the district anything like the layers of different coloured clay; but Professor Ramsay, who saw the section, observed that as it was close to a fault, not much could be said about it. It is possibly stuff *in* the fault. Numerous quarries, most of them small, on White Moor show that the grit is dipping at low angles, 5° to 15° to the south-east; and subordinate features, which can here and there be traced for short distances, indicate thin and inconstant bands of shale. Towards Barnoldswick the dip increases, the beds shown in Moses Lee Clough dipping at angles from 20° to 30°, and among them are several bands of shale. There is a good exposure of the beds in the quarry E. of Moor House, which shows coarse and fine grits, the latter thin bedded and micaceous, with partings of shale, dipping east at angles from 8° to 12°. There is a rather sharp bend in the rocks about here; they strike in one direction nearly parallel to the road which leads by Fanny Gray Inn, and in the other direction towards Barnoldswick Park. There was some indication of a fault here, ranging about north 25° W.; but the evidence was not sufficient to justify me in putting it on the map. Behind Higher Park are quarries showing beds of grit, dipping S. 30° E. at angles of 20° to 30°, and striking at the Lower Yoredale Shales and Limestones seen in the Barnoldswick railway-cutting. This is one of the proofs of the existence of the great Barnoldswick fault, which here throws down the beds on the south against beds more than 2,000 feet below them. Large quarries near Park Close exhibit sections in rather fine evenly bedded grit, generally light gray or whitish. One quarry near the house has a capping of shale. The dip is here east by north 10° to 12°.

Shales between Yoredale, and Kinder, Grits.—These are seldom seen. Some shale is seen in the road between Offa Hill and Stank Top, and along the road leading to Brown Hill. The shales are here probably about 200 feet thick, but they must increase in thickness eastwards, as in the eastern side of 92 S.W. they are much thicker. Muddy shales with thin sandstones were seen north of Foulridge and west of the canal in a little stream east of Hey Fold, and shales are seen in the canal bank near Hollinhurst Bridge.

W. G.

LOWER CARBONIFEROUS ROCKS, EAST OF GISBURN.

The Carboniferous Limestone, east of Gisburn, consists of the main mass of Limestone north of the fault running from Bracewell to Broughton, and that of the Thornton anticlinal. The country about Horton and Bracewell is very drifty and affords few sections. Stock Beck gives no sections between Horton Bridge and Barnoldswick Mill, and no rock is seen in the county north and east of the beck for a width of more than a mile, except in a quarry near Stock Bridge, which has a thick capping of Boulder Clay. West of Bracewell there are some quarries near the brook which comes from Wedacre.

There is a good exposure of beds in Marton Scar, north of Gledstone. Here for a length of about three-quarters of a mile, and for a width of 500 yards, we have, what is very unusual in this district, a stretch of almost bare rock, with several parallel, somewhat broken, ridges, having their scarped sides to the west. These ridges have a curved outline, the dip at the south end of the scar being north of east, in the middle about east, and at the north end south of east. It is from 12° to 15° on the east side, but increases westwards to 25° and 35°. The beds exposed here must be as much as 500 feet in thickness, which is probably nearly all limestone, though there may be some subordinate bands of shale. The limestone is mostly of a light gray colour, though some of it is rather dark, and the bedding is generally regular and well defined. South of Gledstone there appears to be a synclinal, the axis of which ranges about north east and south-west. East of this in a band of country some 400 or 500 yards in width, stretching from West Marton through East Marton quarries to Clint's Delf, N.W. of Cobber Hill Plantation, we get a fair number of sections in limestone, with a pretty steady dip somewhat to the west of north. The limestone seen in the Marton quarries is mostly dark, rather thin bedded, and has in some places shale partings. There is a good exposure of beds at Clint's Delf. The lowest beds seen in the scar are whitish massive limestones; the beds which lie above, seen in the quarry, are generally thin and evenly bedded; they vary in colour from grey to black and have partings of black calcareous shale. Fossils are plentiful in the upper beds.

The country for a long distance north of this band is almost entirely obscured by drift.

The boundaries of the Limestone of the Thornton anticlinal are well seen at Rain Hall Rock and at Gill Rock, where the beds above the limestone are black shales, and the centre of the anticlinal is seen in the beck, north of St. James's Church, Barnoldswick, where the turn over of the beds is shown, though the boundaries about here are obscure. The course of the Limestone is clear about Thornton, from Thornton Hall Quarry to beyond Fence End, but the other side of the anticlinal is obscured by drift, and the only sections which offer themselves are in quarries at South Field Bridge, and at Langber. On both sides of the anticlinal the dip is very high, being at Rain Hall Rock from 50° to 75°, about Thornton from 50° to 70°, and at Gill Rock about 60°. The limestones are nearly always regularly and rather thinly bedded, but rather massive limestone is seen in Gill Rock quarry, and in a quarry near the canal, west of Rain Hall Rock, where the beds are vertical. No doubt there are bands of shale, though they are not so much seen as the limestones. The contorted limestones and black shale seen at the beck side, east of Gill Church, are probably some of the lowest beds anywhere seen in the country under description, brought up by the fault which runs through Gill Church. It is difficult to estimate what thickness of beds is brought to the surface in this anticlinal, owing to the minor contortions, and to the want of a continuous section, but it is probably quite as much as 1,000 feet.

The long and deep quarry of Rain Hall Rock, where the beds are followed along the strike, shows beds of blue and grey limestone resting on shale. Several "backs" or "horses," some of which slightly throw the beds, are met with, and in these calc spar and heavy spar are found.

In a small quarry near Coate Flat fossils are pretty numerous.

In the entrance to Gill Rock we get a section illustrated by the following diagram, which is roughly copied from the one in Phillips's Geol. of Yorkshire, Vol. II.

Fig. 6.

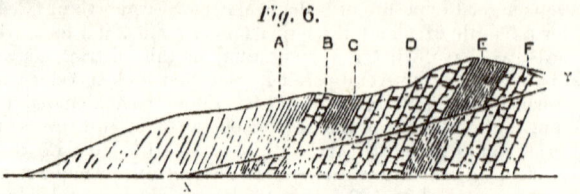

The uppermost beds are black shales, not well seen. These are marked A in the diagram. Below them are thin limestones marked B, and then come other black shales C. The limestone D has been taken as the top of the Carboniferous Limestone, though it is by no means certain that it is the same bed as that quarried at Rain Hall. It is thin bedded, and has been worked to some extent. Below this comes more black shale E, and then comes more limestone F, which is thicker bedded and has been extensively worked along the strike.

A fault which shifts the beds a few yards is seen in the section. Prof. Phillips considered the line X Y to be the inclination of the plane of the fault to the horizon, and he accounted for its low hade (about 11°) and its reversed throw, by supposing the fault to have been formed when the beds were horizontal; the hade would have been then about 50°, and the throw regular.

I think the above explanation unnecessary because I do not consider the line X Y to be the inclination of the fault. If it were so, it should appear in the beds exposed on the western side of the road, which are quite as high as those in which the fault is seen. But there is no trace of it on the west side, and if we go round behind F we find the fault coming nearly perpendicularly down the face of rock, so that the fault is really a vertical one, or nearly so, and nearly in the plane of the paper.

It must be remembered that the face of the section is far from being vertical, the beds below the line X Y being nearer the observer than those above the same line, which is nothing more than the line down to which the beds have been quarried *on one side* of the fault.

West of Barnoldswick the limestone has been largely worked at Limekiln Delf. It is bounded on the west by the Barnoldswick Fault, which here throws down the Lower Yoredale Grit against the shales immediately overlying the Carboniferous Limestone. The limestone here appears also to be bounded on the north by a fault, for there does not seem room for the whole of the limestone seen, to rise, turn over, and dip under the shales seen west of Hollins, and near Cow Pasture.

Lower Yoredale Shales and Limestones.—In the rudely triangular piece of ground, between the Thornton anticlinal and the Stock Bridge Fault, no rock is to be seen except on the western side of the area, between Hollins and Fools Sike. Here in the little streams, north and east of Kirk Clough, dark shales, and thin limestones, often impure, are seen dipping at very high angles, in other places vertical, or sharply contorted, with a general strike parallel to the limestone boundary, west of Brogden Hall.

It may be as well to give here the evidence for the line of fault running from Bracewell to Broughton, which I have just now called the Stock Bridge Fault, and for that of the branch fault running by Crickle House.

If we take a section east of Gisburn from near Moor Lathe to the north end of Kirk Wife Wood, a horizontal distance of 6,000 feet, we get limestones dipping steadily northwards at angles averaging 25° to 30°, and showing a thickness of 2,500 to 3,000 feet. This great mass of beds does not appear to turn over till we get south of a line between Moor Lathe and Yarlside, so that the limestones in the quarries near the B in *Bracewell*, and at Stock Bridge, must naturally be a long way below the upper beds. But unless there is a fault, they must be the continuation of the upper beds seen N.W. of Brogden Hall. Therefore a large fault must exist somewhere hereabout. Then again the undoubted upper beds of the Thornton anticlinal appear to dip under the beds about West Marton and East Marton, which are certainly a long way down in the series, and the anticlinal with its beds dipping steeply N.W. and S.E., strikes at beds which are all dipping one way slightly W. of North. Thus a large fault is wanted between Marton and the Thornton anticlinal, throwing down on the south as at Stock Bridge. The fault is nowhere seen, but the beds seen in Broughton Beck, in places between Broughton Mill and Dancliff Plantation, are a good deal disturbed, and the sulphur wells at Broughton and Crickle Spa may have some connexion with the fault. Professor Ramsay thought that the shales and impure limestones seen near the village of Broughton could not well be put into the Carboniferous Limestone, so the supposed fault going by Crickle House was drawn to separate them from the undoubted Carboniferous Limestone of Clint's Delf. Whether the limestone between these two faults at Broughton Fields is the upper part of the Carboniferous Limestone, or one of the Yoredale Limestones, it is not easy to say.

The Lower Yoredale Shales and Limestones S.E. of the Thornton anticlinal are best seen near Barnoldswick. Some of the lower beds (though not the lowest) are seen in Ousel Dale, and in the road going up to Hill Top, where there would appear to be very little limestone. The shales and limestones are seen in the cutting of the Barnoldswick Railway, north of Barnoldswick Park, and also east of the canal. The band of impure limestones seen in the latter cutting runs by High Close Hill and Keyfield Plantation. The dip is everywhere high from 35° to 60°, or even more, and the direction pretty steady to S.E. and S. by E. In the cutting of the railway near Barnoldswick Park the beds were found to be all turned over at top and ground up, as by a force coming from the north. This, in a country known to be glaciated, seems most likely due to moving ice. The apparently anomalous dip into the hill in the road by Thornton Hall, and again in the road leading from Thornton to the railway station, is perhaps of this character, merely a surface bend. In the cutting of the Barnoldswick Railway, north of Salterforth, a section of which the accompanying woodcut is a sketch,

Fig. 7.

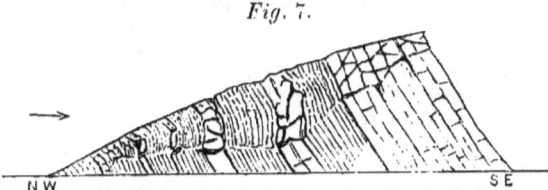

Section in the Barnoldswick Railway, showing Surface Disturbance. From a sketch by Mr. R. H. Tiddeman.

showed the shales and limestones bent at top, looking very much as if they had been forcibly bent upwards, and afterwards by pressure, perhaps that from the weight of the hill above, partly bent back again.

In the country north-eastwards as far as to Elslack, we get scarcely any exposure of these beds. In the lane east of Thornton, leading down to Brown Beck, a calcareous sandstone or very sandy limestone is seen, which cannot be far above the top of the Carboniferous Limestone. It is unusual to find such a bed on this horizon. In Broughton Beck, between Elslack Bridge and Broughton Mill, several bands of limestone and shale may be seen. There is a quarry in good limestone S.E. of The Grove, and limestone much contorted may be seen south of this in the railway cutting. It is compact, grey and dark limestone, with shale partings in places. Encrinites and shells are to be seen, but are difficult to extract. Owing to the few sections and the contortions, it has been found impossible to trace these limestones, but there is no doubt they belong to the beds under description.

The Lower Yoredale Sandstones can be well traced east of the Barnoldswick Fault, from the railway by Bawmier nearly up to the Gill Church Fault. The lower bed is best seen, being that which appears in the railway cutting, and which has been quarried east of Bawmier. Shales above this are seen in the railway cutting, and in a little stream near the o in "Keyfield Plantation," where also may be seen impure limestones and mudstones interstratified with the shales. A little further eastward than the latter place fine grey grit is seen, which is part of the upper bed of the Lower Yoredale Grit. Further eastward we see no more of the Lower Yoredale Grit, except in two places between Earby School and Booth Bridge, where we get small sections of it ; and it probably dies out, so that about Brown House we appear to get a series of shales and limestones, from 3,000 to 3,500 feet thick, between the Carboniferous Limestone of Thornton and the Upper Yoredale Grit of Elslack Moor.

It may be as well to give here the evidence for the great Barnoldswick Fault, which is a marked feature in the structure of the country. In the brook east of Kelbrook the beds are much disturbed, and more than one fault is seen, though which of them is the actual line of the great break cannot be determined. At Kelbrook we have the shales clearly seen to lie above the Upper Yoredale Grit thrown against the lower beds of the same grit, and the shales of the synclinal valley of Salterforth Moss abut against beds which are much lower in the series, and are all dipping one way. The outcrop of the Lower Yoredale Grit dipping at angles of from 30° to 40° is shifted a mile and a half from Salterforth to Limekiln Delf. The fault is seen in a little stream 300 yards S.W. of Barnoldswick Park, and near the quarries at Higher Park, on the line of fault, a mineral vein, probably of iron pyrites, has been struck. Near Hill Top the beds are vertical, and so they are between Springs and Lower Calf Hall. Near Limekiln Delf, on the line of fault, is Dark Hill Well, a very powerful spring, the water of which is important enough to have been the subject of a lawsuit. The evidence for the continuation of the fault westward has been given. Between Salterforth and Limekiln Delf the throw of the fault cannot be less than from 2,000 to 2,500 feet.

On the west of the railway between Sough Bridge and Earby Station is a grit quarry, which from its dip and position seemed evidently to be a part of the Upper Yoredale Grit, separated from the main mass to the east of it by a fault, which I suppose to run east of the quarry in a N.N.W. direction. Where this line crosses the beck to the southward disturbed grit is seen, and the line continued northwards runs east of Rain Hall, where the boundary of the limestone seems shifted, and on by Gill Church, near which the beds are much contorted.—W. G.

MILLSTONE GRIT SERIES.

The subdivisions of this series have been clearly determined. The rocks of this division form two sides of the triangular district now being described with the apex at Colne, and the base of (for the most part) the same beds brought up along the low arch which I have called "the Rossendale anticlinal." Following the order of description hitherto adopted, we shall commence with the basement beds of the series and proceed upwards.

Fourth, or Kinder Scout Grit.—This rock consists of two or more beds of grit, separated by shale. The grits vary in coarseness from massive conglomerate to ordinary sandstones, and over the top of the uppermost bed there occurs sometimes a thin coal-seam. The pebbles of this conglomerate consist of white saccharoid quartz, like vein quartz, but sometimes like the rounded grains of granitoid or gneissose rocks. The sandstones as a whole are largely composed of felspar, sometimes sub-crystalline, and small flakes of mica are plentifully scattered throughout, so that the whole resembles a granitic rock which has been pounded up and reconsolidated. Impressions of plant stems and pieces of carbonized wood are to be found in most of the beds, often in a procumbent position, as if drifted from a distance and imbedded. This is the case with all the grits of the Millstone Series.

The Kinder Scout Grit occupies a large area at the south-west corner of our district, stretching northward in the form of an extensive moorland, intersected by valleys from the northern slopes of Winter Hill to the banks of the Roddlesworth near Tockholes, and embracing Anglezark Moor, Withnell Moor, and Bromley Pasture. This area is bounded by two large faults; one on the west side passes by Rivington Hall, the foot of Stronstrey Bank, and the valley below Withnell, and may be seen in the gorge of Dean Wood near Dean Head Lane; also in the brook courses near the old Anglezark lead mines, and in the brooks and road-cutting east of Brinscall Row. Opposite Brinscall Hall the vertical displacement amounts to the entire thickness of the Millstone Grit Series, as the bottom beds of the lower Coal-measures are brought down on the west side against the base of the Kinder Scout Grit on the east, and probably falls little short of 4,000 feet. From this point northward the throw of the fault decreases.

The fault on the eastern side of the area occupied by the Kinder Scout Grit is the same as that which ranges through Bolton and along the Irwell Valley into Manchester, and known as "the Irwell Valley fault."[*] This dislocation passes below the bridge which crosses the brook at Belmont, where it may be seen; as also in a ravine which descends from Counting Hill and crosses under the Belmont and Bolton road. North of Belmont the fault runs in the centre of a wide valley, and its position can only be approximately inferred in the absence of sections till we reach the Roddlesworth, where its position may be determined by the up-tilting of the beds. At this place it joins the Anglezark fault, forming with it the apex of a triangle.

The basement beds are visible on Anglezark Moor, Stronstrey Bank, and Millstone Edge, consisting of coarse massive yellowish grit and conglomerate, current-bedded, and containing plant remains; above these are two or three beds not so coarse, and separated by bands of shale,[†] which are quarried at Hill Top, Belmont, Bromley Pasture,

---

[*] See "Geology of the Country around Bolton-le-Moors." Mem. Geol. Survey.
[†] In one of which I found *Goniatites* and plant remains in the brook above Higher Hempshaws, on Anglezark Moor.

Hoar Stones Brow, Coal Road Delf, and Coomb Farm. The coal seam which marks these beds as the upper limit of the Kinder Scout Series may be seen at the northern base of Rivington Hill in Dean Brook. There appear to be two beds, the lower a foot thick, the upper consisting of bass and coal 18 inches. Above this are a series of dark shales passing into grey, and becoming sandy upwards; these form the flank of Winter Hill, and lead up to the base of the "Third Grit" of the Millstone Series. The same coal may again be seen at the top of Grange Brook near Belmont; it is here 14 inches thick, resting on six feet of clay and shale, and was formerly worked.

The Kinder Scout Grit forms a long band of rising ground, in some places rocky and serrated, from the fault west of Hoolster Hill to the Nick of Pendle, passing by Mellor, Whalley Nab, and Wiswell Moor. The line is remarkably straight, trending E. 38° N., and all along this line the beds dip towards the south-east at angles varying from 25° to 60°, with an average of 35°. Sections are abundant.—E. H.

The most westerly point where this grit is well seen along the Clitheroe anticlinal is at Hoolster* Hill, two miles north of Hoghton Tower. There, near the top, is a quarry in coarse grit dipping south at 25°. Lower down the southern flank is another quarry in finer grit dipping at 55°. Further west these beds are probably broken by the same fault as that which is seen in the Darwen at its junction with Arley Brook, and which most likely also brings down the Trias against the carboniferous rocks in the Ribble near Alston Hall. Still further west, in ascending Hole Brook from the Darwen, we find a section in very disturbed beds, which is somewhat puzzling; it consists of shale with two or three very thin impure cherty beds of limestone containing fossils, *Encrinites*, *Brachiopoda*, *Goniatites*, &c. These rest on about 20 feet of fine hard thinly bedded grit, which is seen at first to dip south, then, turning round, west, and finally to the north-west at 70°, where it is terminated by a fault, bringing down black shale. At first sight these beds may be thought to have a very Yoredalean aspect, but considering their position below the Third Grits, and the fact that similar shales and limestones occur above the Fourth or Kinder Grit at Sales† Wheel in the Ribble, it is more likely that the sandstones are fine beds at the top of that division. Their arrangement here probably represents the turning point of the Clitheroe anticlinal.—R. H. T.

To the east of Hoolster Hill there are sections again at Ramsgreave Hall, and on crossing the valley of Knotts Brook we find the grit again on the hill above Hollowhead Farm, thrown forward by the fault in the valley. At Parsonage Farm there is a quarry in the rock which here contains veins of barytes. We find it rising in little rocky ridges, from Billington Moor to Whalley Nab, often as a massive conglomerate, and at Nab Wood it is opened out in a large quarry. On the north-east side of the Calder it occurs at Clerk Hill, and may be traced by numerous sections to the foot of Pendle. Its diminishing importance north-east of the Calder relatively to the Upper Yoredale

---

* Spelt Holster on the one-inch map.
† This obsolete word, like many others of North Lancashire, has been preserved by Spenser in his "Shepheards Calender" (for December), and is thus used :
  " To entrappe the fish in winding sale "
and in the contemporary "Glosse" thereon, we find " Sale or sallow, a kinde of " woodde like Wyllow, fit to wreath and bynde in leapes (baskets) to catch fish " withall." So that there is little doubt that this wheel or whirling pool, one of the best netting fisheries on the river, owes its name to the use in it of willow nets or fish traps. The Romans stationed at Ribchester hard by may have used "Salices" for that purpose.

Grit has been already commented on (p. 25). North of Sabden it is seen very clearly to consist of two principal beds, but this division does not seem to persist far along the strike in either direction. A very good section is to be seen in Cock Clough above Sabden Fold, the beds dipping from 30° to 40° to the S.E., and giving a thickness of 550 feet, the underlying shales being 100 feet thick. Beyond this the grit is let down between two north-westerly faults. About Newchurch-in-Pendle sections are very numerous. In two quarries to the S.W. of that village we find coarse massive reddish grit with much felspar, and higher beds consisting of coarse thick flags, and on the ridge to the north-east of it the grit is as coarse as any I have seen in the district containing many pebbles one inch or more in length.—R. H. T. & E. H.

Immediately west of Offa Hill the Kinder-Scout Grit is hidden by drift. Between Offa Hill and Bank Ends the rock is exposed in numerous quarries. It dips at angles of 40° to 50° to the south-east, and is almost everywhere a very coarse yellow-brown grit, in many places conglomeratic. A band of shale in it is seen in the road just behind Offa Hill, and above this shale occurs a red conglomerate. A small fault ranging with the strike of the beds was seen behind Hollin Top. Blacko Water gives no section in the grit, the Boulder Clay hiding everything. We get the grit exposed again in numerous quarries on Blacko Hill, where the dip is near S.S.E. at angles of 25° to 40°. A fault here seems to cross the ridge, which is probably the same as the one seen near Admergill Pasture, and in a quarry near the ordnance station is a fault shown in the accompanying sketch.

Fig. 8.

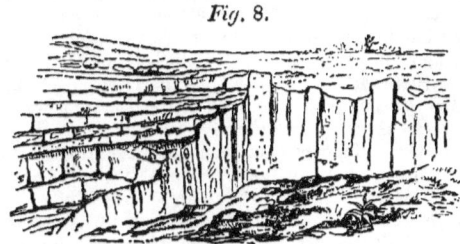

Fault in a Quarry on Blacko Hill.

East of Blacko Tower the Kinder-Scout Grit is again entirely hidden by drift, and when we again see it there are two very distinct beds of coarse grit with a thick band of shale between. These shales are well shown in a little stream at a point on the 1-inch map, just north of the second *c* in the word *Cocker Hill*. The section shows sandy and muddy shales dipping slightly east of south, at angles of 40° to 45°. About 125 feet of shales are seen, and there are probably more of them, as the junction with the underlying and overlying grits is not visible, and I estimate them to be from 200 to 250 feet thick further eastwards, near Foulridge Hall. In the little beck, south of the latter place, shales are seen for some distance dipping a little west of south, at 35°. The upper bed of coarse grit and conglomerate is well seen at Cocker Hill, and the lower grit bed between Foulridge Hall and Dobers. The lower bed I estimate to be 350 feet thick, the upper 200 feet; so that the total thickness of the Kinder Scout Grit is here from 750 to 800 feet. At Blackfield is a curious ridge, which as shown by the shading on the 1-inch map seems to run obliquely to the strike of the beds. It is not, I believe, a drift mound, for one or two poor sections showed grit which seemed in place.

One would think that the trend of this grit ridge certainly indicated the strike of the beds, but from what I could see in the sections it does not. If this were here the strike of the beds, there must be a fault along the western end of the ridge, at the bend in the road, and the base of the Lower Kinder Grit would be thrown nearly against the middle shales seen to the west in the beck. As there was not good evidence for a fault, and the steady dips showed there could be no sharp bending, it was mapped as shown, with Prof. Ramsay's concurrence.—W. G.

Sabden Valley Shales.

Overlying the Kinder-Scout Grit is a thick series of shales, with sometimes bands of grit. On the northern flanks of Winter Hill these shales have a thickness of about 350 or 400 feet, but in the Pendle range it is considerably greater. On the borders of Yorkshire they form the flanks of Way Stone Edge.—E. H.

These shales have been already alluded to as being seen in Hole Brook, west of Hoolster Hill; they may also be seen in the river Darwen, below Samlesbury Mill, where they show a thickness of 625 feet without coming to a base. The entire thickness is probably double this at least. At the mill they are dipping at about 40° to the south. A little lower down the river is a band of ironstone nodules containing fossils well preserved, and forcibly reminding one of a band similar in character and position at Roughlee, east of Pendle. These shales may be seen at intervals all along the Darwen Valley, which is excavated along their strike, nearly as far as Roach Bridge, but there they are covered up by Permian rocks. They seem throughout to contain fossils, *Goniatites*, *Posidonomya Gibsoni*, fish, and in places plants. Thin beds of hard fine cherty grit occur here and there, as in the second northerly reach of the river from Samlesbury Mill, and again at the bend of the Darwen, west of its junction with Hole Brook, and there they form a sharp anticlinal, with some traces of disturbance. They are not seen again north or west of this until we come to the Ribchester synclinal at Sales Wheel. There they may be well seen resting on the Kinder Grit on both sides of the river, and contain many fossils and thin Limestone bands made up of encrinites. They appear again with ironstones under the bridge at Ribchester, on the north side of the river, but are not well exposed. At the bend of the Ribble, south of Ribchester, where it turns to the west, we find black clayey and sandy shales, with ironstones, *Posidonomyæ*, *Goniatites*, and *Calamites*. They seem much disturbed, but the general dip indicates a synclinal trending E.N.E. This is the most westerly point where they are seen.

Returning to the south side of the Clitheroe anticlinal we find them, in dry seasons, exposed in the bed of the Calder, near Whalley, just north of the M of Moreton Park, where there are clayey shales with thin limestones, and a little further south with ironstones. The dips are high, from 40° to 70°, and they cannot here be less than 1,900 feet in thickness. Shales near the base of this series are again seen, but obscurely, in a small watercourse, half a mile due east of Wiswell Moor Houses. The underlying Kinder Grit is dipping at 55° to the S.E., the shales themselves are disturbed, and do not give a true dip. Between the site of Wiswell Mill and Sabden many sections in the higher part of these shales are seen along Sabden Brook; they consist of sandy shales with a few beds of sandy plate and sandstones. Ironstones also occur in places, but I was unable to find any fossils, nor have they that general appearance which elsewhere leads one to expect them.

The lower part of the shales is again seen at the side of Thorneyholme Millpond, near Barley. They are dipping south-east at 55°, and contain fossils and ironstone nodules. At Roughlee, east of the Old Hall, they are again found in the brook, beneath a high bank with a bed of grit running along it. These must be near the top of the series. They contain *Pecten* (which I have not seen elsewhere on this horizon), *Goniatites*, *Orthoceras*, and several species of Gasteropoda. Mr. John Aitken, of Bacup, F.G.S., with whom I first had the pleasure of visiting them, has made a good collection from this spot.*—R. H. T.

Only two sections were seen between Rough Lee Water and the road from Colne to Foulridge. Near the western end of the large canal reservoir, in the road immediately north of Wauless House, some coarse grit is seen, and a little further north shale dipping nearly south at an angle of 30°. Also small sections in grit and shale are seen a little south of the same reservoir, near the top of the letter *k* in *Back of Colne Edge*. The whole thickness of these shales, with subordinate grit beds, must be here about 2,000 feet. They seem to have been cut through in making the Foulridge tunnel, nearly a mile long. The only reference to this I know of is in Phillips' Geology of Yorkshire, Vol. II. p. 74–5, where it is said that the *Yoredale* shales were found in the tunnel. But this is certainly a mistake, the beds cut through in the tunnel must lie *above* the Kinder Scout Grit of Cocker Hill, and not *below* it, as Prof. Phillips thought, and as given in the map accompanying his work. —W. G.

Third Grit.

The shales are surmounted by the sandstones and conglomerates of the *Third grit* itself, in two, sometimes in three, beds separated by shales. At Belmont they are well shown along the brook below the reservoir, and, under the name of "the Ratchers," occur as massive cliffs of conglomerate. Further down the stream, under the Hall Wood, the lower bed of rock may be seen dipping southward beneath the shales which separate it from the upper. The uppermost beds are again shown in the river Roddlesworth, below Tockholes, in the form of fine grit and flagstones, and also below the embankment of the new reservoir, where they are seen to be traversed by a fault which is marked on the map.— E. H.

East of the N.N.Wly. fault which runs past Chorley and Whittle-le-Woods the third grit appears in a few places along an anticlinal which is a continuation of that of Clitheroe. The upper bed only is visible: this is first seen at Radburn, in a small brook above the canal. It is dipping at an angle of from 5° to 10°, and consists of fine grit. It is also seen in Gorton Brook, at the same low angle; and at Seed Lea, N.N.E. of the canal terminus. In the lower part of Gorton Brook, near Duckworth House, the beds are seen turned over to the N.N.W. at an angle of 20°; E. of this a fault intervenes, which throws down the overlying shales, and somewhat shifts the axis of the arch. The next brook (Mill Brook) has just cut down to the top of the grit again, S. of Jack Green. It is again thrown up E. of this by another small fault, but does not show any clear section for some distance. In Mill Brook the shales above it contain ferns, goniatites, and other fossils.— R. H. T.

Along the Pendle Range the Third Grit consists of two and sometimes three beds, divided by shales, the lower of which is often very massive

---

* See Trans. Manchester Geol. Soc. part 2, vol. vii., On an excursion to Clitheroe and Pendle Hill.

and coarse, passing into conglomerate, and current-bedded to a greater degree than the upper.

There are good sections in the river Darwen, south-east of Salmesbury Bridge (see Fig. 9). The beds consist of grits of various degrees of

Section in "Third Grits" at Salmesbury Milldam.

a Hard massive grits contorted. b Principally shales contorted and faulted.
c Grits passing under shales.

coarseness, and of colours varying from grey to yellow and purple. Near the junction with Arley Brook the dip is reversed owing to a large fault which ranges nearly north and south, and has already been alluded to as probably crossing the Kinder Scout Grit, west of Hoolster Hill (see Fig. 10). Other sections are shown in the banks of the Arley

Fig. 10.

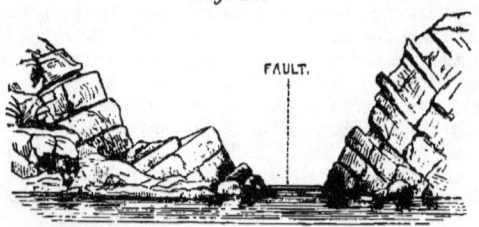

Fault in the Darwen, near Arley Brook.

Brook, in Woodfold Park, in the lower bed; while the upper forms a fine cliff, partly artificial, called Alum Scar, resting upon a thick series of dark blue alum-shales, from which this mineral was formerly extracted. It dips south at 15°, and contains nodules of ironstone with shells.* Sections in the upper bed may be seen in the wood behind Pellmell, and in the lower at Shorrock Ilcy, and further east in the brook by the road side above Yew Tree Inn. Here the dip is S.E. at angles ranging from 35° to 45°.

In the cutting of the Clitheroe and Blackburn Railway near Brown Hill Toll Gate the lower bed of the Third Grit is again laid open, and is seen to be violently contorted (Fig. 11, p. 39). The beds consist of coarse and fine grits irregularly bedded, sometimes concretionary, with bands of shale. Some portions appear to have been pushed laterally over others. This disturbance is doubtless owing to the proximity of a large fault, that traverses the beds transversely from N.N.W. to S.S.E., but which is nowhere actually exposed to view in sections.

---

* As I am informed by Captain Aitken, F.G.S., of Bacup.

Proceeding towards the N.E. along the line of the strike, we find a ridge of the Third Grit at Bank Hey, dipping S.S.E. at 50°; also at

Fig. 11.

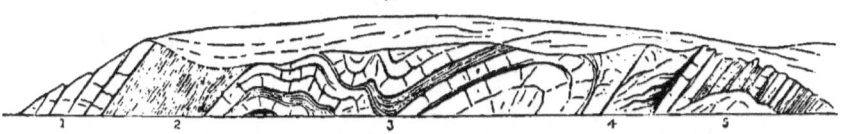

Contorted Beds in Clitheroe and Blackburn Railway.

1. Hard grits.   2. Shales.   3. Contorted grits and shales.
4. Position of fault.   5. Grits and shales.

Height House, Windy Bank, Top o' th' Heights, Black Low. Here it is thrown further north by a fault which brings the upper bed nearly in contact with the Rough Rock. North of Harwood both of the beds form prominent ridges, dipping S.S.E. at 50°, terminating along a fault. The lower bed is a conglomerate. There are good sections also in Sabden brook, above its confluence with the Calder. The lower bed breaks out in the form of massive coarse grit below Read New Bridge, dipping to the S.E. at 45°, under shales with flags and tilestones and two or more thin coal-seams. The upper forms a bar of coarse grit crossing the brook below Cock Hill.—E. H.

The course of these grits may be traced along the southern flanks of the Sabden valley. At Black Hill the lower bed is very massive, thick, and coarse; and at some distance above it occurs a very fine hard siliceous grit, with a coal seam overlying it. This grit is used for road-mending, and is sometimes erroneously called "gannister." It may be seen in the above section in Sabden Brook, where it occurs in two beds with a 3-inch coal seam between. The floor of this seam appears to be variable, for a little further on, in the road between Cobcar Nook and Tewit Hall it presents a section as follows :—

	ft.	in.
Clayey shale.		
Hard fine grit	5	0
Coal	0	5
Fireclay	0	2
Grit.		

Following it along the strike in a wood half a mile distant, we find the seam to have thickened out to 18 inches, and here it seems to rest upon the grit itself.

This grit must not be confounded with a similar bed which occurs *above* the upper Third Grit at Lower Spen, resting upon a 6-inch coal seam, with a floor of clay and shale.—R. H. T.

Between the Calder valley and that of Barrowford Beck the Third Grit has been traced, chiefly by Mr. Tiddeman. It may be seen in section at Hill Top, the Rig of England, Spen, and Ridgaling, where it is crossed by a fault. East of this fault the grit forms a fine cliff overlooking the valley of Rough Lee Water. It again appears at Stone Edge and Colne Edge, where it is a conglomerate, and projects in bosses which have somewhat the appearance of *Roches-Moutonnées*. The dip is here due south, as the beds here form the northern rim of the Burnley basin.—E. H.

Along the eastern flanks of the basin the grit appears on the western slopes of Boulsworth, crossing Thursden Brook at Copy Bottom. The lowest Third Grit seen in Black Clough (the upper course of the

Don about Robin Hood's House) is very coarse and massive, and in some places quite a conglomerate; the upper beds, however, are finer and somewhat flaggy. It dips steeply to the west, and must be several hundred feet thick. The finer top beds of this grit are again seen at the county boundary near Widdop Cross. Above this come beds of sandy shale, seen in the road near Widdop Cross, in Birkin Clough, and in Black Clough; and above these shales comes a bed of sandstone, fine towards the base, where are several shale bands, but coarse at the top, east of Copy Bottom. This band of grit is scarcely traceable southwards beyond Birkin Clough Head, passing apparently into tiles and sandy shales. Above this comes shale, in places black, with flaggy micaceous sandstone in it. These are seen in the little stream east of Copy Bottom, and in the road leading from Thursden to Widdop, opposite the bend in Black Clough, where also may be seen a coal seam 6 to 8 inches thick. The next Third Grit above is fine at bottom, with a considerable shale band some distance up in it. The upper part, seen at Hanson Fold and east of Cockridge, is coarse. It may be traced southwards by Hazel Edge towards the Vale of Cliviger, and northwards y Antly Gate to Oaken Bank, where the coarse upper beds have ecome fine and flaggy, so that the whole is uniformly pretty fine. At Oaken Bank it is thrown down by a fault, seen at the junction of the becks S.E. of Alder Hurst End. In the shales between this and the Rough Rock a thin bed of hard fine grit, with a coal at top, is seen in Swinden Clough.—W. G.

The upper bed may be traced south along the escarpment of Hazle Edge, dipping W. at 30°. Then still further south across Stiperden Moor, crossing Pudsey Clough at Stiperden House. Along this clough good sections are shown, both below and above the grit up into the Rough Rock. The general dip is to the south-west at 35° to 45°

All along the western base of Black Hambledon, the lower of the two beds of grit has been concealed by the great anticlinal fault, which ranges through Tormorden. But on approaching the valley it again appears, forming the fine cliff of Hartley Naze, where it is quarried. The base of the cliff is traversed by the great Cliviger Valley fault, which here forms a junction with the anticlinal fault of Todmorden, and having an upthrow to the south, thrusts forward the Third Grit for a distance of about a mile further up the valley on the south side. From this point southwards the Third Grit, consisting in the main of two beds, assumes an importance in the geology of the district nowhere else accorded to it. Bounded on the eastern side by the great anticlinal fault, and traversed by several longitudinal and transverse fractures, which have the effect of repeating the same beds, once, twice, or thrice; cut deeply into by the long canal-like valley of Todmorden, with its branches stretching into the ridges on either side, the Third Grit gives rise to a series of rocky escarpments, ridges, slopes, and valleys seldom surpassed in picturesque effect; and presenting to the physical geologist a problem as regards identification requiring all his skill to solve, and which indeed could only be solved by the process adopted by the Geological Survey of tracing out in detail each separate bed of rock and shale.

The lower bed is very massive and coarse, sometimes a conglomerate, but not so truly a conglomerate as the beds of the Kinder Scout Grit of Walsden Moor and Blackstone Edge, with which it has sometimes been confounded. It is, however, a good building stone of a white or light brown colour near Todmorden, and yellowish or reddish further south. It forms the cliffs of Robin Wood, Mellings Clough, or Gorpley Wood, and by a fault which passes by Gorpley Mill is brought in again at Gauxholme. On the east side of the valley it is quarried at Knowl Wood.

The same bed again occurs at New Bridge, Warland, and forms the escarpment of Reddyshore Scout.

The upper bed is not so coarse as the lower, and is sometimes separated into two by a thin bed of shale, as at Clough Foot in Dules Gate Valley. Near the top of it there is often a thin seam of coal overlaid by black shales, which may be seen at Clough Foot and Ramsden Clough a foot in thickness. This is the rock which forms the cutting and entrance at the southern end of the railway tunnel north of Littleborough, and which presents a good example of a system of parallel curved joints, ranging N. and S. The rock is cut off by a large downcast fault a short distance south of the tunnel.

The Third Grit again occurs in Rossendale, forming the lower part of the valley from Waterfoot downwards nearly to the junction of Balladen Brook, where it is terminated by a large fault. There are two beds, separated by shales, and on the top of the upper bed there is a little coal, which may be seen in Cowpe Brook, along the banks of which the grit is brought up by the fault above referred to. Sections may be seen at Lumb Holes Mill, Fall Barn Bridge, and at Lower New Hall Hey, both in the railway cutting and in a quarry.

In Scout Moor Brook the top of the grit with its accompanying coal-seam, here having the appearance of cannel, is brought up along the side of the fault which descends along the valley.

The Third Grit occupies the bottom and both sides of the river Irwell from Holden Wood Reservoir to Ramsbottom, where it is again thrown down, and concealed by a large fault which passes under Fletcher Bank Quarry. At Holden Wood Bridge the following section is exposed to view at the weir :—

Section at Holden Wood Bridge. Third Grit.

	ft.	in.
1. Irregularly bedded fine grit	8	0
2. Coal-seam, inferior quality	1	0
3. Shale and stony clay	8	0
4. Irregularly bedded grit	15	0

At the "Trippet of Ogden" massive grit is seen resting on dark shale, the dip being N. at 5°.

The lower bed is very massive, often coarse grained or conglomeratic, and produces an excellent building-stone. In colour it varies from light-red and grey to white, and does not stain on exposure. Quarries are open in it at Fletcher Bank, Bank Lane, Stubbins, &c. It has been employed with success in the restoration of Manchester Cathedral.

The upper bed spreads over a large tract of ground at Shuttleworth Moss and Harden Moor, and along the bank of Cheesden Brook the little coal has been worked.* Here it has a black shale roof, and varies from six inches to one foot in thickness. At Deepley Hill the beds assume the form of a sharp anticlinal broken by a fault along the axis.

The same beds are again shown in Holcombe Brook with three overlying coalseams, called by Mr. Binney "the Holcombe Series." The old workings of the coals are to be found along the banks of the brook.

---

* Mr. Binney, F.R.S., says of these beds, "Immediately above this rock occurs a group of shales of a dark colour, containing generally three thin beds of coal, the lower of which is worked at Cheesden Bridge, where it is about 15 inches thick. The roof of the highest of these seams contains an abundance of shells of the genera *Aviculo-pecten*, *Goniatites*, *Posidonomya*, &c., together with scales of fishes."—Trans. Geol. Soc., Manchester, vol. ii., part 7.

The series as given by the above author is as follows, in descending order :—*

	ft.	in.
Black shales, with *Aviculo-pecten, Goniatites, Posidonomya*	111	0
Top coal - (nearly)	0	6
Black shale	6	0
Middle coal	0	8
Black shale	24	0
Shale, with layers of stone	18	0
Bottom coal -	1	3
Dark shale	12	0
Millstone Grit	-	

A fine section in beds of this series is shown in the banks of the Irwell at Summerseat with one of the above coals. They are again seen in Bradshaw Brook below Turton reservoir.

The Second Grit or Haslingden Flags.—The shales which overlie the upper bed of the Third Grit, and which contain one or more thin coal seams in their lower part, pass upwards into grey sandy shales with or without flagstones. When they *do* occur they constitute the Second Grit Series or "Haslingden Flags."

These beds consist of greyish fine-grained micaceous flagstone and freestone evenly bedded, and sometimes ripple-marked. Their surfaces are also frequently impressed with annelid markings and carbonaceous matter. They are very inconstant, passing often into sandy shale. Sometimes this change is so gradual that it is difficult to tell whether the beds ought to be called " shales " or " flagstones ;" as an instance of this I may adduce the beds which underlie the Rough Rock above Newchurch in Rossendale. While on the north side of the valley they assume the indeterminate character of " strong sandy shale," on the south side they form the excellent flagstones of Cowpe Moss.

These flags are very largely quarried in the central part of our district, along the flanks of Rossendale Valley, where they occur in two thick beds. The upper one of these forms the isolated plateau of Cowpe Lowe and the escarpment of Cowpe Moss; while the lower is quarried at Whitaker Pasture at the base of the hill, and shows the following section :—

	feet.
Grey shale -	30
Sandy shale and flags	30
Grey flagstone	35

The total thickness of shale between the two beds of flagstone there is about 150 feet.

Along the Pendle range north of Blackburn we generally find these flagstones lying about 15 or 20 feet below the bottom of the Rough Rock. They may be seen at Billinge End quarry dipping S. 20 W. at 40°, and again quarries in the hill side from Little Harwood to Harper Clough. In many of these sections the ends of the beds are curiously turned over in a direction opposite to the dip. They may also be seen in a quarry in Read Park, to which reference will again be made. From this point, however, towards the north-east in the direction of Colne these flags are either absent or but sparingly represented.

On the eastern borders of the Burnley Basin we find feeble representations of the flagstones along Thursden Brook just under the Rough

---

* " The Lancashire and Cheshire Coal-field, &c."—Trans. Geol. Soc., Manchester, vol. i. p. 79.

Rock, and again in a quarry west of Hazel Edge, on the north side of Cant Clough Beck; but in the section along Swinden Water, which is intermediate between these points, there is no trace of these beds. This fact is sufficient to show their inconstant character in this district.

Further south we find the flags quarried on the north side of the Portsmouth Valley at Red Water Clough, but all along the base of the Rough Rock from Thievely and Calder Head southward to near Littleborough they are either altogether absent or but feebly represented. This I attribute not to any wedging out of the Rossendale or Haslingden flagstones, but to an actual change of these beds into sandy shale.

The two beds of flagstone are well developed in the Whitworth Valley north of Rochdale and on Ding Moor. I have already described them at Cowpe, and from thence we trace them southward to the western slopes of Scout Moor, but here the lower bed disappears and the upper becomes of less importance.

At Haslingden the lower bed alone appears except at Pike Lowe, and we trace this bed at intervals on both sides of Ogden Valley, joining on the southern sides the plateau of Musbury Heights and Tor Hill. We also find both beds, though sometimes composed as much of shales as of flagstones, outcropping around the western and southern flanks of Holcombe Moor.

There are large quarries in the upper bed at Edgworth, and sections are opened up in both along the railway north of Entwistle station.

North of Rivington the Haslingden flags are very well developed, and are well shown in the valley below Anglezark Lead Mines and on the western and southern slopes of Rivington Pike. They are also seen at Tockholes and in the river Roddlesworth.

First Grit or Rough Rock.—This is the uppermost member of the Millstone series, and forms an easily defined margin for the whole area of the Coal Measures. It is often called "the Sand Rock," from the facility with which it can be ground down or weathers into sand, and it often has a thin coal-seam lying on its upper surface or imbedded in the upper portion of its mass. This coal is known as "the Sand rock coal."*

The Rough Rock consists of a coarse-grained grit passing into conglomerate of white quartz pebbles which reach, but never exceed, the size of a bean. The particles of quartz are held together in a base of felspar, with which flakes of mica are mingled. The rock is often soft and crumbly (though not universally so), and is traversed by planes of oblique lamination, or current-bedding. It contains fragments of carbonised plants and trees. In colour it is generally yellowish, but sometimes reddish or white. It is often quarried for building purposes.

All along the southern slopes of the Pendle range, from Whittle-le-Woods to Colne, the Rough Rock rises at high angles from beneath the Lower Coal-Measures, forms generally a ridge or rocky bank; but in a few places it is of so soft a nature that it produces no feature; and so far from obtrusively thrusting itself on one's attention, requires to be diligently sought after. This is the case north-east of Fence.

At Hoghton Towers the Rough Rock rises into the form of a lofty hill, overlooking the country both to the north and south, and terminated on the east side by a deep ravine, through which flows in a rocky channel the river Darwen. It consists of massive yellow and purple

---

* Also called "the feather-edge coal," a name given it by Mr. J. Hall, and adopted by Mr. Binney, descriptive of the feathery fracture which the coal exhibits in some districts, but not always.

grit, which can be extracted in blocks of the largest size, and in the quarry at the eastern end of the cliff shows a face of rock of about 150 feet in height. The dip is 15° E. of S. at 20', and the rock attains a total thickness of about 400 feet;—a thickness surpassing that in any part of Lancashire, or probably the north of England.

On the southern side of the Blackburn trough the Rough Rock reappears, rising toward the south-east at Pike Low, Withnell, and Stanworth Edge, crossing the valley of the Roddlesworth at Stanworth Wood. At Finington Brook it is crossed by a fault having a downthrow to the east.

The course of the Rough Rock may be traced along the heights above Pleasington, Billinge Hill, and Revidge, where it presents the appearance of a sloping wall of conglomerate at the head of Blackburn Park. Here the dip is S.S.E. at 40°, while further on at Royshaw Hill the angle is 45°–50°. On the east side of Harwood Valley the rock may be seen in a quarry above Peacock Row dipping under grey shales of the Coal-Measures. Along the ridge above Harwood it may again be seen in the little rills which descend the hill, or in quarries which are worked principally for the sake of the Haslingden flags. At Harper Clough quarry the "sand-rock" coal appears; it is of impure quality, and rests on a bed of fire clay which is used for the manufacture of bricks and tiles. This coal is again seen at Harwood Edge.

At the eastern end of Read Park the late Mr. Fort opened out a very interesting section in this rock and the underlying beds, having cut a road through the cliff of Rough Rock into the underlying flagstones. This section, of which a sketch is given, is particularly interesting, as it shows the base of the Rough Rock resting on an eroded surface of the shales.

Fig. 12.

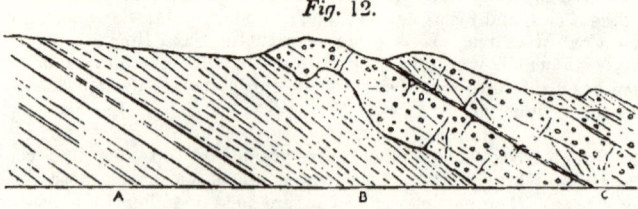

Section in Read Park.

A. "Haslingden flags." B. Sandy shales.
C. Rough rock or first Millstone Grit.

At Ridgaling the Rough Rock is terminated against a fault which throws it down on the S.W. Nor does it appear behind Higherford owing to the presence of a large fault which brings the Coal-Measures against the Third Grit. On crossing a large fault which ranges nearly north and south we find the rock again in the valley west of Colne, forming the ridge on which that ancient town is built. A good section is shown in the brook. South of the town, and below Winewall Bridge, the junction of the Rough Rock with the Lower Coal measures may be seen.

The Rough Rock is quarried at Winewall, and may be traced at intervals southward forming the ridge above Stag Hall, until broken off by a fault at Oaken Bank. By this fault the beds are upheaved to the south, and crop out along the natural cliffs of Deerstone Moor. The dip is here S.S.W. at 10°, but at Willy Moor the dip changes, and the beds dip to the W. along the ridge of Red Spa Moor, crossing Thursden

Brook at New Bridge. There they meet with a fault by which they are thrown down on the south-west about 40 yards, and they may now be traced in a due southward direction forming a rocky ridge across Entwistle Moor, Wether Edge, Sheddon Edge, and Stiperden Moor, as far as Portsmouth Valley, where they terminate in a fine wooded cliff forming the western bank of Red Water Brook.

At the foot of this cliff the Rough Rock is terminated against the Cliviger fault by which it is upheaved on the south, and forms the cliffs along the side of Thieveley Scout. The throw of this fault opposite Cornholme is probably not less than 400 yards.

From Thieveley Scout southwards the Rough Rock may be easily traced over the moors by Heald Moor, Flown Sear Hill, Lumb Foot in Dules Gate (where it contains "the sand-rock coal" eight inches in thickness), across the upper part of Howroyd Clough, over Oatley Hill, Weather Hill, and Shore Moor, and is ultimately terminated at Long Clough, north of Littleborough, against a north and south fault passing by Clough Mill. This fault may be seen in a little dell west of Dearden's Pasture. All along this easterly outcrop the Rough Rock has an average thickness of 70 feet.

Over the centre of the district the Rough Rock is spread over considerable tracts owing to the horizontal position of the beds along the Rossendale anticlinal arch. It forms the flanks of Whitworth Valley, spreads over the plateaus of Brandwood Moor, Coupe Moss, and Knoll Moor. Over this area it is accompanied by the sand-rock coal, which is sometimes workable, as at Tonacliffe, where it is 17 inches thick, at Woodhouse Lane, and north of Ashworth. On the other hand we find it thinning down to four inches on the southern bank of Fern Isle Brook about a mile west of Whitworth.

At Bacup the coal occurs near the centre of the Rough Rock instead of on the top, as at Knoll Moor near Rochdale. It varies in thickness from 12 to 18 inches, resting on a rough fireclay; the coal is of inferior quality. Sections may be seen in Sheep House Clough, and quarries both on the east and west sides of Bacup Valley.

The sand-rock coal has been rather extensively worked on Scout Moor above Edenfield. It occurs about the centre of the grit which forms the crest of the ridge, and has been got both by tunnelling and pitwork. On the west side of the moor the roof is coarse grit, and the outcrop may be seen at the head of Gate Brook. Further east, however, at the top of Grane Rake the roof is shale, and forms the base of the Lower Coal-measures. The average thickness of the coal on these moors is about 15 inches.

The Rough Rock forms the upper escarpment and part of the table-land of Holcombe Moor, and near its centre we find the sand-rock coal, which may be seen cropping out in the cliff above Holcombe village, where the following section is shown :—

		ft.	in.
Rough Rock	Coarse-grained grit	several.	
	Coal	1	4
	Stony under-clay	3	0
	Coarse yellowish grit	20	0
	Fine-grained greyish flagstone	15	0
	Gray sandy shale		

The outcrop of the coal is obscured all along the western escarpment, but it may be seen in several of the watercourses which descend from Wet Moss in a north-easterly direction. At Fall Bank Clough the coal is two feet thick and has a sandstone roof.

The grit is well shown at the southern end of Entwistle tunnel, forming in one part the face of a fault. It also forms a band round the base of Turton Heights, and along the eastern and northern flanks of Darwen Moor. We find it again at Bunker's Hill, south of Blackburn, where the sand-rock coal has been worked, and from this it may be traced southwards by Tockholes Fold, Higher Hill Delf, to Ryal. We also find the same rock forming the escarpments of Haslingden Moor and Cribden Moor, and running up the valleys of Goodshaw and Baxenden, where it is terminated by a downthrow fault which crosses the valley north of the quarry above Baxenden railway station, bringing in the Gannister coal. For further details regarding the position and nature of this rock at various points along its course I must refer the reader to the Geological Maps themselves.—E. H.

The Country on the Eastern Side of Sheet 92 *S.W., bounded by the Great Boulsworth Fault on the south, the Trawden Valley Fault on the west, and the Valley in which runs the East Lancashire Railway on the west and north.*—By W. GUNN.

This district comprises two anticlinals and two synclinals, ranging in a general south-west and north-east direction. Commencing on the north, we have first the Carlton synclinal, between Carlton and Kelbrook, comprising the Yoredale Grit of Carlton, Elslack, and Thornton Moors, rising to an elevation of 1,274 feet above the sea at the beacon on Pinnaw. The extensive lowlying moss between Salterforth and Foulridge may be considered a detached part of this synclinal shifted westwards by the great Barnoldswick Fault. Next we have the Lothersdale anticlinal, the axis of which ranges from Roger Moor by the limestones of Dowshaw and Raygill Delfs in Lothersdale to Park Head Quarry and Carlow Beck. The Reedshaw Moss synclinal lies between Trawden and Ickornshaw, and is but an easterly extension of the Burnley basin detached by the Trawden Valley Fault. The remaining anticlinal may be called the "Watersheddles anticlinal," and lies between Combe Hill and Crow Hill. It has not so well marked an axis as the others, the country being much disturbed and broken by faults. Still, if a line be drawn from a point somewhat north of the ordnance station △ 1,501 on Crow Hill to a point somewhat south of the Wolf Stones, △ 1,454 on Combe Hill, we shall find it will separate nearly all the north and west dips from those which point to the south and east. This axis is certainly a north-easterly extension of that of Heyslacks Clough, on the southern side of Boulsworth, and it may be considered as the termination of that long line of elevation marked by the anticlinal fault which has been traced from beyond Leek in Staffordshire ; for in the country to the northward the anticlinals and synclinals are clearly traceable to the Pendle system of disturbances. It is noticeable that both the anticlinals referred to above correspond to valleys, while the Reedshaw Moss synclinal is a physical basin and the synclinal of Carlton Moor is hilly ground.

The Carlton Synclinal.

This is made up entirely of Upper Yoredale Rocks, the main mass of hilly ground being occupied by the Upper Yoredale Grit, with the Bowland shales and a band of limestone cropping out on its flanks.

The *band of limestone* is nearly on the same geological horizon as the Lower Yoredale Grit in other places, lying at the base of the black Bowland shales. It is always a thinly bedded, generally impure, limestone, much mixed with shale in some places, and in other places very cherty, *e.g.* in the brooks near Yellison House. It is generally of a grey colour. Sections in it are given by most of the brooks between Elslack and Carlton, and there are some quarries in it, though it does not seem to have been burnt for lime. A quarry by the side of the road between Elslack and the Free School will show the thinly bedded character of the limestone, here mixed with shales and a good deal contorted. The upper and under lines of this limestone band as drawn on the map do not probably keep everywhere to exactly the same geological horizons; but they mark out a strip of country where the limestones in the shales are more than ordinarily abundant. This band corresponds to the Upper Limestone of Lothersdale, which will be noticed further on.

The Bowland Shales.—These are well shown in most of the becks that run down from the Yoredale Grit moors between Earby and Carlton, the best sections being in the beck north-east of Thornton Wood and in the two branches of the Ridge Beck, west of Yellison House. They are dark, often thinly laminated, shales, and contain in places towards their base mudstones and impure limestones. About Thornton Wood and Elslack they must be as much as 1,000 feet thick, but nearer Carlton the thickness is considerably less. This may be but an apparent decrease, however, due to a somewhat conventional base line, as mentioned just now. The upper part of the shales and their junction with the overlying grit are well seen in the road leading from Booth Bridge to Mount Pleasant. In some places, as in the becks near Yellison House, the shales immediately below the base of the grit are full of large lenticular ironstone concretions. South of Earby these shales are entirely hidden by drift and alluvium.

The great spread of *Upper Yoredale Grit* between Kelbrook and Carlton has a well-defined boundary on its north-west side all the way round from Earby to Carlton, and its junction with the underlying shales may be seen in the becks south-east of Elslack, in addition to the places mentioned above. The south-east boundary of this synclinal of grit is not so clear, being in several places obscured by drift, and the actual junction of grit and shale is not seen. With the exception of the south-west corner, high dips as a rule prevail all round this area, and S.S.W. of Carlton, along Park Lane, the beds dip at angles varying from 60° to 90°. There are many undulations of the beds in the interior, especially east of Earby. The grit is generally thick bedded and moderately coarse; some very coarse is seen in the old turnpike road a quarter of a mile north-east of Standrise Plantation. Several bands of shale, some of considerable thickness, are interstratified with the grit. One of these is well shown in the beck opposite Carlton Lane Side, where 30 to 40 feet of shale is seen, interbedded with thin sandstones, and this may be the same as that which has been traced by Higher Scarcliff and Rushbank round to near Elslack Mill. This alternation of thin bands of shale and thicker sandstones gives rise to marked features in the ground east of Thornton High Wood and to the south-east of Earby. But these shale bands are always sandy, and often pass very rapidly into sandstones, so that it is not easy to trace them for any distance. They have therefore been coloured as sandstone on the published maps.

Several faults, most of them small ones probably, are seen in the beck west of Carlton; the one seen near Carlton Lane Head may have a considerable throw, and it is not unlikely that it is prolonged southwards across Carlton Moor and that it passes to the westward of Kirk Sikes, as the grit boundary there appears not to accord with the strike of the beds.

The whole thickness of the Upper Yoredale Grit in this synclinal must be at least 600 or 700 feet, and is probably more.

The Lothersdale Anticlinal.

Under this heading will be described the limestones of the valley of Lothersdale, and all the beds above on the south side up to the base of the Kinder Scout Grit. The following is a general section of the beds in this district:—

		Thickness in feet.
Kinder Scout Grit of Sweet Brow and Surgill Rough.		
Upper Yoredale Shales of Surgill	{ Shales Sandy shales and tiles Shales }	about 800
Yoredale Grit, with subordinate shale-bands, of Roger, Hawshaw, and Tow Top Moors		quite 800
Black Bowland Shales	{ Black shales Dark shales with thin limestones }	about 750
Limestone of Hawshaw Slack Delf and Park Head Quarry	{ thinly-bedded with shale-bands }	about 200
Shales, bluish and grey	{ Shales Shales and thin limestones Shales, some sandy }	quite 1,000
		3,550

Limestone of Dowshaw and Raygill Delfs, at least several hundred feet.

It will be seen from the above that there is no Lower Yoredale Grit in this valley. There is a trace of grit at the top of the Limestone in Dowshaw quarry on the south side; but this cannot represent the Lower Yoredale Grit of Twiston Moor, and it must be much nearer the top of the Carboniferous or Scar Limestone.

The lowest beds seen in this dale are the two detached masses of Limestone which have been so largely quarried for lime at Dowshaw Delf and Raygill Delf. They may possibly be identical with the Thornton Limestone, but one cannot be certain about the matter. Between the Thornton Limestone and the Yoredale Grit at the top of Thornton High Wood there appears to be quite 3,000 feet of beds; while in Lothersdale, between the Limestone and the Yoredale Grit, there appears to be not more than about 2,000. Still neither of the sections is perfectly clear, and the Lothersdale one may be considerably thicker. The Thornton Limestone is supposed to be the equivalent of the Scar Limestone of Clitheroe, so that the Lothersdale Limestone may also be the equivalent of the same.

The limestone is of a light blue or grey colour, and the bedding is well marked. Professor Phillips, in the second volume of his Geology of Yorkshire, mentions these light-coloured laminated limestones of Lothersdale, and notices that they hold strings of dale spar and veins of sulphate of barytes, termed locally *cawk* veins.[*] There are three of these cawk veins at Raygill Delf, varying in thickness from 6 inches to 6 feet. The two principal ones dip to the south, and range nearly W. by N. and E. by S., parallel to a considerable throw which crosses the valley from a little north of Surgill Rough, goes past Raygill Clough, and between Green Hill and Dowshaw Delf, and throws down to the south. The most southerly of these two veins is well exposed in Raygill

---

[*] For information about these cawk veins, I am indebted to P. W. Spencer, Esq., of Raygill House, Lothersdale.

quarry, the beds on either side being disturbed and apparently dolomitized. In the work above mentioned a sketch is given of this altered limestone. The northern vein has been long worked, and has been followed underground to a depth of 200 feet. The third vein, which is thinner, connects the other two, and ranges W. by S. and E. by N. Some cherty bands are seen at Dowshaw, where also a thin grit and dark shale may be seen above the limestone at the top of the quarry on the south side. In both quarries the turn over of the beds of this sharp anticline is well seen.

The Shales which come above the limestone may be seen in a little stream east of Dowshaw Delf, and in Hazel Gill, east of Raygill Delf, and the higher thick shales, sometimes calcareous, with bands of dark grey limestone, may be seen in a little stream below Hawshaw Side, and in one that joins Raygill Clough from the south at the *ll* of the words *Raygill Delf* on the 1-inch map. There is not much seen of these beds on the northern side of the dale, owing to the drift.

The Upper Limestone is thinner bedded than the lower, and often contains bands of chert. It is best seen at Hawshaw Slack Delf and at Park Head Quarry. The latter quarry is on the centre of the anticlinal, and the beds seem purer than they usually are, and have been much worked for lime. A small fault ranging W. 35° N., and throwing down to the N., is exposed in the quarry. In several places, as at Hawshaw Slack Delf and W. of Calf Edge, the limestone is contorted. The limestone is shifted from Proctor Height to Smith Hill by a considerable fault which joins the Raygill Clough fault before mentioned near Dowshaw Delf. There is probably a fault between Dowshaw Delf and Smith Hill, for the two limestones are much nearer to one another here than usual, and it is not improbable that the Raygill Clough fault goes on after its junction with the Proctor Height fault, but with its throw reversed. There is possibly a small patch of this limestone in Harden Beck, south of the fault passing by Brown Hill.

The Black Bowland Shales are well seen S.E. of Raygill Clough and near Spen House, and at the latter place the thin limestones which come some way down in the shales may be seen. The Black Shales may also be seen at Higher and Lower Burnt Hill, where the Yoredale Grit is thrown down against them by a considerable fault, and at the bend of the road near Bleara Lowe; and in a gulley between this and Proctor Height the ironstone balls which are so common just below the base of the Yoredale Grit may be seen, as well as at Kitchen on the south side of the dale.

The Yoredale Grit.—This grit is of the usual character, and there are subordinate shale bands which give rise here and there to minor features. The grit is very coarse indeed at Hare Law and in a few other places. The trend of the grit is very clear from Roger Moor to Street Head, excepting a small portion of it east of Ayneslack. The dip is high and increases from about 20° on the west side to upwards of 60° at Dale End.

Several large faults all throwing down to the south shift the outcrop of the grit near Surgill Head, Hawshaw Delf, Ayneslack, Scald Bank, and Copy House. The coarse grit of Hare Law seems faulted on both sides. The bend round of the top of the grit towards Kelbrook is clearly seen, and is interesting when we consider that the Kinder Scout Grit above strikes straight across the valley at Foulridge.

Upper Yoredale Shales.—The best sections in these are in the upper part of Surgill Beck, at Nelly Hole near Scald Bank, in Moss Houses Beck near Hare Law, and in Lancashire Gill.

Thin beds of sandstone are not unfrequent, and these may be seen in Moss Houses Beck. About here the general succession seems to be—

Kinder Scout Grit :
Shales
Shales and flags
Coarse grit with shale partings } Upwards of 600 feet.
Shales -
Shales and tiles -
Yoredale Grit.

Reedshaw Moss Synclinal and Watersheddles Anticlinal.

This will include all the rest of the country under description from the base of the Kinder Scout Grit up to the Lower Coal Measures of Reedshaw Moss.

The *Kinder Scout Grit* forms a well-marked ridge from Foulridge eastward, excepting near the Lancashire-Yorkshire boundary, and N.E. of Stone Gappe, where it is obscured by drift. It is of its usual coarse character, often a conglomerate, and is specially well seen on Noyna east of Foulridge, where it is about 800 feet thick, and includes one or two well-marked shale and flag bands which are seen in Moss Houses Beck. These appear to die out towards Piked Edge, and the whole thickness of the grit is less. But in the country east of Stone Gappe beyond the limits of this Memoir (in 92 S.E.) these shales set in again and swell out the grit to near the thickness above mentioned. It is crossed by four out of the five large faults spoken of as breaking the Yoredale Grit, and two of them near Oliver and Black Lane End shift the outcrop of the grit much more than the width of its outcrop at the surface. The large fault near Ayneslack does not appear to break the Kinder Scout ridge, and is supposed to be cut off by a strike fault put on the map in broken lines.

Sabden Valley Shales.—These, which must be from 1,500 to 2,000 feet thick, include several beds of grit, which are most conspicuous in the lower half of the shales. These beds of grit cannot be traced west of a north and south line through Laneshaw Bridge owing to thick drift, and even east of this line they cannot be traced continuously. They appear in places to pass rapidly into sandy shales, and in others to be wedged-shaped masses of grit surrounded by shales. As they are all below the base of the Third Grit and some of them in the country further east coalesce with the Kinder Scout Grit, it has been thought advisable in colouring them as grit to give them the Kinder colour. The fault which crosses these beds near Cowling is seen in a little stream E. of Over House ; but the course of the fault past Oliver, between that house and Stone Head Beck, is very uncertain, though it is clear a large fault must cross somewhere about here. The best sections in the upper part of the shales are these :—In Shawhead Beck near Knarr Side, black, brown, and purple shales are seen, with ironstone and occasional sandstones. *Goniatites* may be found in places. On the Yorkshire side of the boundary in Black Scars Beck dark shales with ironstone bands and nodules are well exposed, and some of the thin sandstones in the shales may be seen in the upper part of Stone Head Beck east of Warley Wise. In the shales further down Stone Head Beck, opposite Westfield, some shells were obtained, and the faults drawn on the map may be seen. Near Gill some lead has been obtained from a vein, which is probably one of these faults.

Third Grit.—This forms a ridge or series of ridges, lower, and not so marked as that of the Kinder Scout Grit, but generally parallel to it. However, on the eastern side of the area about Ickornshaw the two part company, owing to the decrease of dip in beds above the Kinder, and to a large fault which shifts the outcrop of the Third Grit from Middleton to Hallan Hill.

It is composed of several beds of grit separated by shales. The shale bands have been put on the map—at least the more important of them—where it was possible to trace them with any approach to accuracy; but they are not always the same in number, nor are they always on the same geological horizons, and it must be understood that the portions coloured as grit are merely those where the grits are in excess and the shales are subordinate, or where it is impossible to say from the want of sections whether there is shale or not.

The lowest bed is generally very coarse and massive, and is best seen at Knarrs Hill. It is of a red colour north of Salt Pie.

The following details of borings made near Laneshaw Bridge will show the numerous alternations of grit and shale in these beds :—

Boring No. 1.		ft.	in.	Boring No. 2.		ft.	in.
Soil		2	0	Soil		1	6
Blue marl	} Drift	34	0	Stony marl	Drift	37	6
Gravel and sand		54	9	Dun metal		18	0
White rock		3	9	Dark do.		7	9
Blue metal		48	0	Coal		0	10
Dark do.		24	10	Grey rock		3	0
Coal		0	2	Dark metal		1	5
Warrant earth		3	0	Coal		1	0
White rock		12	6	Blue metal		21	9
Dun metal		18	0	Grey rock		7	6
Grey rock		4	7	Blue metal		4	0
Dark metal		41	8	Brown rock		1	8
Light do.		5	3	Dun metal		6	6
Dark do.		6	5	Grey rock		3	4
Iron band		0	4	Blue metal		1	6
Dark metal		0	8	Grey rock		4	7
Dark rock		4	6	Grey linsey		3	10
White do.		16	6	Grey rock		3	8
Dark metal		4	7	Light metal		3	10
Light do.		3	2	Grey rock		4	0
Linsey		4	3	Linsey		1	4
White rock		16	4	White rock		5	11
Linn and wool		21	0	Linsey		2	6
Iron band		0	8	White rock		16	2
Linn and wool		28	5	Dark metal		1	0
Dark metal		15	0	Coal		0	2
Linn and wool		3	0	Blue metal		11	6
Iron band		0	9	White rock		13	6
Linn and wool		1	4	Grey rock		4	8
White rock		25	4	Dark metal		3	7
Linn and wool		3	0	White rock		3	5
Light metal		3	0	Linsey		2	3
White rock		55	6	White rock		9	0
Blue metal		18	6				
White rock		4	7	Total of solid strata		173	2
Total of solid strata		398	7				

The first of these borings was made by the beck side east of Laneshaw Bridge, and includes in its upper portion part of the beds between the Third Grit and Rough Rock. The second boring, if I am rightly informed, was made considerably north of the village; and it would be in the beds about the middle of the Third Grit; but as in the beck near the beds are seen to be dipping at angles of 60° to 70°, the real thickness of the beds is considerably less than is shown in the boring. The beds of the first boring are no doubt all higher than those in the second boring. The total thickness of the Third Grit series here is probably quite 1,000 feet. I was told that near Ball Grove a coal 1 ft. to 1 ft. 6 in. had been cut in making a drain, and this would seem to be about the horizon of the coals near the top of boring No. 2.

The coarse massive grits of Barn Hill and Steeple Stones south of Combe Hill probably belong to the Third Grit series, as also do those of Broad Head Moor, but the identification of the grit beds about here is very uncertain.

The Shales below the Rough Rock may be seen in the beck north of Monkroyd and at Lumb, south of Ickornshaw, and we often get a flaggy band near the top of these which may be the equivalent of the Haslingden Flags. This is seen at Lumb, and south of Combe Hill, but as further east it coalesces with the Rough Rock, which also it appears to do on Crow Hill, we have given it the same colour as the Rough Rock.

There is an old coal pit near Emmott Hall where coal was got about 70 years ago. I was informed by Mr. Wright the proprietor that the seam worked was 1 ft. 6 in. thick and 30 yards below the surface, and that another coal 4 ft. 6 in. thick had been bored to at a depth of 56 yards; but Mr. Townsend of Laneshaw Bridge, to whom I am indebted for the borings above given, informed me that he had never heard of the thicker seam, so that the information about that may be incorrect. Some few years ago an attempt was made to get up a company to reopen the old pit, but it was never formed. Mr. Townsend made several borings with the object of finding the seam elsewhere, but was not successful. This seam must lie in the shales under description.

The Rough Rock can be well seen at Winewall and Raven's Rock and up the river Laneshaw. It laps round the small basin of Lower Coal Measures in Reedshaw Moss, and spreads out over a large area north of Combe Hill, where it includes some bands of shale. The two beds of grit at Raven's Rock and Bank's House seem both referable to it.

The Lower Coal Measures of Reedshaw Moss are seen in the river Laneshaw to consist principally of shales, but one bed of grit is seen south of Laneshaw House which may represent the Woodhead Hill rock. There are two coals cropping out in the beck which have both been worked; one of these seen opposite Laneshaw House is about two feet thick, and I was shown that the outcrop of this on the north side of the basin, which outcrop is usually concealed by clay.

The other seam cropping out at the main bend in the stream has a bastard Gannister floor; it was 1 ft. thick; the distance between the seams I could not well ascertain, but was probably not more than six or seven yards, and both seams appear to be much nearer the Rough Rock than are the mountain mines at Colne.

In the river Laneshaw, south of Laneshaw House, a coal 8 inches thick may be seen resting on the Rough Rock.

There is a small patch of Lower Coal Measures south of Crow Hill, adjacent to the Great Boulsworth fault.

CHAPTER III.

THE BURNLEY COAL-FIELD.

BY EDWARD HULL, F.R.S.

INTRODUCTORY.

ALTHOUGH the structure of the Burnley Basin has been only slightly investigated till within the last few years, yet several very valuable memoirs on different branches of its physical structure and palæontological treasures have been published since the establishment of the Geological Society of Manchester in 1840. That very little was known of the district before this time may be gathered from the appeal which the authors of the "Outlines of the Geology of England and Wales" make to the resident geologists of Manchester and its neighbourhood to investigate the geological structure of South Lancashire ; and although in the map and section which accompanies that work the trough-like form of the Burnley coal-field is indicated, the coal-field itself is passed over in silence, and the authors proceed from the description of the Manchester or "South Lancashire Coal-field" to that of Ingleton, which they designate as the "North Lancashire Coal-field," ignoring that of Burnley altogether.*

---

* In the following pages I have availed myself of the assistance of previous writers on the geology of the district, amongst whom may be specially named, Mr. Elias Hall, who in his "Key to the Geological Map of Lancashire" (1836) gives an estimate of the resources of the Burnley coal-basin.—Mr. E. W. Binney, F.R.S., "On the Marine Shells found in the Lancashire Coal-field," *Trans. Geol. Soc. Manchester*, Vol. I. (1840).—"On Fossil Shells in the Lower Coal-measures," *Ibid.*, Vol. II. (1860).—Mr. T. T. Wilkinson, F.R.A.S., "On some Fossil Trees discovered at Burnley." *Trans. Hist. Soc. Lanc. and Chesh.*, Vol. IX. (1857).— Messrs. J. Whitaker and T. T. Wilkinson, "On the Burnley Coal-field," *British Association Report*, 1861, and "Abstract of the Principal Mines of the Burnley Coal-field," by the same authors. *Trans. Hist. Soc. Lanc.*, Vol. XIV.—Mr. T. T. Wilkinson, "On the Drift Deposits near Burnley." *Trans. Geol. Soc. Manchester*, Vol. IV. (1863).—Mr. George Wild, "On the Fulledge Section of the Burnley Coal-field," *Ibid.*, Vol. IV. (1863).—Captain J. Aitken, "On the Gannister Coal and Higher Foot-mine of Bacup." *Ibid.* Vol. V.(1864).—Mr. J. Dickinson, F.G.S., "Sections of the Coal-strata of Lancashire." *Ibid.* Vol. IV.—Mr. T. T. Wilkinson, "Additional Bed of Coal and Cannel in the Burnley Coal-field." *Proc. Lit. and Phil. Soc. Manchester*, Vol. III. From the authors of some of the above papers my colleague Mr. Tiddeman and myself have also received much personal assistance, and we are also indebted to the following gentlemen for ready help and information—Mr. James Eccles, F.G.S., Springwell House, Blackburn ; Mr. Dixon Robinson, of Clitheroe Castle ; Colonel Towneley, F.R.S., Towneley Hall, Burnley, and his agent, Mr. Storey ; Mr. Towneley Parker, Cuerdon Hall, near Preston, and his agent, Mr. H. Jobbling, Fulledge Old Hall ; Colonel Every Clayton, of Rowley Hall ; Mr. B. Chaffer and his son, Walverden House, Marsden ; Mr. T. Simpson, proprietor of Oswalwistle Collieries, and his manager ; the Rev. P. Graham, Turncroft, Darwen ; Mr. H. H. Bolton, of Baxenden and Rossendale Collieries, and to the former proprietor, the late Mr. Jonathan Hall ; Mr. T. Beswick, Shorey Bank Colliery ; Mr. Walsh, proprietor, and A. Taylor, manager, of Hey Fold Colliery.

COAL-MEASURES.

Lower Coal-Measures.

The Lower Coal-measures, or Gannister Beds, form the largest portion of the country under description. The tract which they occupy has a triangular shape, with its apex at Colne, its base the Rossendale anticlinal, along which the Millstone Grits rise to the surface, its north-western side the Pendle range, and its eastern, the Millstone hills of the Yorkshire borders. In its lap it holds the Burnley coal-basin, and it sends from its south-eastern angle a long arm southwards to form a junction with the South Lancashire Coal-field in the direction of Rochdale.

The beds consist of fine-grained grits and flagstones, often rippled and micaceous, with intermediate beds of shales of divers colours and textures. There are also several coal-seams in the lower part of the series, known as "Mountain Mines," from the fact of their being found generally amongst the hills; and one of these seams, which in the neighbourhood of Rochdale and Bury has a peculiarly hard siliceous floor, called "Gannister" or "Galliard," is known as "the Gannister Coal." The series has as its base the Rough-rock or uppermost bed of Millstone Grit, and its upper limit the *Arley Mine*.

A complete section in the Lower Coal-measures may be obtained between the outcrop of the Arley Mine at Cliviger Colliery and the Millstone Grit at Redwater Clough, on the northern side of the Portsmouth Valley, and again on the hills south of Dineley, from the outcrop of the Arley Mine at Easden Wood to the Cliviger fault at Dineley. The general section as here presented is as follows:—

General Section of the Lower Coal-Measures, Cliviger.

	ft.	in.
Arley Mine Coal	4	0
Underclay	4	0
Grey shale and flagstone (Riddle Scout Rock)	70	0
Grey evenly-bedded flagstone with partings of sandy shale (Old Lawrence Rock)	115	0
Grey, sandy, and blue shale	145	0
Fine-grained, thin-bedded, sharp-grained grits with shales, passing downwards into tilestones (Dineley Knowl)	290	0
Shale	40	0
Irregularly-bedded grit	30	0
Black Clay Coal, in 2 beds (6 in. to 24 in.)	1	0
	695	0

Darwen; Mr. J. Place, proprietor, and Mr. Harwood, manager, of Hoddlesden Colliery; the proprietors and manager of Turton Moor Colliery; Captain Fishwick, proprietor of Spotland Colliery, near Rochdale; Mr. Whitehead, mineral surveyor, Rochdale; Mr. Charles Bradbury, manager of Church Colliery; the late Mr. Fort, M.P., of Read Hall, proprietor of Great Harwood Collieries; also Mr. Jillett, mineral viewer of Derby, the head manager, and J. Redfern, underlooker of the above collieries; Captain Le Gendre Starkie, Huntroyde Hall; Sir J. P. Kay-Shuttleworth, Bart., F.R.S., Gawthorpe Hall; also his son, Mr. Robert Kay-Shuttleworth, and manager, Mr. Kingston; Messrs. Green and Collinge, proprietors, and Mr. J. Jobbling, manager, of Cliviger Colliery; Messrs. Brooks and Pickup, proprietors of Towneley and Hambledon Hill Collieries; Mr. Witham, proprietor, and Mr. J. Hargreaves, manager, of Altham Colliery.

It is also right that I should state that in the survey of the neighbourhood of Rochdale I had the assistance of my colleague, Mr. A. H. Green, F.G.S.

LOWER COAL-MEASURES.

	ft.	in.		
Fireclay, shale	20	0		
Flagstone and hard grit				
Grey sandy shale	270	0		
Black shale with ironstone				
Foot Coal with Gannister and fireclay floor	1	0		
Fireclay, middling quality	3	0		
Sandy shales	18	0		
Coal, 40 yards, or Upper Mountain Mine	2	0		
			1,009	0
Fireclay, middling quality, passing into shales	24	0		
Coal	0	2		
Fireclay	3	0		
Black shale with *Goniatites*	96	0		
Gannister Coal or *Mountain Mine*	3	10		
			1,136	0
Shales	30	0		
Coal, Lower Foot Coal	0	10		
Fireclay, passing into shale	50	0		
Flagstone, passing downwards into hard grit (Woodhead Hill Rock)	120	0		
Dark shales (not seen at this spot)	56	0		
Coal	0	4		
Fireclay	2	0		
			1,395	2
Coarse grit (Rough Rock).				

In constructing the lower portion of the above section I was much assisted by Mr. Matthew of the Hane Brick and Tile Works. Several of the above figures of thickness are only approximations, and are probably under-estimates. The thickness of the entire series varies from 1,500 to 2,000 feet in the direction of the Pendle Range.[*]

Flagstones.—The flagstones of the Lower Coal-measures are of good quality, and held in high estimation. Some of the beds, however, in the same quarry come out only as ashlar building stone. and in this case are less profitable. The presence of ripple marks or annelide tracks is prejudicial to their quality as flagstones.

The *Old Lawrence Rock* of Mr. Binney, which is the uppermost flag-rock of the series, preserves a very uniform aspect over the whole district; being compact, of a greyish colour weathering brown, micaceous and evenly bedded, with partings of grey sandy shale. The following section taken at Mr. B. Chaffer's quarry at Catlow, near Burnley, will give a general idea of this rock[†] :—

Section at Catlow Quarries.

	feet.
1. Local drift clay and gravel	6
2. Fine-grained hard ashlar and flagstone	12
3. Grey sandy shale	10
4. Fine-grained brown and grey micaceous flag-stone, ashlar, and tilestone	33
5. Soft shaley rock called "rag"	6
6. Ashlar rock	3
Total	70

---

[*] *See* Mr. E. W. Binney's general section of these beds in Trans. Geol. Soc. Manchester, Vol. I., p. 75–78.
[†] I have to express my obligations to Mr. Chaffer for his personal assistance in the survey of this neighbourhood.

This rock is also extensively quarried on Entwistle Moor and Worsthorn Moor, where it is known as "the Worsthorn Rock." Along the southern margin of the Burnley Basin these flags are often worked. They occur at Longshaw Quarry, south of Dineley; at Crown Point, Brown Head Moor, Habergham; at Hapton Park Scout, and Hambledon Scout. These flagstones form the crest of the escarpment, to the north of which they roll over, and dip under the Burnley Basin at Hapton and Henfield. Sections are shown along Castle Clough, Altham Clough, and at Oakenshaw. The same beds are again brought in on the downcast side of the large fault which crosses the moors north of Bacup, between Deerplay Hill and Slate Pit Hill at the head of Dule's Gate.*

Having thus described the range and character of the uppermost flag-rock, it is scarcely necessary to allude to the others, which follow in a position more or less regular; but it may be as well to say a few words regarding the "Woodhead Hill Rock," which lies near the base of the Lower Coal-measures, and in an intermediate position between the Millstone Grit and the Gannister Coal, or Mountain Mine (*see* general section above, p. 55). This rock differs from the others of the series in its variable qualities, though it is always to be found in its proper place. Thus at Bacup we find the rock occurring on the west side of the valley as an evenly-bedded flagstone; but on the east side as a highly current-bedded rubbly stone, fit only for rough building purposes. In other places, however, it produces a building material of excellent quality and colour, and I would particularly refer to the quarry at Wensley Fold, Witton near Blackburn, and to another south of Ridgaling, Lowerford near Colne, as examples. West of Blackburn this rock must be of unusual thickness, judging from its position with reference to the Rough Rock and the high angle at which it dips. Sections may be seen in it along the valley of the Darwen at Fenniscowles, and for some distance downwards.

COAL-SEAMS.

Gannister Coal, or Mountain Mine.—This is the most valuable seam of the Lower Coal-measures, and has been worked at intervals over the entire district. It varies in thickness from 18 inches to 4 or even 5 feet amongst the hills east of Bacup. It is generally of excellent quality, though soft, with a laminated structure, and produces a coke which is highly esteemed for foundry purposes; large quantities of which are annually produced from the small coal or slack at the Baxenden Collieries.

The 40-Yards, or Upper Mountain Mine.—This seam varies in thickness from 14 inches to 3 feet at Darwen and Oswaldwistle. It is usually accompanied by a valuable fireclay floor, which at Littleborough is largely used for fire-bricks and pottery. The same clay is also employed at the large works of Messrs. Brooks and Picup, at Towneley near Burnley.†

Darwen Coal-field.—This little coal-field forms a strip, running southward out of the Blackburn basin. It is bounded on the east by a large fault, which passes southwards by Oswaldwistle, Pickup Bank

---

* Captain Aitken seems to think that one or two points, such as Lowe Hill and Tooter Hill, are capped by this rock.

† The best fire-bricks made at these works are manufactured from a bed which is obtained on the moors north of Bacup, amongst the shales and flags which lie above the Upper Mountain mine.

and Edgworth Moor to Quarlton; and on the west, by another large fault, which ranges in a N.N.W. direction along the western side of Darwen Moor.

It is also divided into two longitudinal segments by a large fault, parallel to the others just described, which runs along the bottom of the valley in which the town is situated. The change in the physical features of the ground on each side of the fault are very remarkable. On the west or upcast side, the Millstone series, capped by the beds of the Coal-measures, are elevated into high moorlands, intersected by deep and wooded dells, and bounded by rocky cliffs and wooded banks. On the eastern or downcast side, the surface presents merely the form of a slope gradually rising from the bottom of the valley to the long ridge east of the town along which the old Roman Road to Ribchester was carried. The position and effects of the fault speak to the eye of the observer with strong force as he stands on the edge of the moorland, at a height of 600 or 700 feet above the town; for near him is the *outcrop* of a coal-seam which is worked at a depth of several hundred feet below the bottom of the valley. Taking a line across the fault through Over Darwen Church, I have calculated that the throw of the fault amounts to more than 400 yards at this point.

In the Darwen district there are two seams principally worked, the Upper Mountain Mine—here called the Yard Seam—being three feet in thickness, and the Gannister Coal, or Half-yard Mine, the thickness being only from 18 inches to 2 feet.* This latter lies 70 yards under the former. At 12 yards above the Yard Seam is a little coal of 10 inches, resting on a true Gannister floor. This seam may be seen cropping out in the deep cutting near the entrance to the railway tunnel. Over this are shales, and then flags and tilestones of good quality, which are worked in pits belonging to the Rev. W. Graham, at Turncroft.

As it is stated by Mr. J. Dickinson, in his paper on the Coal-strata of Lancashire, the strata which intervene between the 10-inch and the Yard seams at Darwen gradually thin out in the direction of Hoddlesden, and at Old Hoddlesden Colliery the two seams come so close as to be worked together. In this position they continue for some distance northward, but divide at Beltham, when the upper part is called " the Half-yard," and the lower " the Little Coal."†

The Yard Mine, as well as the Half-yard, are now nearly exhausted over the Darwen district. The former is, however, still worked in Mr. Walsh's colliery at a depth of 114 yards, and here, as well as under the tract of ground about Turncroft and Sough Fold, the lower seam has not yet been got. On Darwen Moor the Half-yard Seam is extensively worked by the Messrs. Shorrocks by tunnelling into the hill, and the coal is conveyed by rail and steep inclines down into the town. At the Old Lyon's Colliery. the Yard Mine, which forms a little outlying patch, was worked, but is now entirely exhausted. It was also worked on Turton Moor, on the downthrow side of a fault.

Besides the above, there is also a coal-seam one foot thick, lying 18 yards above the " Half-yard " mine at Darwen.

At Turton colliery we find the following section‡ :—

---

* Trans. Geol. Soc. Manchester, Vol. IV., p. 163.

† Messrs. Shorrocks, the proprietors of these mines, allowed me every facility for the survey of the district, in which I received much assistance from their intelligent manager.

‡ Supplied to me by Gabriel Garbet, the manager, with the sanction of the proprietors.

		ft.	in.
Gravel and shale	·	51	0
Bin Coal		0	10
Rock ·	-	54	0
Shale, &c.		54	0
Coal		0	10
Fireclay	·	6	0
Coal, called "Half-yard Mine"	·	1	8

It is very difficult to correlate these beds with those of the Darwen district. They are probably not identical

Quarlton and Turton.—The Coal-measures of Quarlton lie on the downcast side of a large fault, which ranges in a southerly direction from Pickup Bank by Wickenlow Hill. The downthrow of this fault is on the west, and opposite Holcombe Hill; it must amount to about 400 yards. Both the Upper and Lower Mountain Mines have here been worked in pits from 70 to 80 yards in depth, and I was informed by John Sedden that the fault had been struck in the workings of the upper seam at Quarlton Colliery.

The general dip is eastward, at 15° to 20°. The outcrop of the upper seam is visible in a little dell north of the farm called "Top of Quarlton," and that of the lower or Gannister Coal occurs in a brook-course at Hey Head and at Quarlton Fold. The roof is shale, and the floor Gannister rock. Further west, and lower down the valley, the Rough Rock may be seen in quarries at Knotts and Edgeworth.

The Lower Coal-measures are again thrown in by a fault ranging in a north-westerly direction through Turton Bottoms. Its position can be well determined in the railway cutting at Over Houses, Spring Bank, Quarlton Vale House, and Wauves Reservoir, at the western extremity of which the beds may be observed standing on end. The coal-field of Turton is merely a narrow belt, bounded on both sides by faults. The fault at the opposite side may be observed in a ravine, at its junction with Bradshaw Brook, to which the name of "Jumbles" has been given, probably by the miners, with reference to the dislocated state of the strata. The fault at this point brings down the Gannister Coal against the Rough Rock, and the throw of the fault is about 50 yards.

Littleborough and Wardle.—The Coal-measures of this district are bounded on the north-west by the uprising of the Millstone Grit of Ashworth Moor, Knoll Moor, Tunnicliffe, Whitworth, and Shawforth. Here the coal-field is contracted to the breadth of about half a mile, formed by the high ridge called "Trough Edge" (1,480 feet), which justifies its name in being literally a physical trough, on both sides of which the Mountain Mines, with the underlying sandstones, shales, and grits, crop out.

From the eastern base of Trough Edge the Coal-measures are bounded by a narrow strip of Rough Rock, skirting the flanks of Weather Hill, Ramsden Hill, House Pasture, and spreading over the top of Shore Moor. Here the boundary becomes a fault, which is visible at the junction of Brook Holes Clough with Horse Pasture Clough, and ranging in a S.S.E. direction by Fox Stones Hill and Calderbrook, crosses the Todmorden Valley, and continues its course southward along the western slopes of Stormer Hill, Draught Hill, and Whitaker Moor to Longden End.

The strata over the whole of this district have a general southerly dip, and are traversed by numerous faults, which for the most part range in a north-westerly direction. Along the southerly part of the district the mines are worked in collieries of considerable depth, but northward the seams rise to the surface and crop out along the flanks of the moorlands, where they are worked by adits, and their outcrops can be traced with great facility. Not only are the Upper and Lower Mountain Mines here worked, but also occasionally the little seams which lie above and below the Gannister Coal, called the "Upper" and "Lower Foot Mines." This is indeed often a matter of necessity, as the main seams are almost exhausted over a considerable tract, as stated by Mr. Dickinson.* The highest seam that is worked is one called the "Bassey Mine," which, according to the same authority, lies about 30 yards above "the 40-yards" or Upper Mountain Mine. As already stated, this latter seam has a floor of valuable fireclay, which is worked at Littleborough. I shall now trace the course of the Mountain Mines from west to east, noting some of the spots where the outcrop may be seen.

Near Spotland Mill the Gannister Coal is worked in a colliery belonging to Captain Fishwick at a depth of 130 yards; it is about 2 feet in thickness. The outcrop of the "40-yards" Mine occurs at Lower Fold, close to a fault of 20 yards downthrow to the S.W. Higher up the valley of the Spodden at Dunnisbooth Wood a fine section showing the outcrop of the Gannister Coal is shown in the banks of the brook. This is one of the best spots I know of for examining the peculiar floor of this coal. The lower part consists of an inferior sandy fireclay, and over this is laid the Gannister Rock, with an extremely irregular floor, varying in thickness from one to two feet. The rock is as hard as flint, and contains numerous casts of *Stigmariæ ficoides*. Upon this floor, with a rather uneven surface, rests the coal, about 2 feet thick, and with a dark gray shale roof. These beds are thrown out by a fault which crosses the river at the new railway bridge, and brings up the Millstone Grit to a higher level. The outcrop of the Gannister Coal may be seen at Toncliffe Brow, and it is here accompanied by the "Upper" and "Lower Foot" Mines.

At Copy Nook Coal-pit the crop of the Lower Foot Mine and the Gannister Coal may be seen. The Gannister rock is there a yard in thickness. The Upper Mountain, or "Half-yard" Mine, as it is here called, may be traced almost step by step along the upper slopes of Trough Edge, Middle Hill, and Rough Hill (1,425 feet). It is there 2 feet in thickness, and has a thick floor of fireclay.

The Wardle Valley, which is traversed by Higher Slack Brook, forms an anticlinal, as the beds dip into the hills on both sides. The centre of the valley is formed of "Woodhead Hill Rock." On the eastern side the Gannister Coal may be observed cropping out at Scotch Cote Spring, where it is 2 feet 3 inches in thickness. It may be traced southward by Further Barn, Bank Hey, High Barn, and the brook-course a little above Wardle, where it dips to the south and disappears beneath the river bed. The outcrop is well shown again at Clough House Brook, with a fine exposure of the Gannister rock, which is here very siliceous, weathering with the appearance of White Quartz rock. The outcrop of the Upper Mine is again seen in the banks of Ryding's Dam, with a shale roof, in which *Goniatites*, *Aviculo-pecten*, and *Orthoceras* are abundant. At Small Bridge Colliery the Gannister Coal, less than 2 feet in thickness, is worked at a depth of 120 yards, and again at a depth of 126 yards at Cleggs Wood Colliery, by Messrs. Knowles. On the upcast side of a fault which passes along the east of the pits, the Upper Mine is brought to the surface, and crops out along the crest of Cleggs Wood Hill, where the fireclay of the floor is extracted.

---

* Trans. Geol. Soc. Manch., Vol. IV., p. 161.

The outcrop of the Upper Mine may be traced along the side of Starring Hill and East Hill, and that of the Lower near the base of East Wood. At Higher and Lower Shore Hamlets the Woodhead Hill Rock rises to the surface in very massive proportions, dipping beneath the Gannister Coal to the westward, and is broken off by a large fault on the rise, which brings in the sandstone of Rake Foot Quarry, which occupies a position some distance above the Upper Mountain Mine. The appearance of the rocks on each side of the fault being so like, and the dip nearly the same, one might easily suppose them to be identical.

The outcrop of the Gannister Coal may be observed at Ealces Wood above Lane Foot, and may be traced southward by Whitaker to Syke Farm. Along this outcrop the dip is eastwardly, but the beds rise again, crop out, and dip to the west on approaching the great fault before described as passing along the western slopes of Whitaker Moor. My colleague, Mr. A. H. Green, who assisted in the survey of this neighbourhood, gives the following section at the outcrop of the Gannister Coal at Whitaker Coke Kilns :—

	ft.	in.
Flaggy sandstone	2	6
Grey shale	2	6
Coal	1	10
Gannister floor	2	0

The outcrop is again visible at Lydgate, in the brook, and at the top of the bank; that of the Upper Mine, with its yellowish fireclay floor, and the crop of both seams, may be traced along the side of the brook to the banks of the canal at Pike House Mill.

On the north side of the valley of the Roch the Upper Mine can be traced from Whitfield to Handle Hall Colliery, where it is worked by an adit and is 2 feet to 2 feet 3 inches thick with a fireclay floor and a black shale roof. At Ringing Pots Hill the outcrop of the Gannister Coal may be seen in two places in close proximity to a large fault with an eastwardly dip which would seem to bring it in under Gauder Hall.

At Long Clough there is a coal-seam only a few feet above the Rough Rock, and therefore near the base of the Lower Coal-measures. It lies about 55 yards below the Gannister Coal which crops out on the bank above the brook. The coal is about 18 inches thick and of inferior quality. It is probably the seam known in some places as the "Lower Yard Mine," and called by Mr. Binney in his tabular view of the coals of Lancashire " the Bassy Coal."*

Bacup Coal District.—This district is connected with that of Littleborough by the narrow neck of Trough Edge. It extends along the northern side of the Valley of Rossendale as far west as Goodshaw and Wolaw Nook, and crossing the hills along the line of the Thievely Fault at the northern base of Dirplay (or Deerplay) Hill, it follows the line marked by the fringe of Rough Rock which crops out along the southern side of Portsmouth valley southward to Weather Hill.

The outcrop of the Lower and Upper Mountain Mines may be observed on the western side at Shackleton Holmes, both being 2 feet in thickness and 67 yards apart. Here the Gannister floor is on the point of disappearing. Continuing on the westerly crop we find both seams at Hogshead Colliery, with shale roofs; but about half a mile further north a change takes place in the roof of the Lower Mine, in that the shale roof is replaced by irregularly-bedded sandstone, which according

---

* "On the Lancashire and Cheshire Coal-field." Trans. Geol. Soc. Manchester, Vol. I., p. 77.

to Captain Aitken,* who has very carefully investigated the geology of this till lately inaccessible† district, attains a thickness of 60 feet and upwards. At the entrance to Oaken Clough Colliery the sandstone roof may be seen resting on an eroded surface of the coal, here 32 inches in thickness, which is its average thickness around Bacup. At High Houses and Old Hey on the western edge of Reaps Moss the outcrop of the Upper Mountain Mine may be seen. The roof is black shale with bands and nodules of ironstone. The distance between the two seams is about 55 yards, and at 12 yards above it is the Upper Foot Mine, to which I shall recur when I come to speak of the eastern and northern outcrop.

The outcrop of the seams may be traced either by workings or by other indications along the sides of valleys of the Irwell and Greave Brook to near the heads of these dells, when they are cut off and thrown down on the north-east side by a fault which may be called "the Irwell Springs Fault," as it has doubtless some influence in producing the springs which are the sources of the Irwell along the base of Dirplay Hill.

At Grime Bridge Pit, one of the Baxendell collieries, the Gannister Coal is worked at a depth of 33 yards; the coal is about $2\frac{1}{2}$ feet thick and crops out at Carr, Lower Dean Head, Nab Plantation, Nab Rough, and Bent Hill adit. The fault which traverses the district in a north-westerly direction from behind St. John's Church, Bacup, to Meadow Head in Goodshaw Booth, was laid down on the maps with the assistance of Mr. H. H. Bolton, of the Baxenden and Rossendale collieries.‡

At Gambleside there is a fine opportunity of observing the outcrop of both the Upper and Lower Mountain Mines. The Upper caps the hill, being only covered by a few feet of fine grit resting on shale, and has a fireclay floor. Below this there is a series of shales about 50 yards in thickness, with a little flagstone in the centre, and then the Lower Mountain Mine, 3 feet in thickness, with a Gannister floor; the section occurs in the lane leading from Gambleside to the reservoir. The seam here contains numerous pyritous and earthy nodules which require to be picked out before the coal is sent to market. These strange and unwelcome visitors are not uncommon over this part of the country.

Returning to the eastern side of the Bacup district, we find a remarkable change in the thickness of the Gannister Coal caused by the junction of the Upper Foot Mine with this seam, and a corresponding increase in the thickness of both these seams, so that the union of a seam usually less than a foot along with another of an average of 2 feet produces a solid seam of 4 or 5 feet in thickness. The manner and phenomena of the junction of the two seams are now well understood, owing to the circumstantial account thereof by Captain Aitken, published in the Transactions of the Geological Society of Manchester,§ accompanied by a diagramatic section taken in the workings of the coal under Tooter Hill. The line of contact lies in a direction nearly N.W. and S.E., passing under Tooter Hill to the north of Foul Clough Colliery on the eastern side of Trough Edge.‖ The junction of the two seams is

---

* "On the Gannister and Higher Foot mines of Bacup &c. Trans. Geol. Soc. Manchester, Vol. V., p. 186.
† I mean till the railway was made.
‡ To whose kindly assistance I am much indebted. The late Mr. Jonathan Hall also rendered me much aid in the survey of the district about Haslingden.
§ Vol. V., p. 185.
‖ At this adit the thickness of the seam is 2 feet 3 inches before the contact; but further north, at the "Wreck Beds," the seam at its outcrop is 5 feet after the contact of the upper and lower seams.

very abrupt, and takes place appareutly along the side of a sudden depression of one part of the coal-bed, or the elevation of the other. The hollow part has then been filled to a level with the sandy matter which forms the roof of the lower seam, and over the surface of this plane the upper foot mine passes across to unite with the lower thick seam. At the point where the depression occurs the coal is traversed by numerous slickensides and joints running at an angle of 45° to the board or cleavage of the coal, having their surfaces coated with a thin layer of oxide of iron, and called by the miners "gagantails." "After the union of the two seams," the author of the paper quoted above remarks, "a marked " difference takes place in the thickness and quality of the coal, the " character of the roof, and in other respects. The coal is not so good " in quality as the Yard Coal (the name applied to the Gannister Coal " at Bacup), and is consequently better adapted to the purpose of gene-" rating steam than for domestic use, and distributed throughout the " entire mass are found numerous balls generally named 'bullions.'" The roof is a strong black shale, in which are found flattened nodules known as "baum pots."*

Besides the spots already mentioned, the outcrop may be well seen at Dules Gate brickworks, where the coal is 5 feet thick. About 70 yards above it is the Upper Mountain Mine, 18 inches in thickness, with a fireclay floor of 4 feet used at the brick and tile works. In the black shale roof of this seam *Aviculo-pecten* occurs, and in the roof of the underlying 5-feet seam *Goniatites, Aviculo-pecten,* and *Posidonia.*†

From Dules Gate the outcrop of the Gannister Coal may be traced along the eastern side of the moor by springs, and occasional old workings, to the picturesque and precipitous ravine called in the upper part Green's Clough, and in the lower "Beater Clough." Along Beater Clough the brook is precipitated over several cascades formed by ledges of grit and flagstone down to the Portsmouth Valley.

On ascending the valley we find the beds dipping to the S.W., and at an elevation of about 550 feet above Portsmouth we meet with the Rough Rock containing "the Sand-rock Coal;" a little above this "the Woodhead Hill Rock," producing a cascade, then shale surmounted by "the seat rock" of the Gannister Coal. A rare and very beautiful plant is here to be seen. Just below "the seat rock" a tunnel has been driven into the side of the hill through shale for the purpose of working the coal. This tunnel is probably many years old, and its sides are covered with an extremely minute moss, which when seen at a certain angle with the light presents a delicate green iridescent lustre similar to that of the wing of some birds. The lustre can only be seen under the shade, and the plant itself is almost imperceptible to the naked eye. It is known as "the shining moss." Seldom does the outcrop of a coal-seam offer so great an attraction as at this spot.‡

Higher up the brook course we come to the outcrop of the Upper Mountain Mine, 2 feet in thickness, with a roof of black shale containing fish scales. The outcrop of both the seams may again be found on the moor above Thievely Scout, and near to the spot where they are terminated against the fault at Black Clough. The crest of the escarpment of Thievely Seout is here formed of the Woodhead Hill Rock.

---

* *Supra cit.*, p. 187.
† Mr. E. W. Binuey, "On the Marine Shells of the Lancashire Coal-field." Trans. Geol. Soc. Manchester, Vol. I., p. 87–88. I have also myself seen them at South Graine in the roof of the 5-feet seam.
‡ I am indebted to Mr. Joseph Sidebotham of Manchester for the name of this moss, *Schistostega pennata.* (*See* Buxton's Botanical Guide, p. 169.)

Accrington and Blackburn Districts.—The tract overspread by Lower Coal-measures between the northern outcrop along the flanks of the Pendle Range, the southern outcrop on the moors near Haslingden, and having its eastern and western limits at Wholaw Nook and Hoghton, has an area of about 40 square miles, the greater part of which contains the *Gannister Coal*, and a smaller portion the *Upper Mountain Mine*. This tract forms the south-western limit of the Burnley basin, the offshoots from which towards the south have already been described under the head of the Darwen and Bacup Coal-districts (pp. 56 and 60).

Over the greater part of the Accrington district the strata lie in a position approaching the horizontal, as at the Baxenden and Oswaldwistle collieries. A very low but appreciable arch or anticlinal may be traced in a direction nearly due east and west, branching from the main axis of the Burnley Trough at Blackburn,* and ranging eastward by Cowhill Moss, Wolfenden Syke, to Church, where it coincides with the line of fault which passes by Hyndburn; from this it ranges by the Huncoat Hall, the "Top of Barley," and Hapton Lower Park to Wholaw Nook, beyond which it gradually disappears. All along the southern side of this line there is a southerly dip, as may be observed at the quarries on Brandy House Brow, Blackburn, Lunch Barn, and Stanhill, the river section at Foxhill Bank, and the escarpments of North Rake and Hambledon Scout. The reversed dip, on the northern side of the axis, may be observed in quarries at Audley Higher Barn, Blackburn, Cowhill Fold, Dunkenhalgh Colliery, and in sections along Hyndburn Brook, Whinny Hill, in the railway cutting at Old Accrington, in Altham Brook, Huncoat, Castle Clough Wood, Park Gate, and Mickle Hurst Brook. It is, in fact, by the northerly dip of the strata along the line here indicated that the southern rim of the Blackburn and Burnley Trough is formed.

The district is also traversed by several faults, most of which have been laid down on the maps with the friendly assistance of the proprietors and managers of Oswaldwistle, Baxenden, Church, and Hapton collieries. Others have been traced on the ground.

The highest seam is the Foot coal, a 10-inch seam with a fireclay floor, which is worked at Gaulkthorn Pottery in Oswaldwistle. On the southern outcrop, along the edge of the Burnley Trough, this seam may again be found in the brook course in Dunkenhalgh Park, Matt Bridge, Oswaldwistle, and Brocklehurst Wood, east of Accrington.

At a distance of 12 yards below this is the "Upper Mountain Mine," which, from its thickness, is called the "Yard Mine" at Oswaldwistle. The outcrop of this seam occurs a few yards above Broadfield Colliery, and again at Iccanhurst. On the western side of a fault of 40 yards downthrow to the west, passing by Broadfield Moor, this seam forms a trough with a southern outcrop by Town Bent, Duckworth Hill, and the northern outcrop by Lottice Brook to Foxhill Bank. In some places it is very irregular in thickness, and even entirely absent.† It may be found cropping out in a brook course, at Higher Cattle Plantation, on the north side of Oswaldwistle Moor. The seam is here 3 feet thick, and is brought up at a steep pitch near a fault. On the northern outcrop, the Upper Mountain Mine (as is supposed) was formerly reached in a pit 80 yards deep near Cob Wall Bridge in Little Harwood. The seam was here about 30 inches in thickness.‡

---

* The exact line of the change of dip at Blackburn lies along a fault which was observed in a quarry on the south side of the canal, east of the bridge at Grimshaw.

† As I was informed by the manager of Oswaldwistle Colliery, J. Howarth, who gave me much assistance in the survey of this district.

‡ Information kindly afforded by Mr. Armisted, the proprietor of the colliery.

The *Lower Mountain Mine*, or *Gannister Coal*, is of good quality, and produces a coke, but is of variable thickness over this district. Thus at Baxenden Colliery the average is 26 inches; at Broadfield Colliery, Oswaldwistle, 18 inches ; Cabin End Factory, Knuzden, 26 inches ;* Whinny Edge, Blackburn, 18 inches ; Mill Hill, 27 inches ; Feniscowles, 20 inches ; and along the southern flanks of the Pendle Range the thickness augments considerably, as the thickness at Higher Cunliffe Colliery was found to be 4 feet.† At Church Colliery, in the centre of the district, the thickness is 28 inches ; and at Fulledge, 4 feet.

The roof of the Gannister coal is grey or dark blue shale, and the floor a fireclay of inferior quality. The outcrop of the seam may be observed at Cat Hall Level and Wood Nook Level, Accrington. In the neighbourhood of Blackburn it may be seen cropping out along the top of the cliffs of flagstone at Brandy House Brow, where it was formerly worked ; we there obtain the following section :—

Outcrop of Gannister Coal, south side of Blackburn.

		ft.	in.
1. Brown shale -	-(about)	10	0
2. Black shale	-(,,)	3	0
3. Gannister coal	-(,,)	1	6
4. Fireclay -	-	5	6
5. Shales and tilestones	-	15	0
6. Fine grained, evenly bedded, greyish flagstone		20	0
7. Grey and yellow current-bedded grit ("Woodhead Hill Rock")		15	0

The extreme westerly point to which the Gannister coal extends is Feniscowles, where it was formerly worked to within a few yards of the church. In the river banks below, and in the cliffs behind Feniscowles House, a reddish sandstone—sometimes massive, in other places flaggy—may be observed, which is undoubtedly the Woodhead Hill Rock, which underlies the coal, as shown in the above section.

At the Cherry Tree Firebrick Works the proprietor kindly furnished me with the following section of the coal-series, but I am unable to identify the seams further than to suppose that the lowest is the Gannister. Mr. W. Livesey, of Mill Hill, also gave me an account of the seams near the factory worked by the late Mr. Turner, and I place the two sections side by side. I believe there is a fault between the two spots, which is probably the prolongation of the Darwen Valley fault. The dip of the strata is northerly.

Sections in Lower Coal-measures near Blackburn.

	Cherry Tree Works.		Mill Hill.	
	ft.	in.	ft.	in.
Drift and strata	120	0		
1st coal	- 1	4	210	0
Strata	30	0		
2nd coal	1	8	1	10
Strata -	45	0	60	0
3rd coal	- 2	3	3	0

In the following pages we shall follow the course of these "mountain mines" along the southern flanks of the Pendle Range to Colne, and then southwards to the Cliviger Valley.

---

* Proved by boring made by Messrs. Pilkington & Co.
† As I was informed by Mr. J. Swarbrick of the Little Harwood Brick and Tile Works.

LOWER COAL-MEASURES.

Harwood, Huntroyde, Marsden, and Worsthorn Districts.—Along the southern slopes of the Pendle Range, from Great Harwood to Wheatley Lane, the Lower Coal-measures, although of an absolute thickness of about 2,000 feet, form but a narrow slip interposed between the Millstone Grit and the outcrop of the Arley Mine. This is owing to the steepness of the dip along two thirds of the width of the belt which may be taken at an average angle of 45° towards the S.S.E.

Fig. 13.

Section through Little Harwood Hall, showing the sudden change of dip along the foot of the Pendle Hills.

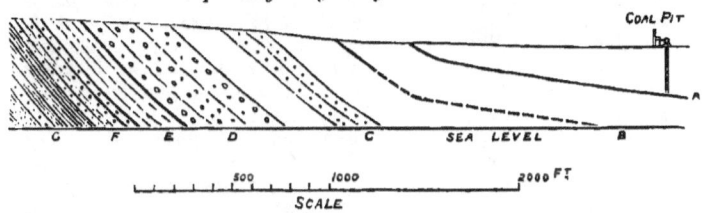

A. Upper Mountain Mine (coal).
B. Lower   do.   do.
C. " Woodhead Hill Rock."
D. " Rough rock " (Upper Millstone Grit).
E. Grey sandy shales.
F. " Haslingden Flags."
G. Chiefly shales.

On the opposite side of the Burnley trough these same beds, although somewhat reduced in aggregate thickness, spread out to at least double the breadth they occupy on the Pendle side, owing to the more gentle inclination of the beds towards the axis of the trough. The change from the high to a low dip, on descending from the Pendle hillside into the plain, is remarkable for its abruptness; and this is the case from Blackburn to Lowerford, a distance of 14 miles. The suddenness of this alteration in the pitch is indicated in the woodcut above (Fig. 13), and also in a great degree conforms to the change of slope in the surface of the ground. The line or axis along which this change takes place, from an angle say of 40° to 5° or 10°, may be traced as follows :—Commencing at Finiscliffe Bridge, at the west of Blackburn, we may draw the line through the Market Square to Little Harwood Hall ; from thence through Side Beet, Lower Cunliffe, and along the south side of Harwood Lower Town. Crossing the river Calder, west of Dunkirk Farm, the line then ranges by Simonston Hall, south of the Padiham reservoir, High Whitaker, Fence Church, Wheatley Lane, and Higherford; south of this line the angle of dip is seldom greater than 10°, north of it seldom less than 40°.

The outcrop of the Mountain Mines along the slopes of the Pendle Hills is generally indicated by refuse heaps and pits of a past generation. None of these are deep, for the coal all along its outcrop, in the language of the country, is " a rearing mine," rising to the surface or dipping beneath with such rapidity, that with the appliances of the last century it could not be unwatered at a greater depth than 30 or 40 yards from the surface. The difficulty and expense of working a seam with such an inclination was also almost prohibitive. Henceforward, however, with our improved machinery and modes of working, these natural obstacles will be overcome; and Mr. Fort's colliery at Great Harwood is the first of several which may be expected in course of time to open out on an extended scale the Gannister Coal along its northern limits.

The northern outcrop of the Gannister Coal at Blackburn is not known, as far as I could ascertain; but from the position of the Woodhead Hill Rock at Wensley Fold and Shear Brow, I should judge that it must be near the position of the park gate on the Preston road. At Little Harwood (as already stated) the Upper Mountain Mine has been reached at 80 yards, and the outcrop of the Gannister Coal must be a short distance north of the Hall. Further to the north-east it has been worked at the outcrop in fields north of "Blow-up."* At Higher Cunliff the seam has been worked from the outcrop to a depth of 60 yards at the canal side; the dip is S.S.E. at 45°. From this point as far as Harwood Colliery the outcrop has not been ascertained; but at Pit No. 3 the seam is reached at a depth of 80 yards, and has a thickness of 3 feet 9 inches; the dip is S. 20 E. at 18°, and the outcrop is a few yards south of St. Bartholomew's Church. About a mile to the N.E. of this is Pit No. 2, which is the deepest and best-constructed colliery working the mountain along its northern outcrop.† The depth is 169 yards to the coal, which is nearly 4 feet in thickness and dips to the S.E. at 50°. The outcrop of the seam, which is concealed by Till, is at a distance of 300 yards on the rise of the engine-house. The seam crosses the river Calder between the two bends, Cock Wood Bridge, and has been traced through Read Park, where it was formerly worked from the outcrop. The Upper Mountain Mine has also been proved here to be 20 inches in thickness. The Gannister Coal crops out at Haugh Head Farm, and below Trap House, where the dip is S.S.E. at 50°. North of Huntroyde it is broken through by a large fault, ranging in a N.W. direction, with an upcast to the N.E. This I infer from the relative position of the Rough Rock, and old pit works on both sides, and also from the fact that it is the line of "the Hargreave Fault," which has been fully proved further south.

The outcrop of the Gannister Coal above Cuckoo Hall is well marked by old pit heaps of black shale, containing *Goniatites* and *Aviculopecten*. These were pointed out several years since by Sir J. Kay-Shuttleworth to myself and the members of the Geological Society of Manchester, on the occasion of an excursion to that neighbourhood. The shells above named are those which generally overlie the "Upper Foot Mine," or "bullion coal;" but we may suppose that (as at Bacup) the two seams are here united.‡ At Northwood the outcrop may be easily ascertained, from the position of "the Woodhead Hill Rock," in the quarry at Hollin Brow. This is a fine-grained irregularly-bedded sandstone, full of stems of trees and plants.§ The dip is S.S.E. at 40° or 45°, and from borings which have been made we get the following series‖ :—

---

* So called from the bursting of the boiler of the engine belonging to a little coal pit formerly at that spot.

† Opened out under the direction of Mr. Jillett, mining engineer of Derby, on the property of the late R. Fort, Esq., M.P., of Read Hall.

‡ Such seems to be the opinion of Mr. Binney, F.R.S., and Mr. Dickinson, F.G.S. Trans. Geol. Soc. Manchester, Vol. II., p. 81.

§ A collection of which has been made by my friend Mr. Robert Kay-Shuttleworth, who accompanied me in the survey of this neighbourhood.

‖ These and several other sections were furnished with the permission of Sir J. Kay-Shuttleworth, by his surveyor, Mr. Kingston, to whom I am indebted for much friendly assistance.

Borings at Northwood Farm, Padiham Heights.

No. 1.	ft.	in.	No. 2.	ft.	in.
Drift clay, &c.	9	8	Drift clay, &c.	19	7
			Shale	13	5
Shale	1	8	Rock and shaley flagstone (Rag)	48	4
Rag (flagstone)	37	8	Shale	6	1
Shale	4	9	Rock and Rag	48	4
Black shale	0	6	Shale	5	8
Coal	2	2	Coal	2	0
Seat	5	1	Seat	5	10
Rag	26	10	Rag	25	1
Black shale	4	10	Black shale	3	5
Coal	1	0	Coal	1	5
Seat	6	11	Seat	1	9
			Rag	9	9
Rag	4	11	Shale and Rag	11	5
Shale	19	8	Shale	7	5
Black shale	2	0			
Coal	0	6	Coal	0	7
Seat	5	9	Seat	6	2
Rag	5	9	Rag	8	6
Shale	69	7	Shale	3	1
			Rag	9	3
			Shale	53	5
Coal (Gannister seam)	4	6	Coal (Gannister seam)	3	11
Seat	1	1	Seat stone	2	10
Rock and Rag	13	2			

The thicknesses of the strata and coals as given above ought to be reduced in the proportion of 3 : 4, in order to arrive at the true measures, owing to the high dip of the beds, which varies from 40° to 45°.

Continuing our course towards the N.E. we find indications of old coal-workings by the side of a brook course about 400 yards north of Fence Church; then again along the hill side above Boggart Hole and Blackwood farms at Wheatley Lane. Above Rushton Thorn Inn we lose sight of all indications of the outcrop, owing to the overspread of Drift Clay, till we reach the meadows south of Lower Fulshaw, where there are several old pits. Here (judging from the proximity of the coal to the Third Millstone Grit, which forms the ridge overlooking Rough Lee Water) the beds must be terminated by a large fault, with an upthrow to the north, which is marked on the map as ranging in an east and west direction to the north of Higher Ford.

On the south side of Colne Water, at Swinden Hall, the outcrop of the Gannister Coal may again be observed in several places between that and the Burnley Road. The roof consists of grey and black shale with *Goniatites*, and large concretionary cement stones, called "bullions." I was unable to make out the thickness of the seam, but from appearances it must be considerable, probably 4 feet. The general dip is westerly at a small angle, and it is probable the beds are thrown down to the south by a fault, as the flagstones and shales between Bolt House and Marsden Hall must be considered to overlie the Gannister Coal, although they appear to dip under it.

Along the flanks of Caster Cliff Hill the outcrop of the Gannister Coal may be traced by lines of old hollows and pits. Indeed, the whole hill seems to have been covered with coal-pits. The roof is black shale, which may be seen in the dell below Hill Head, and the top of the hill is capped by the Upper Mountain Mine, the fireclay floor of which may be seen in the bend of the road at The Nook. The general dip is towards the N.W., and the beds are traversed by a large N.W. fault, which passes by Woolpack Row, and may be seen in the river bed below Colne at Walk Mill. By this fault the outcrop of the coal is thrown farther down the hill, nearly to the river side, and the beds are thrown out by another in a parallel direction, which ranges by Park Laith and the entrance to the adit of Fox Clough Colliery. On the east side of this fault, which is an upthrow to the north of about 35 yards, the outcrop of the coal may be traced from Far Laith to the top of the river bank west of Winewall, where it terminates against another upcast fault, which may be seen in the bed of Trawden Brook.

Returning to Caster Cliff, and the outcrop of the Gannister Coal at Gib Clough Head, we find a little south of this by the lane-side a quarry in hard siliceous grit, dipping N.W., which I have no hesitation in referring to the position of the "Woodhead Hill Rock," underlying the Gannister Coal. Between this quarry and those of Crawshaw Hill and Catlow there must be a large downthrow fault, as the flagstones of Catlow are altogether a different kind of rock, and lie at a considerable distance *above* the same coal-seam. The exact position of this fault I was unable to ascertain, but from a comparison of the sections in the brooks near Marsden Hill, I am of opinion that it probably ranges in a N.W. direction by Gib Hill.

The coal-seams which lie under the Catlow flagstones have been proved and partially worked at Clough Head. The coal-pit is now closed, but I am enabled, through the assistance of Mr. B. Chaffer,* to give the following condensed account of the measures passed through:—

Section at Clough Head Coal-pit, Marsden.

		ft.	in.
Drift clay and various strata	-	120	0
1st Coal ("Black Clay Coal"?)	18 inches to	2	1
Various strata	-	150	0
2nd Coal (Upper Mountain Mine)		1	5
Various strata	-	150	0
3rd Coal (Gannister seam)	from 4 feet to	6	0

Of the Gannister Coal from 18 to 14 inches were cannel, the remainder coal. It was also stated that a fault was found to run W.N.W. and E.S.E., with a downthrow on the north side of 14 yards. Mr. Gunn supplies the following paragraph:—

[The Lower Coal-measures of the Trawden Valley form a wedge-shaped mass let down between two faults, which may be called the Cowfield Fault and the Trawden Valley Fault respectively. The first of these, which bounds the Coal-measures on the south, is probably a continuation of the fault which throws down the Gannister Coal near Colne, and which is seen in the river at Walk Mill. It may be seen in the small plantation east of Cowfield; it indicates its presence by a high dip of 70° about 300 yards east of Draught Gates, and may again be seen in two places near Alder Hurst End. The Trawden Valley Fault may be seen at the waterfall near Lumb in the brook that comes down from Beaver Cote, and again in the little stream near Stunstead, and it shifts the outcrop of the Rough Rock from Winewall to Lidget. The strata in this wedge everywhere dip to the north or north-east, and they may be seen, *e.g.*, north of Lodge Holme, dipping at an angle of 30°, as

---

* Mr. Chaffer was kind enough to obtain an account of the sinking from J. Sagar, the foreman at the colliery.

if to pass under the Rough Rock, while that grit close by is lying horizontal, or even dipping north-east. Thus the existence of this line of fault is incontestably proved. The part of the valley opposite Alderbarrow is thickly covered with drift, so that no sections are to be had, but opposite Far Wanless a thin coal is seen in the beck, with shales and sandstones. South of the Corn Mill the beds are vertical and a north-west fault is seen. Beech Beck, which comes down from Cowfield, gives a pretty good continuous section, which as we go up the brook appears to be certainly a descending series, though several faults ranging north-west cross the stream and make it difficult to make out the beds. We have first flaggy sandstones, afterwards sandy shales, and opposite the church dark shales. Just above a small fault is seen, and a little higher up another fault, with vertical beds on one side and horizontal beds on the other. Higher up we get flags, tiles, and shales, and a little above the junction with the little stream coming from Draught Gates is seen another fault, which I suppose to throw down on the west. It is perhaps the same as that which ranges east of Fox Clough. I was told that near New Laith a coal seam of a foot or so in thickness had been dug into in draining. About 100 yards further up the beck than the fault last mentioned a fireclay is seen in the bank, with a coal the thickness of which I was unable to ascertain. Higher up we get alternations of sandstones, flags, and shales, but the section is not continuous. A coal 5 inches thick is seen in a quarry about 300 yards north-west of Trawden House, on the western side of the road leading to Colne. The details given will show that it is not easy to say whether workable coal seams underlie this valley].

The Gannister Coal (as I am informed) has been bored to in Shuttleworth Pasture, but here the strata are deeply buried under drift gravel, and for want of evidence I have been obliged to sketch in dotted lines the uncertain outcrops of the coal-seams. At Broad Bank Hill there are some traces of the outcrop of a coal-seam which I take to be that of the Upper Mountain Mine. From this point southward the easterly rise of the beds is very rapid, from 25° to 50°, so that we pass from the outcrop of the Upper Mountain Mine to the Rough Rock within a very narrow space; and as there are few natural and no artificial sections the outcrop of the coal-seams are not often exposed to view. These outcrops, however, can be very well made out in the valley of Swinden Water, on the southern slope of Extwistle Moor. The Upper Mountain Mine has here been worked to a small extent in a pit marked on the ordnance 6-inch map. The roof is black shale, with ironstone, and the dip is W.S.W. at 25°. I could not obtain a sufficient section to make out the thickness of the seam. The outcrop of the same coal may be seen again on the south side of the valley, between which and the former there seems to be a small fault. About 100 yards N.E. of this latter outcrop the Gannister Coal, about 4 feet in thickness, may be observed rising to the surface at an angle of 30°. The roof of the coal is dark blue shale, and the floor, clay resting on fine gritstone. The dip is west. Further up the valley we find the Millstone Grit in the form of conglomerate. The Gannister Coal may again be seen on the northern slope of Worsthorn Moor, where an opening has formerly been made into it. The roof is blue shale, the floor white under-clay, and the thickness 4 feet. The dip is west at 40°. Further south the Gannister Coal may again be seen cropping out on the southern banks of Sheddon Clough; the shale roof is here fossiliferous. From this point there is no difficulty in tracing the outcrop southward to the Portsmouth Valley along the banks of Coal Clough and Barn Edge Pasture, where it has been worked along a line of shallow pits; the dip is here W. 15°

S. at 25°, and in the shale roof may be found the usual fossils, *Aviculo-pecten papyraceus*, and *Goniatites Listeri*.

In the neighbourhood of Burnley at Fulledge the thickness of this coal-seam, there known as "the Bullion Coal," is 4 feet. Mr. George Wild, who has left us a minute account of it as found here, states that it often contains large quantities of rounded pyritous nodules in the upper part, together with impure cannel. The roof of the coal is dark blue shale, containing large rounded nodules of hard calcareous ironstone (cement-stones or "bullions") lying on the top of the coal and often jutting into it. These nodules contain species of *Nautilus*, *Goniatites*, and *Orthoceras*, also *Aviculo-pecten*, one nodule containing remains of a large fish, along with a very fine specimen of the tooth of *Cladodus*.*

We have now traced the outcrop of these lower workable coals all round the Burnley Basin. There can be no doubt they form an important source of supply for future use. The quality of the Gannister Coal is generally good; the thickness varies from 3 to 4 feet, or even more, and as yet it cannot be said that around the Burnley Basin, in consequence of the rapid descent of the beds from the outcrop, anything more than the shavings of this great block of coal have been carried away by mining. Nor is it difficult to ascertain the area of the Gannister Coal still remaining. Taking the area bounded on the north by the outcrop from Little Harwood to the fault east of Barrowford, then along the line of this fault to the outcrop in the Thursden Valley, then southward along the outcrop to Portsmouth, and from this point back again to Little Harwood, along the line of the Thieveley and Hambledon faults, we have an area of nearly triangular form, over which very little of the Gannister Coal has been worked, of about 50 square miles, or 32,000 acres in extent. Taking as an average 3,500 tons per acre, and deducting one tenth for the quantity already extracted and for loss, we have a residue of (in round numbers) 100,000,000 tons remaining for future supply from the Gannister Coal.

It is not too much to assume, considering the value and thickness of the seam, that the whole of it is within workable depth, the greatest depth under the Burnley Basin being little over 700 yards.†

MIDDLE COAL-MEASURES, BURNLEY BASIN.

The Middle Coal-measures embrace all the series of strata, including sandstones, shales, and clays of various descriptions, with beds of coal from the *Arley Mine* upwards. The Upper Coal-measures which occur in considerable force in the neighbourhood of Manchester are not represented in the Burnley district. In the following account of the Burnley Basin, I shall avail myself of the information contained in the valuable papers of Messrs. T. T. Wilkinson, F.R.A.S., and J. Whitaker,‡ and of Mr. George Wild.§

---

* Trans. Geol. Soc. Manchester, Vol. IV., p. 189.

† As the fossils of this district have not as yet undergone an investigation by the Geological Survey, I here give the names of those stated by Messrs. T. T. Wilkinson, F.R.A.S., and Whitaker in this paper on the Burnley coal-field. "The roof of the "Upper Mountain Mine (Spa Clough Top Mine) contains rays of *Gyracanthus*, teeth "of *Rhizodus*, *Megalichthys*, *Holoptychius*. That of the Lower, *Buccinum*, *Catillus*, "*Bellerophon*, *Pecten*, *Goniatites*, and *Orthoceras*." Trans. Historic Society of Lancashire and Cheshire, Vol. XIV.

‡ Read before the Brit. Association meeting at Manchester, 1861, and published with additions by Mr. Wilkinson in the Trans. Historic Soc. Lanc. and Chesh., Vol. XIV.

§ "*On the Fulledge Section of the Burnley Coal-field.*" Trans. Geol. Soc. Manchester, Vol. IV.

The Burnley Basin.—Understanding by the term Burnley Basin that part of the coal-field enclosed within the outcrop of the Arley Mine, its general shape may be described as that of a pear, lying with its footstalk pointing towards Blackburn and its head towards Colne. Owing, however, to the fracture of its southern side by two large faults, this pyriform shape is disfigured by a large protuberance extending along the Cliviger Valley, and embracing in fact the mining ground of the Cliviger Colliery Company. These faults may be called the Cliviger Valley and the Worsthorn faults; they have a general north-westerly direction, but towards the south-east gradually approach each other, though without actually forming a junction. In fact the Worsthorn fault appears to die away along the south side of Warcock Hill. These faults, by causing a downthrow of the Arley Mine between them, have added considerably to the area of this important coal-seam.

The coal-field is traversed by a series of large faults ranging in a north-westerly direction, some of which have been traced right across the trough from side to side. The effect of these faults, which for the most part throw down the beds on the north-east side, is to break the continuity of the southern outcrop of the Arley Mine, and give the margin of the basin a zigzag outline. To these faults I shall have occasion to refer hereafter. I shall now give the general section of the Middle Coal series, as proved at the Fulledge and Cliviger Collieries near Burnley, and on the Gawthorpe estate further to the westward.*

Coal-series at Fulledge and Gawthorpe, near Burnley.

FULLEDGE SECTION.

Name.		ft.	in.
Surface soil, clay, &c.		10	6
Boulder clay		27	0
Grey rock and strong shale with plants		14	0
Coal (Doghole)		5	0
Seatstone and blue shale		16	0
Coal (Charley)		1	0
Soft blue shale		2	9
Coal (Kershaw) { Coal, 0/8; Shale, 0/9; Coal, 2/0 }		3	5
Seatstone		4	6
Shales		24	9
Sandstone		9	0
Rag (sandy shales and flags)		22	2
Shales, with stems of trees and plants		10	8
Coal (shell bed) { Coal, 0/8; Shale, 0/10; Coal, 2/4 }		3	10
Blue shale with plants		10	2
Coal (Burnley 4-foot) good quality		4	0
Soft floor		1	0
Grey sandstone		24	0
Coal (Old Yard) inferior quality. { Coal, 0/7; Shale, 1/8; Coal, 2/10 }		5	1
Grey and dark shale		117	6
Bad cannel		0	4

GAWTHORPE SECTION.

Name.	ft.	in.
Lightenhill Park.		
Strata	21	0
Coal (supposed to be the Doghole)	6	0
Mary Anne Pit.		
(There is some uncertainly about the identification of these strata with the Burnley series above the "Lower Yard" Coal.)		
Strata	121	0
Coal	3	0
Strata	39	0
Coal (two seams with parting)	5	0
Strata	58	6
Coal (Mary Anne seam)	5	0
Strata	15	0
Coal	2	0

---

* Furnished by Sir J. Kay-Shuttleworth. The Cliviger section includes the five lowest seams, and was furnished by Mr. J. Jobbling, manager.

SECTION—continued.

Name.	ft.	in.	Name.	ft.	in.
Shaley sandstone (Rag)	8	4	Underclay and shale	39	0
Hard sandstone	5	6	Sandstone	4	6
Rag and soft metal	16	2	Coal, 6 inches, shale	12	6
Dark shale	13	0	Coal-smut	4	0
			Shale	8	6
Coal (inferior quality)	1	2	Coal-smut	3	0
Dark shale with *Anthracosia robusta* (?)	13	0	Strata with four thin coal-seams	183	0
Shale and coal	3	0	Shale	18	0
Coal (Lower Yard) { Coal, 1/1; Clay, 0/8; Coal, 1/2 }	2	11	Coal	1	9

Cornfield Pit.

Name.	ft.	in.	Name.	ft.	in.
Soft seat and shale	17	10	Soft seat and shale	19	0
Sandstone and shaley flagstone	36	0	Rag (flaggy sandstone)	7	2
Blue and black shale with thin coal-seams containing shells and fish remains	21	10	Shale	28	5
Blue shale with plants and *Spirorbis*	2	0	Roof	0	8
Coal (Low Bottom) (steam coal)	3	6	Coal (four-foot mine)	4	3
Seatstone	2	6	Seat and shale	16	4
White rock and rag (Tim Bobbin Quarry)	15	5	Rag and white rock (Ightenhill)	42	4
Dark grey shale (large *Anthracosia*)	4	5			
Inferior Cannel (fish remains)	2	3	Shale	7	0
Dark blue shale (*Anthracosia ovata, Modiola*)	4	3			
Rag and rock	18	5	Rag and Rock	16	3
Blue shale	39	5	Shale	4	8
Black shale with fish remains	0	4			
Coal (Fulledge thin bed)	2	10	Coal (Yard Mine)	3	0
Coal and shale	7	8			
Seatstone	4	0			
			{ Bing or clay, thinning out to a few inches at Padiham, and causing the union of the yard and 6-feet seams, then called "11 foot mine" }	9	7
Blue shale, &c. (*Anthracosia*)	32	4			
Black shale (fish remains, rare)	0	3			
Coal (Great Mine) { Coal, 2/4; Shale, 1/0; Coal, 1/7 }	4	11	Coal (6-foot mine) with partings	6	2
Soft seat-earth	0	6			
Rock (sandstone)	30	6	The strata below the 6-foot have not yet been sunk through at Gawthorpe, but have probably been pierced by boring between two upcast faults, 200 yards S. of Ightenhill Manor House, where a seam of 3 feet 3 inches was reached at 100 yards.		
Shale, with ironstone bands	22	0			
Slaty coal	1	8			
Strata supposed to be about	30	0			
Coal	0	5			
Rock	10	0			
Grey shale with ironstone	21	0			
Coal (China bed)	2	0			
Seat	2	0			
Rag and rock	15	8			
Blue and black shale	26	1			
Shaley Cannel (crackers)	2	6			
Black and blue shale	8	0			
Rag and rock (Bragget Hay Pasture)	18	7			
Black shale	4	1			
Rag rock	5	9			
Black shale	2	8			
Coal	1	0			

SECTION—*continued.*

Name.	ft.	in.	Name.	
Seat	5	7		
Coal and shale	1	9		
Ring rock	0	9		
Coal and clay	2	3		
Coal (Cally Mine, or Dandy bed)	2	9		
Seat	7	0		
Sandstone	22	0		
Blue shale and flaggy rock	40	0		
Grey shale, with six ironstone bands	17	0		
Black shale with ironstone balls	5	6	Dark grey shale	⎫
Grey shale roof	3	6		⎪
Coal (Arley Mine, Cliviger, 4 feet)	4	0	Coal (Arley Mine)	⎬ Section below Higham.
Seat clay, with two bands of coal	5	0	Seat clay and shale	⎪
Sandstone (Riddle Scout)	9	0	Flagstone	⎭
Black shale with ironstone	3	6		
Hard brown sandstone and flags	19	6		
Grey shale ("Grey layers")	25	0		

The *Arley Mine*, also known as the "Habergham Mine," and the "Marsden Four-foot," is, without doubt, the most valuable of all the seams in the Burnley Basin, as it is also the most extensive. It is available for household, gas, and coking purposes, and fetches a higher price than any coal in the district. It is generally admitted to be identical with the Arley Mine of Wigan, and the Royley of Oldham and Rochdale, and this opinion is confirmed on the grounds of the quality and thickness of the seam, its position in the coal-series, with reference to the Mountain Mines, and the nature of the roof, together with its fossil remains. Taking all these things into consideration, it is impossible to doubt the identity of the seams in these places. Mr. Wild remarks, with reference to the Arley mine at Burnley, "This coal is 50 inches thick, and the "upper half of it, with the exception of a few inches at the top, is a "strong cubical stone coal, giving off a very large quantity of gas (car-"buretted hydrogen). There is no distinct parting in the seam, yet the "lower half is a soft, dull-looking dust coal, free from shale, but containing "very much of what is locally called 'rotten coal' or 'lamp black.' "The upper half, or stone-coal, deteriorates in quality as we approach its "north-eastern outcrop near Marsden Heights, at which place, and at "Marsden, in some instances it is nothing but a bastard cannel."* In the black shale roof the remains of fishes are plentifully distributed, and amongst others fine specimens of *Megalichthys, Rhizodus,* and *Diplodus* have been found.

The coal is extensively worked at the Cliviger, Habergham, Marsden, Altham, and Hapton Collieries, while a large proportion of the seam has already been extracted from the outcrop inwards. Under the whole of the Gawthorpe estate, however, it still lies undisturbed.

Outcrop of the Arley Mine.—The extreme westerly point to which the Arley mine extends along the centre of the Blackburn and Burnley trough is Oakenshaw. Here we find in Banks Wood numerous old hollows and shallow pits, where the coal was worked in former times. At this point, however, it would appear that before the coal actually

---

* Trans. Geol. Soc. Manch., Vol. IV., p. 187-8

crops out in the bed of Hyndburn Brook, it is thrown out by a fault the position of which may be ascertained opposite the Print Works by the change of dip shown by certain flagstones in the river bank at Fiddlers Wood.

At Clayton the Arley mine was formerly worked from the centre of the trough to both its northern and southern outcrop.* It was here from 4 to 5 feet, and even more, in thickness, and was partly cannel. According to the statement of "Old Sagar" much of the coal still remains to be recovered. At Altham the outcrop of the coal may be very clearly traced in Altham Clough; on the north side of the canal aqueduct the dip is N. at 10°, and the coal is thrown out to the westward by a fault of 50 or 60 yards. The underlying flagstones are shown in a quarry on the top of the bank between the canal and railway. At Shuttleworth Hall the depth is only 35 yards, the dip is N. at 6°. At a distance of 350 yards to the east of the Hall the coal-field is traversed by a fault of 70 yards downthrow to the east, which throws back the outcrop to Hapton Hall. The overlying rock is shown in a quarry on the north side of the railway dipping N. 15° W. at 5°. From this point the outcrop takes a N.N.E. direction to Bradley Fold, where it is supposed to meet with another large fault, throwing down the beds to the eastward. I was unable, however, to ascertain the exact position of this fault, owing to the absence of sections in the upper part of Hapton Clough, but the effect is to throw the outcrop of the Arley Mine as far back as Habergham Clough, where it may be very clearly made out by numerous old workings and the character of the accompanying strata. West of Habergham Hall the beds are traversed by a small fault, which is shown on the mining plans to range in a N.N.W. direction to Lower Houses, and from this point the outcrop may be traced by means of the old hollows and pits along the bank of a little dell south of Habergham Hall to Sep Clough. East of this point the coal is again thrown down by a fault which passes in a N.N.W. direction by Appletree Carr.† On the east side of this fault the outcrop is at Wood Plumpton, beyond which it is terminated against the great Cliviger fault, by which it is thrown back to the south-east as far as Calder Head, a distance of 3½ miles.

The outcrop of the Arley Mine on Riddle Scout is well known.‡ The depth to the seam at the colliery at Calder Head is 97 yards, but of this 60 yards was composed of gravel, so that for the purpose of tracing the outcrop the depth can only be taken at 37 yards. A few yards to the west of the pits the great Cliviger Valley fault terminates the coal in that direction.

Although the dip of the beds at the outcrop of the coal at Cliviger is 15° to the west, and we get the three overlying seams brought in before reaching Holme, less than a mile distant, yet the Arley Mine nowhere descends to any considerable depth along this valley, as it is repeatedly brought to the surface by faults, the largest of which has been struck a few yards to the north of the coal pit on Hartley's Pasture. Along Green Clough, above Holme, there is also a reversion of the dip, by which the "Calley," "Crackers," and "2-foot" seams are made to crop out.

---

* In tracing the outcrop in this neighbourhood, I had the assistance of Charles Bradbury, manager of Church Colliery; J. Hargreaves, manager of Altham Colliery; and a miner of the name of "Old Sagar," whom I picked up at Read Hall.

† This, and several of the principal faults about Habergham, were traced on the Survey maps from the mining plans in possession of Mr. Dixon Robinson of Clitheroe Castle.

‡ In tracing the outcrop and faults of the Cliviger district, I had the friendly assistance of the proprietors and managers of the Cliviger Collieries.

Referring to the map for the position of the outcrop of the Arley Mine in this district, I shall merely mention a few spots where it may be traced;—on the top of the cliff at Riddle Scout, on both banks of Green Clough, the river bank near Honey Holme, in brook at Broughton's Wood, in pits at Causeway Side Inn, and in Shedden Clough Brook, a few yards above the ventilating pit, which is itself only 7 yards deep to the coal. On the north side of the brook the coal is thrown out by the Worsthorn Fault, but the coal reappears at Extwistle Hill, and may be very clearly seen at Proctor Cote, and on the top of a cliff on the southern banks of Thursden Brook. Then the dip is west at 5° to 10°.

The outcrop may be traced without much difficulty from Haggate along the northern slopes of Marsden Heights. At the time of my visit I had the advantage of seeing the coal laid open in the excavations for the service reservoir of the Marsden waterworks. Here adits had been driven in the coal from the outcrop. The seam was 40 inches thick, with a black shale roof and fireclay floor. The dip was found to be W. 10° S. at 6°. The outcrop was again found in Marsden churchyard,† and from this point it may be traced along the meadows to Clough Bottom by the old hollows, and the falling in of the ground caused by the extraction of the coal.

Crossing Pendle Water, we find the outcrop indicated by old hollows and pit banks by the side of Raven's Clough Wood, and the coal itself, with its underclay and seat-rock (Riddlescout Rock) is very well shown in the banks of the Clough opposite Old Laund Hall. The dip is here S.S.E. at 10°. At this point the extreme northerly extension of the mine is reached, and the outcrop gradually turns towards the south and west, and follows the line of the Pendle range. All the way to Huntroyde the outcrop is seldom visible, owing to the thick deposit of driftclay and sand which conceals the beds; at the same time we get assistance in our survey at several points. At Pendle Forest coal pit, near Fence, the Arley Mine was worked by means of a "whimsey" at a depth of 60 yards, a great part of which depth was doubtless sunk through drift, the outcrop therefore cannot be far from this spot. Following the brook course, which runs southward from Fence, we have little difficulty in determining the position of the outcrop to be at 300 yards below the road, where the Riddlescout Rock below, and the blue and dark shales with ironstone above, mark the position of the coal, which itself is covered over from view. The dip is here S.S.E. at 10° to 20°. It might have been expected that the outcrop of the Arley Mine would have been found in Moor Isles Clough; but although the beds which overlie the coal at a short distance are there very well shown, dipping at a moderate angle to the S.S.E., the section terminates in a mass of boulder clay which fills the bed of the brook where the coal with its associated shales ought to appear.‡ West of this clough the coal is probably thrown down to the south-west by a fault which may be seen near the junction of the brook with the Calder.

The outcrop of the Arley Mine may be seen in the brook course about 250 yards south of Higham, together with its shale roof, fireclay floor, and the Riddle Scout Rock. The dip is S. 20° E. at 45° to 50°.

---

* July 1867.

† Strange to say, at the time of my visit a grave was being dug out in the coal.

‡ My first visit to this clough was in February 1860, in company with the members of the Geological Society of Manchester, at the invitation of Sir J. Kay-Shuttleworth. An account of this visit is given by Mr. Binney, F.R.S., in the Trans. Geol. Soc. Manchester, Vol. II., p. 49 et seq.

During the excavation for the Padiham waterworks reservoir the Arley Mine was laid bare at its outcrop. I was informed by the man in charge of the reservoir that the coal was found to run in a north-easterly direction, rising at an angle of about 2 to 1, or 25° to 30°, and was 4 feet in thickness. Near the edge of Huntroyde Park the coal must be broken off against "the Hargreave fault," which has been proved at Padiham and Gawthorpe in the workings of the "11-foot Coal." Along the southern side of Huntroyde Park the Arley Mine has been worked, as well as in several pits west of Padiham. In one of these, near Dean Plantation, the depth was 80 yards. The outcrop of the coal occurs in the brook a few yards above Dean Bridge. The coal itself is not shown, but the roof of black shale with ironstone, and the fire-clay floor with the subordinate flagstone, are very well shown. The dip is S. 10° E. at 10°. The overlying sandstones are well shown in the valley south of Dean Bridge.

At Simonston Hall the Arley Mine is thrown nearly a mile to the southward by the agency of "the Simonston fault." This fault is very well shown in the brook course at Clough House, and again in Dean Plantation. On the west side of this fault the outcrop begins near Dean Brook Bridge, and runs through the meadows along the northern bank of the river Calder, under which river the coal lies at a depth of only a few yards. The coal was formerly worked by an adit entering the seam near the outcrop and descending with the dip. The trucks were drawn up the incline by means of an engine driven by water power. West of Dunkirk farm the mine is traversed by a fault, which (as I have already stated) crosses also the southern outcrop west of Altham Clough, on the other side of which the coal has been worked in the collieries belonging to Mr. Lomax of Clayton Hall, as previously described.

From the above description it will be observed that the Arley Mine between the Hyndburn Brook at Oakenshaw and Padiham occupies the centre of the Blackburn and Burnley trough, and has both a northern and southern outcrop. The axis of this trough follows the valley of Hyndburn Brook as far as Clayton Bridge. From this point it may be traced through Syke Side, and thence to the head of the river Calder opposite Eaves Barn. Following the course of the river by Padiham Bridge, it crosses Ightenhill Park south of Gawthorpe Hall, where the original "trough" assumes the form and arrangement of a "basin."

Area and Resources.—The total area of the Arley Mine is rather less than half of that of the Gannister Coal, being about 23 square miles, or 14,720 acres. Allowing 5,000 tons per acre, we have, as the total original quantity of coal in this seam, 73,600,000 tons; and assuming that of this one-tenth part has already been extracted, and allowing for loss and waste, there remains about 65,000,000 tons for future supply.

Great Hambledon.—It is a matter of general observation that the Millstone Grit occupies the highest ground, the Lower Coal-measures the intermediate, and the Middle Coal-measures the lowest in the Carboniferous districts of the north of England; but in the case of Great Hambledon, a hill lying midway between Burnley and Haslingden, we have a remarkable exception to this general arrangement; for here the Arley Mine, together with the overlying seams called "the Dandy" and "China" seams, is found at an elevation of 1,250 feet, the summit of the hill itself being 1,342 feet. The general position and arrangement of the beds with respect to the Burnley Coal-field will be better understood by reference to the following section (Fig. 14) than from a description.

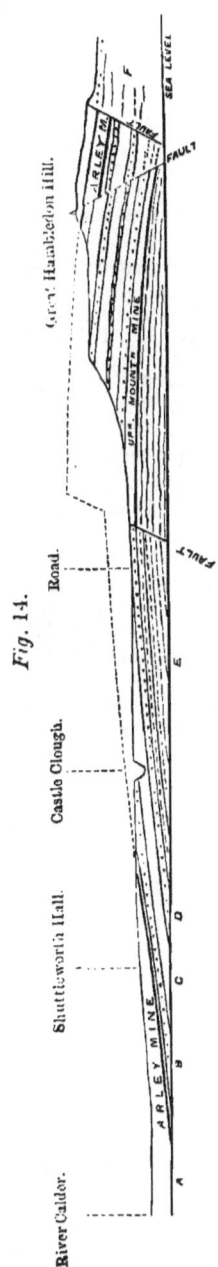

Fig. 14.
Section from the Burnley Basin across Great Hambledon Hill.

The greater part of the Arley Mine lies between two parallel faults which cross the hill from east to west. The western outcrop of the mine is at the "Slate Pits," under Mole Side Plantation, and the whole of Mole Side Moor is covered with little mounds and old pit-heaps, suggestive of the name; one of these pits close to Hambledon Hall is 70 yards deep. Near this pit one of the faults passes, having an up-throw to the north of 80 yards. It crosses the hill close to the Trigonometrical Station of the Ordnance Survey, and brings the Arley Mine on the north against a seam which I take to be the *Cliviger 2-feet* or *China Bed*. This fault continues in an easterly direction north of the Cupola Pit, and seems to be a prolongation of the Thieveley Fault which enters Cliviger Valley. The parallel fault which forms the boundary of the Arley Mine on the south of Hambledon Hill may be seen near the head of Whin Hill Clough close to the outcrop of the Gannister Coal. Here this coal and the Arley Mine are nearly brought in contact on opposite sides of the fault, the throw of which will therefore be about 370 yards.*

The Arley Mine again occurs at the head of Easden Clough on the northern flank of Deerplay Moors. It forms a narrow band about a mile in length along the north side of the Thieveley Fault. The outcrop may be seen on the moor at White Hill.

The Dandy Bed or *Cally Coal of Cliviger.*—This is a thin seam lying at Cliviger 39 yards above the Arley Mine. At Cliviger it consists of two seams separated by 9 inches of clay or "bing," the upper seam being 18 inches, and the lower 34 inches in thickness; it has only been worked to a small extent. In the Gannow Pit, Burnley, this seam lies 45 yards above the Arley Mine, and is about 30 inches in thickness. It is often less than this, and probably an average of 2 feet would not be far from correct. The roof

---

* In the survey of this district I was materially assisted by Mr. Pickup, one of the proprietors of the colliery at Hambledon Hill, and his manager, J. Hudson.

of the coal is generally sandstone of variable thickness, often protruding into it and diminishing its thickness. Mr. Wild states that in such cases the cleavage of the coal is to some extent disturbed or destroyed, causing an impediment to the working.

The outcrop of the Dandy Bed may be seen at the base of the cliff at Bradget Hey, above Cliviger Valley, also at Robin Cross and Greencliffe Nook. At Habergham the coal lies at the bottom of the quarry near the hall, and crops out in Habergham Brook about 450 yards above its confluence with Hapton Clough Brook, also in the road south of Four Lane Ends, and Sep Clough. Its northern outcrop may be seen in Raven's Clough Wood, and the rocky dell above Linedred Lane near Marsden. Its thickness is here 27 inches, with a sandstone roof 10 feet thick, and a fireclay floor.

The China Bed or Cliviger 2-feet Coal.—Lying about 16 yards above the Dandy bed at Cliviger there is a seam of inferior cannel mixed with bituminous shale, and called " crackers," from its crackling in the fire, and 20 yards above this is the " 2-feet seam." This coal is of inferior quality and has not been worked. In the section of the Gannow Pit it is mentioned as 2 feet in thickness, but towards its southern outcrop on Bareclay Hills it thins out. In Cliviger the outcrop is at the foot of the cliff called " The Lowe."

The Slaty Coal.—This seam, according to Mr. Wild, lies 17 yards below the " Great " or " Bing Mine," and consists of a heavy, dull, inferior coal, having a blue shale roof with thin bands of ironstone. In this shale scales and teeth of fishes are found.

The Great or Bing Mine.—This seam forms the base of Messrs. Wilkinson and Whitaker's " Burnley Upper Series." The exact distance of this seam above the Arley Mine at Burnley is only now being directly proved by the sinking of the new colliery to the Arley Mine at Bank Hall, but the distance has been indirectly obtained by calculation at Fulledge Colliery. Mr. Wild, by taking the known depths of the two seams on opposite sides of the fault in Fulledge Meadow, makes the distance 140 to 150 yards, and Messrs. Wilkinson and Whitaker give the distance 160 yards.*

This mine consists of two seams at Burnley separated by bing or clay from 1 to 2 feet in thickness. It may be seen at the bottom of Picup Quarry at Habergham Eaves, and there presents the following section:—

	ft.	in.
Fine-grained compact grey sandstone with partings of shale	25	0
Great or Bing Mine { Coal	1	4
Clay	1	4
Coal	1	6

In the section of Fulledge Colliery the roof is shale, and the coal is as follows:—

	ft.	in.
Great or Bing Mine { Coal	2	4
Shale	1	0
Coal	1	7

---

* Bank Hall Colliery belongs to the executors of the late Mr. Hargreaves, who, I have no doubt, would personally have been perfectly willing to give me any information they possessed for the purposes of the Geological Survey; but, unfortunately, Mr. Waddington having taken a different view, and having objections to furnish any information whatever, I am unable to give the details regarding this and several other collieries around Burnley.

The outcrop of this mine may be observed in the river banks of Pendle Water below the Lodge Farm. The coal is there in proximity to a fault of unknown magnitude, and the following section may be observed:—

		ft.	in.
Grey shale (forming the banks for several feet)	-	8	0
Great Mine { Coal	-	2	0
Dark shale and clay	-	2	0
Coal	-	1	6
Clay, passing downwards into grey shales with sandstones underneath		20	0

The outcrop of this seam at Habergham is near to Green Hill, where the dip is north. It is broken off to the eastward by the Cliviger Valley Fault, and on the downcast side is thrown back as far as Old House near Towneley Tunnel. At Fulledge Colliery the depth is 200 yards.

At Gawthorpe Collieries this seam is extensively worked under the name of the "6-foot" mine. It is 73 yards deep at the Cornfield Pits, and about 60 yards at the Broughton Pit. In the direction of Padiham and along the right bank of the Calder, the overlying "Yard" seam, which at Ightenhill Park is separated from it by 16 yards of strata, gradually approaches the "6-foot" seam by the thinning out of the intervening strata, till at length they almost unite and form "the 11-foot" mine of Padiham.

The western outcrop of this mine occurs in the bed of the Calder about 300 yards above Padiham Bridge; it then skirts the north side of the village below the quarries, and may be traced to the edge of Huntroyde demesne by a line of old pit-hillocks and hollows. At Higher Slade the ground has fallen in in several places. Beyond this point it meets with the Hargreave Fault, by which the outcrop is thrown about 300 yards to the south-east, and on the opposite side it may be traced on the ground from Hargreave Farm northward by old hollows. The coal itself may be seen in the banks of the brook below High Whitaker, but on the opposite bank is concealed by masses of sand and clay. Near the Hollins Farm the coal is probably broken off against a large fault which ranges in a north-north-westerly direction across the river at the Hag plantation. The throw of the fault at this place is probably about 100 yards up to the north-east, and the crop of the 6-foot coal is brought in all probability to the south side of the river. Here, however, the strata are much obscured by drift, but I have drawn a hypothetical line on the map representing the outcrop, crossing the Calder south of Hunterholme, and round by Pendle Hall to the bend of the river at the foot of Moor Isles Clough, where a thick coal was observed by me to crop out, which I cannot suppose to be any other seam than "the 6-foot" seam. At this point the coal is thrown out by the large fault which passes by Royle in a north-westerly direction.

I have already described the outcrop of this seam on the banks of Pendle Water near the Lodge Farm. Here it meets with a fault, between which and Burnley Lane Head the outcrop of this seam is unknown to me, partly owing to the absence of coal mining in that neighbourhood, and partly to the overspread of thick masses of Boulder clay. It is probable, however, that it overlies the sandstone of Burnley Lane Head, and crops against the bottom of the Boulder clay in the flat north of Bardon House and Rake Head Mill. In the neighbourhood of Burnley, the "bing" or clay which separates the great mine into two seams contains *Trigonocarpum* in abundance, as also *Lepidostrobus* and *An-*

thracosia,* and in the roof of black shale about 3 inches thick containing remains of *Megalichthys* and *Pleurodus* (a tooth). Three feet above the coal is a bed of small *Anthracosia*, "very numerous and nearly all closed, with the anterior side downwards."†

Area and Resources.—The Great Mine occupies an area of about 6½ square miles, or 4,160 acres; and if we allow 4,500 tons to the acre, and deduct 6,000,000 tons for quantity already got and waste, we have an available quantity for future supply of (in round numbers) 12,000,000 tons.

Fulledge Thin Bed.—At Burnley this seam is 34 inches thick, with a very thin parting of shale at 12 inches from the top. At Gawthorpe this seam is known as "the Yard Mine," and is there worked. The black shale roof is richly stored with fish remains, amongst which are stated‡ to be those of *Megalichthys*, *Ctenacanthus*, *Gyracanthus*, *Pleuracanthus*, and *Ctenoptychius pectinatus*, and *C. apicalis*.

The southern outcrop of this seam near the point where it is broken off against the Cliviger Valley Fault is near the junction of the Todmorden and Bacup roads. At Burnley the outcrop may be about half-way between Habergham Eaves Church and Pickup Quarry. In the neighbourhood of Padiham, the strata which intervene between this seam and the "Great Mine," measuring at Fulledge 50 feet, thin out, and the Yard and Six-feet Seams unite (as stated above) to form the "11-foot mine" of Padiham. The outcrop, therefore, is the same with that of the "11-foot mine," already described. At the Broughton Pit the depth is only 21 yards. North of Burnley the outcrop is generally concealed by boulder clay, but it probably ranges from Bardon Clough by Burnley Lane and Phesaudford Mill to the banks of the river Brun below Ridge End, where it was pointed out to me by Mr. T. T. Wilkinson.

Cannel Seam.—Lying about 20 yards above the Fulledge Thin Bed, is a seam of cannel, partly of inferior quality, resembling an oil shale, but which would appear not as yet to have been properly tested. As I had no opportunity of seeing this seam *in situ*, or actually mined, I cannot do better than insert Mr. Wild's description of it. He says, "The Cannel, 30 inches thick, is met with nine yards below the Blind-"stone bed. The upper part of this seam is dull, heavy, impure cannel, "while the middle is bright and light, getting rather heavy towards "the bottom. On the top of this bed are found large *Anthracosia* in "abundance, with the valves mostly open and crushed; and imme-"diately below, in the Cannel, remains of fishes, *Diplodus*, *Rhizodus*, "*Megalichthys*, &c."

Low Bottom, or Blindstone Coal.—This seam is "the 4-foot mine" of Gawthorpe. At Fulledge Colliery it was found at a depth of 153 yards, and was 42 inches thick, with a dark brown pyritous band ("blindstone") from one to two inches thick, lying near the roof, but in some places thinning out altogether. The coal is bright and soft, and used principally as steam coal. The roof of shale contains plants, with *Spirorbis* attached to the stems, and Mr. Wild states that he has found a shell resembling a small *Goniatite*. Above this is another "fish-bed, described by Mr. Wild as a conglomerate, containing nodules "of ironstone, smuts of coal, fossil plants, bones, scales and teeth of

---

* Messrs. Whitaker and Wilkinson. *Supra cit.*
† Mr. G. Wild, Trans. Geol. Soc., Manch., Vol. IV.
‡ By the authors above named.

" fishes, held together by a coprolitic cement," and over this comes dark shale, with bivalve shells, and a small univalve.

The southern outcrop of this seam may be found on the banks of a little dell below Copy Wood near Birks House, close to the Cliviger Valley fault. The seam was cut through in the railway tunnel. The northern outcrop may be seen in the banks of the river Brun below Ridge End, and in the overlying shales may be found abundance of *Anthracosia*. The outcrop may be traced still further westward to the canal aqueduct where it crosses Barden Clough.

The Low Bottom Coal has been exhausted under the Fulledge estate with the exception of some pillars, and is still mined at Whittle Field Colliery. It has been worked under Royle, and crops out in the river bed below the bank on which the house is built. The outcrop may also be observed on the top of the sandstone quarry at Ightenhill, and as the same rock again reaches the surface at Tim Bobbin Delf, the crop may be taken at the foot of the steep slope formed of this sandstone. In both these quarries the dip is eastward, but on the north side of Ightenhill Park the beds become nearly flat, and the coal crops out round the summit of the hill at the Hollins, Old Car, and Meadow Head.

The Low Bottom Coal is brought in to the west of the outcrop just described by the large fault which passes across the river Calder at the Hag Plantation, and ranges in a south-easterly direction by Ightenhill Park Old Colliery to Tim Bobbin Inn.* By this fault the coal is thrown down to the bottom of the hill as far as Gawthorpe Hall. At Cornfield Colliery the depth is 30 yards, of which 12 are composed of gravel, and the outcrop crosses the river a few yards above the Wooden Bridge.

It is not improbable that this seam occupies a small area on the north side of the river on the downcast side of the Hargreave fault. The rock of Padiham Quarry may be considered without any hesitation to be identical with that of Ightenhill Park and Tim Bobbin Delf, which immediately underlies the 4-foot coal. At Padiham Quarry it dips N. 15 E. at 6°, and therefore the 4-foot coal ought to set in immediately to the north of the quarry. It may be found, however, that while this is the case, the coal itself has long since been worked out.

The Lower Yard Coal.—The description of the remaining Coal-seams of the Burnley district need only occupy a short space, as their range is extremely limited, and the coal has for the most part been already exhausted. *The Lower Yard Mine* consists at Fulledge of four thin seams of coal in the following order :—

	ft.	in.
Coal, inferior quality	0	6
Clay or shale	0	6
Coal, inferior quality	0	6
Brown stone (bituminous)	1	6
Coal, with large pyritous nodules (workable)	1	1
Shale parting	0	8
Coal (workable)	1	2

In the roof of the uppermost seam occurs a large *Anthracosia*, probably *A. robusta*. This seam has been worked at Fulledge, and under Towneley Park, along the upper part of which it crops out. It also

---

* Between this fault and another, which passes close to the west of Ightenhill Park Quarry, there is an enclosed piece of ground, where the borings to test the coal do not correspond with those on the opposite sides of the fault. A solution of this enigma may be, that " the 6-foot," and all the higher seams, are thrown out over this space, leaving only the Dandy and Arley Mines underneath.

comes to the surface along the northern edge of the Clough which passes by Towncley Hall, and in which the underlying beds may be very well seen; the dip is N.N.W. at 6° to 10°. The northern outcrop crosses the river Calder opposite Crowwood House.

The Old Yard.—This coal was 67 yards deep at Fulledge Colliery, and is in two seams. The upper is 6 inches thick, of inferior quality, and is separated from the lower, which is 34 inches thick, by shale, crowded with plant remains. The lower seam is also of poor quality, having many thin partings of shale in it. This seam has only been sparingly worked.

The Burnley Four-foot.—This is a seam of good quality, but of limited area, and ought in fact to be described in the past tense, as the whole of the coal, with the exception of some pillars, has been cleared out. Its area is bounded on one side by the Cliviger Valley Fault, on the other by the Fulledge Main Fault, the breadth being only 820 yards, and its length between the northern and southern outcrops at Stoney Holme Cottage and Burnley Wood being little more than a mile. The roof of fine blue shale contains *Asterophyllites*.

The Shell Coal, or *Vicarage Mine*, was 51 yards deep at Fulledge Colliery. It is in two seams, the top 8 inches, the bottom 28 inches, with a parting of shale from 4 to 10 inches thick. The shale roof of this seam contains the trunks of fossil trees standing in an inclined position. This coal was formerly worked under, and to the north and south of, Burnley, and is nearly exhausted.*

The Kershaw Coal was 26 yards deep at Fulledge, and 20 yards at the Old Salford Colliery, and consists of two seams, the upper 9 inches thick, the lower 2 feet, separated from the upper by 9 inches of clay. The northern outcrop is at Danes House, the southern at the canal aqueduct over the Calder. The entire seam, which originally had an area of only half a square mile, is worked out where practicable.

The Doghole is the highest seam in the Burnley Basin, and was only 19 yards deep at Fulledge, and 12 yards at the Salford coal-pit. It is from 5 to 6 feet thick and of middling quality, and is all worked out with the exception of some pillars. The sandstone, which is laid bare in the bed of the river Brun, near St. Peter's Church, overlies this coal by a few yards. In the roof, which consists of strong blue sandy shale, are found fine specimens of several Carboniferous plants.

RESOURCES OF THE BURNLEY BASIN.

Having now described the entire series of coal-seams in the Burnley Basin, it remains only to say a few words regarding their areas and resources. In consequence of the general configuration of the basin, and the vertical distance by which the lower seams are separated from each other, the areas of the coal-seams become very rapidly contracted as we ascend in the series from the *Gannister Coal*. Thus (as shown above) the area of the *Arley Mine* is only about one half that of the *Gannister Coal*, while the area of the *Great Mine* is less than one third of that of the *Arley Mine*. After passing the *Low Bottom* Mine in ascending order, there are none of the remaining seams which add any appreciable amount to the resources of the basin. The *Lower Yard* Mine has an area of only three square miles, over which it has been worked to a considerable extent, and owing to the beds of shale by which it is split up is of but little value. The *Old Yard* Mine is of small area, and still more

---

* The old mining plans of this and the Kershaw and Doghole coal-seams are in the keeping of Mr. Dixon Robinson of Clitheroe, who kindly allowed me to inspect them.

inferior quality, while the *Burnley Four-foot*, the *Shell*, the *Kershaw*, and the *Doghole* seams have already been exhausted, as far as they are capable of being worked. In giving a summary of the resources of the Burnley Basin, I shall therefore leave out of consideration all the seams of coal above the *Low Bottom*, or *Four-foot* seam of Gawthorpe, but even with the above qualification it will be observed that the quantity of available coal is still very large.

Area and Resources of Coal-seams, Burnley Basin.

Name.	Area, Square Miles.	Tons per Acre.	Deduct for Loss. Quantity gotten. Tons.	Quantity Remaining, (Round Numbers) Tons.
Low Bottom Mine	4	4,000	3,000,000	7,000,000
Fulledge Thin Mine	3	2,500	2,500,000	5,000,000
Great Mine	6·5	4,500	6,000,000	12,000,000
Arley Mine	23	5,000	7,000,000	65,000,000
Gannister Mine	50	3,500	10,200,000	100,000,000

By adding up the last column we get an estimate of the entire quantity of coal remaining in the basin for future supply to a maximum depth of 700 yards, namely 189,000,000 tons, in which is not included the quantity from the Upper Mountain Mine, which, on account of its thinness, I have thought it safer to omit from the estimate.[*]

FAULTS OF THE BURNLEY BASIN.

On referring to the maps, it will be observed that the majority of the faults which traverse the Burnley and Blackburn trough take a direction west of north, and east of south; thus corresponding with the system of fractures which characterize the South Lancashire Carboniferous District. Those near Rivington, Turton, Darwen, and Holcombe have already been alluded to when treating of those places. I shall therefore confine my observations here to some of the spots where the faults of the Burnley and Blackburn trough may be observed, or the grounds on which they have been drawn on the map.

Commencing near the western extremity of the trough, we notice first the fault which is drawn nearly parallel to the Clitheroe railway across the strike of the beds between Royshaw and Little Harwood. The evidence of this fault is twofold. In the first place no one can examine the contorted state of the Millstone Grits in the cutting of the railway at Lower Wilworth without being satisfied that a line of disturbance is close by; and in the second place, the strike (or direction) of the beds on either side of the valley show that they are shifted from their original line, and have been thrown back on the west side. The throw of the fault at Little Harwood may probably be about 150 yards.

The fault at Harper Clough may be clearly determined at the quarry, where it will be observed that the Rough Rock on the east side is brought down against the Third Grit on the west. The line of the fault between Harper Clough and Dunscar is marked by a prominent knoll of grit, at the base of which a spring bursts forth.

The Oakenshaw fault has already been alluded to as being indicated by the change of dip of the strata in the river Hyndburn; and its position further to the north-west seems to be pointed out by the sudden

---

[*] Mr. Elias Hall in his "Key to the Geological Map," gives the quantity of coal at 70,212,266 tons (1836).

truncation of the ridge of Millstone Grit at Bowley Hill. At the time of my survey this fault had not been reached from Great Harwood Colliery.

Altham Clough fault has been proved on both sides in the working of the Arley Mine; it has a downthrow to the N.E. of 50 or 60 yards.

Hapton Hall fault has also been proved on both sides in the working of the Arley Mine; by its means the outcrop of this seam is thrown to the south from Higher Shuttleworth to Hapton Hall, a distance of nearly half a mile. The flagstone of Castle Clough will, on examination, be found to be a rock of a different character from that which is shown in the railway cutting on the other side of the fault.

The fault which passes by Padiham Green and Bentley Green is inferred from the fact of the outcrop of the Arley Mine being thrown to the southward a distance of a mile from Stone Moor to Habergham Clough, but its precise position I had no means of ascertaining for want both of sections and information. Its position, however, on the banks of the Calder at Padiham was approximately determined. The downthrow is to the east.

The fault that ranges from Long Lane plantation to the west of Habergham Hall by Lower Houses has been drawn from the mining plans of the Gawthorpe and Habergham collieries; it is shown in Habergham Brook very plainly.

The faults which pass under Rose Grove Station, Bareclay Hill, and Rose Hill, have also been obtained from plans of Habergham collieries.

Cliviger Valley Fault.*—This is the most important of all the lines of fracture in the Burnley district, producing marked effects both on the outline of the coalfield and the physical features of the landscape. To this fault, and the denudation which accompanied and followed it, we owe the valley which gives birth to the two rivers—the twin Calders—and presents varieties of scenery unsurpassed in diversity and loveliness in Lancashire. The south-western side of this valley presents the features of a rocky escarpment surmounted by moorlands, and descending into the valley by green slopes. The edge of the escarpment is formed, as at Thievely Scout, by the Woodhead Hill Rock, in other places by the Millstone Grits of various ages, and towards Burnley by banks of Lower Coal-measures. From Thievely northward, where the Lower Coal-measures takes the place of the Millstone Grit, the scenery is not of so bold a character. The fault generally lies close to the base of the southern escarpment; and on the northern side the country descends into the valley along more gentle slopes than on the opposite side. The position of the fault has been determined at the colliery at Calder Head and at Towneley Colliery. Here the fault itself, and a branch which for some distance splits off and then rejoins, may be seen in the sections at Copy Clough and in the brook course at Lower Timber Hill. The fault has been found in the workings of Whittle Field Colliery to pass under Clifton, and seems to pass a little east of Broughton Pit. Here, however, it appears to be on the point of dying out, the downthrow not being greater than 10 to 12 yards.

The Fulledge fault has been proved in the working of Fulledge Colliery. In conjunction with the Cliviger Valley fault it forms a

---

* The position of this fault at Cliviger Collieries has been determined for me by the proprietors and manager, at Towneley Colliery by Mr. Pickup, and at Burnley by Mr. Wild of Fulledge, and Mr. H. Jobbling of Fulledge House, manager of the estates of Towneley Parker, Esq., of Cuerden Hall. In the survey of this locality I was kindly assisted by the brother of Mr. Pickup of Towneley Colliery, proprietor of Copy Clough Quarry.

trough in which are contained the higher coal-seams of the Burnley Basin. The fault passes from Bank Hall southwards along the valley of Towneley Park, but appears to die out before reaching Cliviger Mill, as it has not been crossed in the levels of the Arley Mine near that spot. Its prolongation further north than Danes House is uncertain. At Fulledge Colliery the slope of this and the Cliviger Valley fault was found to be 1 to 2, so that the area of each successive coal seam in descending order became more and more contracted (see Fig. 15); so

Fig. 15.

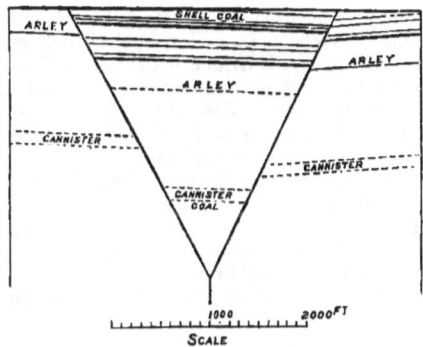

Section at Fulledge Colliery.

that after working all the seams down to and inclusive of the *Great Mine*, the sinking of the pits to the Arley Mine, the most valuable of them all, was abandoned because it was considered by the proprietor, Mr. Cardwell, that there would not be a sufficient breadth of coal to admit of profitable working after the outlay of the sinking and additional plant had been incurred. This will be rendered clear from the following section across the trough.* Mr. Wild cut across the 100-yard fault to the upcast side and worked the Dandy and Arley mines.

The Thieveley Fault.—This fault is of special interest as being metalliferous in several parts of its course, a fact stated by Messrs. Whitaker and Wilkinson† in the paper from which I have already quoted. The fault branches off to the westward from the Cliviger Valley fault at Copy Bottom, ascending the hill by Thieveley farm and the northern end of Thieveley Scout, crossing Black Clough and Easden Clough a short distance below their sources. Here the amount of the throw is exactly the vertical distance from the Upper Mountain Mine to the Arley Mine, or about 1,000 feet. At White Hill it is intersected obliquely by the Deerplay Hill fault, but continues its westerly course, crossing Greenhill Clough near the reservoir, and then, on reaching the western side of the valley, it splits into two branches, and crosses Hambledon Hill, as already described. In the workings of the Upper Mountain Mine the fault has been struck where it crosses the Rawtenstall Road about 100 yards south of the milestone, three miles from Burnley. The fault is a downthrow on the north side.

The traces of the old workings for lead may be seen at Thieveley, and on the moors on both banks of Black Clough. Mr. Wilkinson also

---

* Those who wish for an explanation on mechanical principles of the manner of the formation of trough-faults, will find one in Mr. Jukes' "Memoir on the South Staffordshire Coal-field," 2 Edit., p. 196–7.
† Trans. Hist. Soc. Lancashire and Cheshire, Vol. XIV.

mentions the occurrence of lead ore on Hambledon Hill, at the quarry. The ores appear to have been both galena and carbonate of lead, and the veinstone a fine grit.

The *Deerplay Hill Fault* is one of the most important in the district. Commencing at Ramsden, on the western slopes of the Vale of Todmorden, it ranges in a north-westerly direction to Huncoat, where it appears to pass into a low arch or anticlinal axis. It is everywhere a downthrow on the north side, attaining its maximum north of Bacup, along the base of Deerplay Moor.

The fault has been proved in several places and may be seen in others, amongst which may be mentioned the waterfall in Ramsden Clough, the brook course which descends into Howroyd Clough on the northern slope of Inchfield Moor, South Graine at the head of Dules Gate, where it was pointed out to me some years since by Mr. J. Dickinson;[*] the throw is here about 100 yards. The fault may again be determined by the reversing of the dip at Clough Head Brick Works; again at Scar End Brook, where the Gannister coal is cut off against it; at Irwell Spring Print Works, where the throw is about 250 yards and a maximum; at Whitewell Brook; in the bed of Limy Water at Whorelaw Nook, where it has been proved in the mines; in a brook course east of Porter's Gate, and in Hapton Tower Brook, where the dip of the strata is reversed. In Thorney Bank Clough it seems to cross at the lower end of the wood. The line of fracture is throughout its course remarkably straight.

The faults at *Rowley Colliery* have been laid down from mining plans; one of these passing by the west side of the pit may be seen in the banks of the river Brun; the other, which passes across the Fulledge trough, ranges in E.N.E. direction by Brunshaw and Rowley Hall, and terminates against the Worsthorn fault. It is a downthrow of 22 yards on the north side at Rowley, and is visible in the banks of Swinden Water. This is a well-authenticated example of faults crossing each other.

The fault which runs along the valley of the river Brun, below Hollins, is shown at two points of intersection of the brook course. In the bed of the brook under the house the beds are sharply reversed, and further up the stream the fault itself is seen in the bank, a sketch of which is given (Fig. 16). On the east side of the fault massive sand-

Fig. 16.

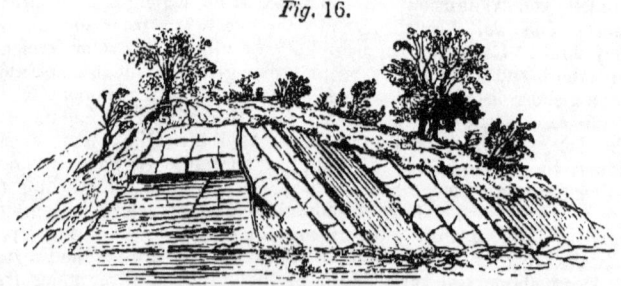

Sketch of the Hollins Fault, in the Bank of the River Brun, near Burnley.

stone is brought up in the river near Ormerod Wood; this rock lies immediately under the Dandy Coal, or about 40 yards above the Arley mine, and again appears higher up the stream at Rock-Water.

---

[*] During an excursion with the members of the Manchester Geological Society in 1862-3.

The *Worsthorn Fault* has been traced from Shedden Heys to the north of Burnley Lane Head in the workings of the Arley Mine. At Cliviger Colliery this seam is thrown out by the fault on the north-west side. The position of the fault may be determined within a few yards above Hurstwood Hall, for the grey shales and flagstones which form the banks of the brook about 50 yards above Hurstwood hamlet are on the upcast side, and belong to the Lower Coal-measures. The fault has been ascertained to pass a few yards west of Worsthorn Church, and its position at Houghton Hay has been proved in the working of Rowley Colliery. Further towards the north-west it has been bared on the upcast side in working the Arley Mine at Hag Gate Colliery and Marsden. On the west side of Pendle Water it passes under Montford, and has been traced almost to the western outcrop of the coal. The downthrow is on the south-west side.

Marsden and Colne Districts.—The principal faults which traverse the districts formed of the Lower Coal-measures and Millstone Grit in the neighbourhood of Marsden and Colne have already been described in the pages in which I treat of these formations, and to which I must refer the reader for further information (p. 44).

Padiham and Gawthorpe.—Several of these faults have already been alluded to when describing the outcrop of the coal-seams, but some require further notice.

One of these is the large fault which passes in a N.N.W. direction at the junction of Pendle Water with the Calder. At its southern extremity from Crowwood House to Royle it has been drawn from plans of the workings of the "Low Bottom" coal. It might be supposed to be a continuation of the Fulledge fault, which has been traced to Dane's House, but I have been unable to connect it therewith, and from the old plans of the "Vicarage mine" works the levels of this seam would appear to have passed across the position of the fault without a break. It is possible, however, by a bend to the northward to connect the faults, and the question as to their continuity will doubtless be settled ere long by actual mining. The position of the fault is indicated by the verticality of the beds in the river bank at Ingham's Farm, and it again gives indications of its presence near the junction of Moor Isles Clough with the Calder, where the beds are much broken and disturbed. The fault is an upcast to the north-east, but to what amount is uncertain as the mineral structure of the ground on that side of it is at present very obscure.

The Ightenhill Fault.—The fault which crosses Ightenhill seems to split into two opposite the Old Hall, each of the arms being downthrown outwardly (*see* p. 81, footnote). The strata, consisting of shales and flaggy sandstones on the east side of this fault, may be seen in the left bank of the river Calder at the Hag Plantation; the ordinary dip is here reversed, being N. at 15°; the opposite side of the fault is filled in with boulder clay, a phenomenon by no means unusual in Lancashire.* The northern prolongation of the Ightenhill fault across Padiham Heights may be very clearly traced by the break in the continuity of the Rough Rock east of Copthurst near Higham, the rock being there thrown to the southward about 100 yards. Owing, however, to the rapid dip of the beds (50°) the horizontal displacement appears to be less than in the direction of the centre of the basin where the dip decreases.

---

* Examples which occur to me, at this moment, are to be seen in the railway cutting between Turton and Entwistle Station, and in the banks of the river Goyt near Disley.

The Great Anticlinal Fault.—By this name my colleague, Mr. A. H. Green, and I have distinguished the line of fracture which we have traced from Wetley and Leek in Staffordshire northward by Disley, Marple, Saddleworth Valley, to Todmorden and the western slopes of Black Hambledon, a distance of 53 miles. Through this distance the beds are uplifted on the eastern side, and the dip is reversed. This reversal of the dip, however, does not always take place exactly where the fault passes, but sometimes a short distance from it on the eastern side, in which case we may suppose that the elevatory force has not been sufficient to overcome the cohesion of the beds, or the force of friction along the line of fracture. One of the instances in point may be observed on the northern flank of Blackstone Edge, where the Kinder Scout Grit forms an arch, the axis of which passes under Blackstone Reservoir (see Fig. 17).

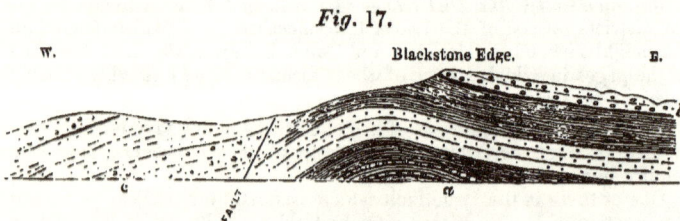

Fig. 17.

W.   Blackstone Edge.   E.

Section across Blackstone Edge.

a. Yoredale Beds.   *b.* Kinder Scout Grit.   *c.* Millstone Grit.

But whatever may be the position of the beds in close proximity to the anticlinal fault, the general change of dip from west to east at, or near, this line is apparent along the borders of Lancashire and Yorkshire. There is also another very evident change which may also be noticed as accompanying the line of "the anticlinal fault," which seems to indicate this fracture as having been the axis along which elevatory forces exerted their highest efforts. All along the western side of the fault, from Staleybridge northward, we find the beds broken by faults, much disturbed, and dipping westward at right angles; but on crossing the fault itself we pass almost immediately into a country where the stratification is remarkably regular, and where the beds are but slightly removed from the horizontal position. This fracture must therefore be regarded as the axis of elevation of the Penine Chain. I shall now state a few points where the fault itself, or its effects, may be clearly observed in our district, referring to previous memoirs for an account of its prolongation southwards.*

Commencing at the southern extremity of Blackstone Edge, we find the beds of shale and grit much shattered in the banks of Longden End Brook at the foot of Little Hoar Edge. The fault itself is shown in a little brook east of Blackstone Edge Fold, again in Light Hazles Clough on Chelburn Moor, in Calf Lee Clough, by lines of springs (Cat Stones Spring and Shaw Stones Spring) along the western slope of Walsden Moor, and the abrupt termination of Langfield Edge. On the north side of the Portsmouth Valley its position is shown by the change of dip between the beds of Kinder Scout Grit and conglomerate above Lower Hartley, and that of the beds which form the prominent cliff of Hartley

---

* "Geology of the Country around Oldham," p. 57. "Geology of the Country around Stockport, Macclesfield, &c." See also "On the Millstone Grit of North Staffordshire, &c.," by Messrs. Hull and Green, Journ. Geol. Soc., Vol. XX.

Nez and those in the brook course to the north of it; the dip in the former case being north at 25°, in the latter west, and west 10° south at 20°.

The position of the fault near Stiperden House may be determined by observing the high S.W. dip of the beds in the Clough below, and their approximately horizontal position at the Hawkstones. The last place I have observed the fault is in Black Clough, one of the upper tributaries of Cant Clough Beck on Worsthorn Moor. Mr. Gunn contributes the following paragraph descriptive of the fault further north :—

" The fault appears to go to the west of the Gorple Stones and Hare Stones bringing up the Yoredale Shales on the east; and though not actually seen, it most certainly crosses the head of the Widdop Valley, as is inferred from the trend of the beds there, for we have the Yoredale Shales of the valley brought up against coarse grit on the west. The fault is supposed to range along the western flank of Boulsworth, for there does not appear room for all the Third Grits to rise and crop out between the well-ascertained Rough Rock of Willy Moor and the Kinder Scout Grit of Boulsworth Hill. It is certain that a fault must pass between the Upper Third Grit of Antly Gate and the beds north of Round Hole Springs, for these latter are dipping at high angles of 40° to 60° to the north-west, and striking right at the Third Grit, which dips at low angles to the west and south-west, and can be traced round to Alder Hurst Head parallel to the Rough Rock. Near Boulsworth the arch of the anticlinal leaves its direction and strikes off to the north-east, up Heyslacks Clough and to the north of Crow Hill, ranging parallel to the axis of the Burnley Basin, and gradually disappearing to the eastward. Boulsworth may therefore be considered to be the northern termination of the Great Pennine anticlinal, and the country immediately to the north which is much faulted, as the common battle ground, so to speak, of the forces which produced both the Pendle and Pennine systems of elevation.*

A fault which throws down in the same way as the anticlinal fault ranges along the Trawden Valley, but it would be a misnomer to consider it a continuation of the anticlinal fault, and it seems rather to belong to the system of N.N.W. faults which are so numerous in the Lancashire Coalfield."

---

* See for similar case, Mem. Geol. Survey, Geology of Country round Oldham, by E. Hull, p. 8.

CHAPTER IV.

THE CHORLEY COAL-FIELD.

(Coppul District.)

THE Coal-measures to the south and west of Chorley form the northern limit of that part of the South Lancashire Coal-field generally called " the Wigan district," as pourtrayed in Sheet 89 S.W. of the Geological Survey Maps, and already described.* This district extends westward from the large fault which I have described under the name of " the Great Haigh fault,"† which has a downthrow to the west of about 600 yards at Kirkless Hall Colliery. This fault ranges nearly northwards, with a diminishing amount of displacement, and is supposed from the results of borings made over the property of Mr. Cardwell, M.P., at Ellerbeck, to range along the east side of the Wigan and Chorley Road towards Yarrow Bridge, in which direction it has been proved in the workings of the coal-seams at Duxbury Colliery. To the east of this fault the Lower Coal-measures are brought to the surface, while on the west side several of the Coal-seams of the Middle Coal-measures, from the Cannel and King coals down to the Arley Mine, inclusive, are known to exist.

The northern portion of the Coal-field, which may be conveniently called "the Coppul district," is for the most part concealed under a deep deposit of boulder clay, sand, and gravel, so that sections are rarely visible even in some parts of the deeply-cut watercourses. It fortunately happens, however, that along the banks of the Yarrow, from Birkacre Printworks up the stream almost to its source, a nearly continuous section can be made out; which, together with the mining operations which have for many years past been in progress, has made the geological structure of this neighbourhood perfectly clear.

The Carboniferous Rocks are terminated towards the west by a supposed large downthrow fault, ranging through Bispham (Sheet 89 S.W.) in a north-easterly direction by Knowles Wood and Eccleston Green. In the brook courses here the Carboniferous Rocks will be found in close proximity to the New Red Sandstone; and although the actual line of junction is not visible, yet the general arrangement of the beds, together with the straight course of the boundary itself, favours the supposition that it is not an overlap of the newer on the older formations, but an actual fracture or fault.‡

The Coal-field to the east and south of the boundary fault above described is traversed by a series of fractures, which seem to be continuous with those which intersect the Wigan district; up to the time of the completion of the survey, about 15 years ago, the connexion of several of them had not been proved by actual mining operations in the Coppul district. These fractures are supposed to terminate against the boundary fault of the Coal-field; the evidence, however, is somewhat of a negative character, as it is impossible to say to what extent the Triassic district is intersected by faults, owing to the paucity of sections.

---

* E. Hull, Geol. of Wigan, 2nd edition Geol. Survey Memoirs.
† Ibid., p. 25.
‡ This supposed fault is referred to in the memoir on the Geol. of Wigan as terminating towards the N.W. the little tract of Permian Rocks shown in Skillaw Clough, p. 27. The continuation of it near Salmesbury has been determined by Mr. Tiddeman.

THE CHORLEY COAL-FIELD.

The Coal-measures are divisible into an upper and lower series; the base of the former being assumed to be on the horizon of the *Arley Mine* coal, although it is questionable whether it ought not to include a thin series of shales which overlie the highest beds of the "Up-Holland flags."

Lower Coal-measures, or Gannister Beds.—These consist in the Chorley district of a series of massive reddish sandstones, micaceous flagstones, and shales, with several thin beds of coal ("Mountain Mines"), two of which are workable, and have been mined to a considerable extent over the districts of Charnock Richard, Chorley Moor, and Ellerbeck. The whole series attains a thickness of 1,800 to 2,000 feet, measured from the Arley Mine down to the "Rough Rock," which forms the highest bed of the Millstone Grit series.

The lowest beds are to be seen in a brook course east of Hurst House Delf, Heskin, where the following section was made out:—

		ft.	in.
1.	Coal with shale roof -	1	0
2.	Hard siliceous floor	0	10
3.	Flagstone (base not seen) - about	5	0
4.	Black shale	4	0
5.	Coal -	0	10
6.	Hard siliceous floor -	0	5
7.	Strata (probably shale) not seen, reposing on Millstone Grit -	10	0

A series of purple sandstones and shales a little higher in position may be seen along the banks of the Yarrow below Bolton Green for some distance, with a dip varying from S. to S.W. These underlie the "Lower Mountain Mine," which is the representative of the "Gannister Coal" of Lancashire. This seam is about 18 inches in thickness, and has been worked in a pit 18 yards deep, a short distance north of Charnock Green. Micaceous flagstones, which lie a few feet above the seam, may be seen in a small quarry by the Preston and Wigan Road, dipping S. at 6°.

The Upper Mountain Mine, being thicker than the Lower, has been more extensively wrought in this district. The quality is good, and it has a thickness varying from 22 to 32 inches, 28 inches being the average. It is overlaid by dark shale, containing bivalve shells (*Anthracosia*), and the floor is rather hard sandy fireclay. The following may be taken as an average section in the district of Charnock Richard:—

Section of the Lower Coal-Measures at Charnock Richard.

		ft.	in.
1.	Hard reddish sandstone, used for building and paving	78	0
2.	Flaggy micaceous sandstones and shales	141	0
3.	THIN COAL	0	7
4.	Potter's clay -	6	0
5.	Sandstones and shales passing down into black bass with shells -	42	0
6.	COAL, UPPER MOUNTAIN MINE (average)	2	4
7.	Underclay, rather hard and sandy -	2	0
8.	Hard flaggy sandstone and shale -	30	0
9.	Bluish shales resting on micaceous flags and shales -	240	0
10.	COAL, LOWER MOUNTAIN MINE (Gannister Coal) -	1	6

---

* Information was very readily afforded by the proprietors and managers of the collieries, amongst whom I could especially mention Mr. Darlington, Mr. J. Smith, and Mr. J. Whalley.

The outcrop of the Upper Mountain Mine, together with the directions of the faults by which it is intersected, have been traced on the map of the Geological Survey, chiefly from the plans of the collieries in which it has been worked. Near the northern end of Spring Wood in Heskin Hall grounds it was reached in a pit at a depth of 72 yards, but in this neighbourhood it does not crop up, owing to the boundary fault at Eccleston Green, against which it strikes ere reaching the surface. It is brought, however, to the surface by a large fault, which ranges N. and S. by Heskin Bridge, and which produces a downthrow on the west of about 100 yards. At Ox Hey Wood, on the east side of this fault, I saw the outcrop of the little coal, which lies 16 yards above the Upper Mountain Mine. The potter's clay, which here underlies it, is 6 feet in thickness.

The coal is considered to be broken off before reaching Red Lane by a large fault in a N.W. direction under the Bottom Mere of Park Hall, and which is considered with every probability to bring down on the S.W. side the Arley Mine, nearly on a level with the Upper Mountain Mine on the opposite side of the fault. The depth from the Arley Mine to the Upper Mountain Mine has been variously estimated from 330 to 360 yards, which represents the amount of the displacement opposite Park Hall. The outcrop of the coal is seen in a lane above Roscoe Wood, east of Charnock Green, but in the valley of the Yarrow it is concealed beneath a thick deposit of Drift gravel. It has been worked around Gillibrand Hall, and from this it stretches in a N.E. direction towards Chorley, cropping out a short distance N. of Chorley parish church, and terminating against the large fault already described as ranging in a northerly direction by Yarrow Bridge Inn.

The displacement owing to this fault is so great, that the outcrop of the coal is thrown to the southward for a distance of two miles, and it again sets in at Halliwell Field, and crops out along the north side of the hill on which stands the old mansion called Hall o' th' Hill. In this district the coal has been worked in pits belonging to the Right Hon. E. Cardwell by Mr. Rowbottom.* A bed of massive reddish sandstone, which overlies the coal, is opened out in the railway cutting south of Halliwell Field.

Middle Coal-measures.—Under this head we include all the beds from the Arley Mine upwards to the Worsley Four-foot Coal, which is probably represented by the seam that crops out on the bank of the Douglas at Fairclough Wood. In Sheet 89 N.W., however, only the lower part of the series is present.

The beds consist of an alternating series of reddish, grey, and yellow sandstones, shales of various characters, fireclay, and beds of coal, which rest upon floors of soft unctuous clay, generally called "warren earth," or "warrant," in Lancashire. The shales often contain nodules, or bands of clay-ironstone, but not in sufficient quantity or thickness as to render them available for economic purposes.

The Arley Mine.—The lowest coal-seam of the Middle Series is the *Arley Mine*. This seam, which in quality and value ranks next to the Cannel Mine around Wigan, has its ultimate outcrop towards the north in the Chorley district. Its thickness varies from 3 feet 6 inches to 4 feet 6 inches, and the position of the outcrop has been ascertained, either by borings or actual workings, from the large fault north of Yarrow Bridge on the east, to that which ranges by Howe Brook House, Heskin, on the west. At Coppul Colliery the difficulty of working the

---

* In a fault at this colliery galena and iron pyrites were found.

Arley Mine is increased by the presence of a roof of white clay between the usual black shale and the coal. At Duxbury Colliery this is separated from the coal by 6 feet of black bass or shale. I shall now indicate the evidence from which the position of the outcrop has been traced on the map, as collected by myself at the time the survey of the district was in progress (1859).

North of Yarrow Bridge the outcrop has been ascertained by the workings of Duxbury Colliery to be a little north of Red Bank Cottage; but a pit which was sunk for some depth at Halliwell's Farm, near the Duxbury Road, and which was intended to work the Arley Mine at its extreme north-easterly limits, was drowned out by running sand. At Lighthurst the coal is traversed by a fault ranging nearly north and south, having a downthrow of 15 yards on the east, and which passes by the west of Duxbury Colliery. On the west side of the fault the coal has been worked to a considerable extent from Burgh Hall Pit, which was 90 yards in depth, and the outcrop of the coal (under a deposit of drift, sand, and gravel) was proved in a field about 180 yards north of Birkacre Colliery.

At Dry Dam Pit the Arley Mine was worked at a depth of 60 yards; and a short distance west of the pit it was found to be thrown out by a large upcast fault which ranges in a north and south direction, and was found further north in the workings of the Mountain Mine west of Gillibrand Hall. West of this fault the position of the Arley Mine is somewhat uncertain, but the outcrop (under the clay and sand) is considered to range by Coppul Railway Station in a south-west direction till it meets the large downthrow fault which ranges by Haydock's Farm, where it is again thrown back as far north as the little colliery which stands by the roadside near this place. In this pit the depth was only 30 yards, and in two others further west on either side of the same road (leading to Chorley) the depth was 12 and 32 yards respectively.

A short way further towards the south-west the coal is cut off and thrown down by the prolongation of the large fault already described as ranging by the east of Park Hall, where it was proved in the workings of the Mountain Mine. This is doubtless the same as the "Great Shevington Fault" described in the memoir on the "Geology of Wigan,"* which produces such marked changes in the position of the strata along its course through the Wigan Coal-field.

The position of the Arley Mine, and some of the overlying coal-seams south-west of the fault, has been pretty well ascertained by borings and shallow pits. One of these latter was sunk near the southern end of the plantation called "Little Wood" west of Park Hall, and borings were put down in position south-west of this pit, so as to prove the course of the seam along the level or strike of the beds. From these borings, and judging by the strike and dip of the Mountain Mine further north, we may infer that the coal trends in a south-west direction by "Cooper's Allotment" with a dip towards the south-east.

To the south of Pyebrook Hall the Arley Mine must be traversed by the north and south fault which has been found at Heskin Collieries, and which has a downthrow of about 100 yards to the west. Beyond this it has been worked near the crop in Pyebrook Colliery at a depth of 40 yards, and was found to reach 5 feet in thickness. This is the extreme limit towards the north-west of this remarkable coal-seam, which has been proved to extend from Prescot and St. Helen's on the south to beyond Burnley on the north, and Oldham, Staleybridge, and

---

* 2nd edit., p. 24.

Poynton on the south-east. Before its area was circumscribed by denudation it must have been much greater in this part of England; while (it may not be out of place to add) it has been attempted with much success to identify it with the "Black Shale" or "Silkstone" coal of Yorkshire and Derbyshire beyond the intervening gap produced by the Pennine Chain.*

The Smith Coal.—This seam lies about 64 yards above the Arley Mine and is of good quality, though only from 2 to 3 feet in thickness. It is worked at Duxbury and Coppul Collieries. At the former the beds assume the form of a trough between the two faults already described, in the centre of which one of the pits of Duxbury Colliery is situated.† The Smith Coal crops out along the banks of the Yarrow a few yards below Yarrow Bridge, and here the dip is towards the W.S.W. The crop at the opposite side of the trough is at Carr Houses, where the dip is E.S.E. From Duxbury Colliery the seam takes a S.W. direction, and crops out in the bed of the Yarrow, about 100 yards above Dry Dam Colliery. Here the roof of the coal consisting of black shale with scales and teeth of fish, and shells of *Anthracosia* may be observed.

The Bone Coal.—This seam lies about 25 yards above the Smith Coal, and has been worked both at Duxbury and Coppul Collieries; in thickness it is about 25 or 26 inches. It derives its name from the presence of splint or "bone" coal along with the ordinary coal of the district. Its outcrop occurs in the bed of the Yarrow near the bend above Duxbury Colliery, and about 180 yards above Dry Dam Colliery, not far from that of the Smith Coal already indicated.

The Yard Coal.—This seam lies about 47 yards above the Bone Coal. It has as yet been very little worked within the Duxbury and Coppul districts, though it is considered in the Wigan field a very valuable seam. The outcrop occurs in the bed of the Yarrow at Duxbury Mill. Some distance above it is a seam of very inferior quality called the "Ravin Mine."

Cannel and King Coals.—The position and thickness of these seams have already been described in the memoir on "The Geology of Wigan."‡ They are both thinner towards their northern outcrop at Duxbury than at Wigan. At Duxbury the thickness of the Cannel seam is about 12 inches, and of the King 18 inches, but at Welsh Whittle the thickness of both seams has increased. The following section was taken at Mr. Key's Colliery:—

Section at Welsh Whittle.

		ft.	in.
Strata (depth from surface) - - -	-	27	0
Cannel - - -	(average)	1	2
Strata - - - -	-	45	0
King Coal - -		2	5

In the roof of the Cannel seam, consisting of black shale, were found scales of fishes, bivalve shells (*Anthracosia*) and fern fronds.

The Cannel and King coals crop out a short distance above Mr. Key's Colliery in Syd Brook, also at Throstle Nest, in the bed of the Yarrow

---

* Hull's "Coal-fields of Great Britain," 3rd edit., p. 241.
† I may here express my acknowledgments to the proprietors and managers of Duxbury and Coppul Collieries, who afforded me every information and assistance in their power during the progress of the survey.
‡ pp. 17–18.

near Throstle Nest, and below Mr. Hargreave's new colliery. The following section of strata both above and below the Cannel seam was made out in descending the Eller Brook to its junction with the Yarrow:—

		ft.
1.	Flaggy sandstone interstratified with shales (about)	15
2.	Grey and dark shales with ironstone bands and nodules immediately overlying the Cannel and King coals	20
3.	Position of Cannel and King coals	—
4.	Seat clay of coal, fireclay	3
5.	Black shale with bands of ironstone	30
6.	Hard cherty sandstone	,,
7.	Grey shales with ironstone	,,
8.	Sandstone and shales interbedded	,,

These are the highest seams of coal which occur in Sheet 89 N.W. Yet, they form but the lower members of the series which are found in the Lancashire Coal-field, and which set in with the dip of the strata in the direction of Standish and Wigan.

E. H.

Coal-Measures, etc.

Since Prof. Hull completed his survey of the Carboniferous tract in the south-east corner of 89 N.W., two or three new sections have been exposed; one of these occurs in a cutting for a railway siding near Ellerbeck, south of Coppul, where the following succession of beds is seen:—

Fig. 18.

SECTION IN A RAILWAY SIDING, NORTH OF ELLER BECK HALL.

NORTH-WEST.   SOUTH-EAST.

Lower Coal-measures, Hard Grey Sandstone.
M.S. A thin bed of Middle Sand.

Another railway section in the Lower Coal-measures occurs near Heapey, east of Chorley; thin bedded sandstones associated with dark grey shales, containing calamites and ferns, are seen for some yards in the cutting east of the bridge.

In the district west of Coppul Station several new collieries have been opened during the last ten years, in fact this tract was at the time of Prof. Hull's survey entirely unworked. To the north the Blainscough Hall Company have several pits, in one of which occur the coal-seams given in the following abstract, for which I am indebted to the manager, Mr. Dickenson.

	ft.	in
Soil	1	6
Sand	10	6
Strong buck-leaf marl	30	0
Quick sand	5	0
Buck-leaf marl	7	0
Loam and sand	5	0
Buck-leaf marl	8	0
Quick sand	0	6
Buck-leaf marl with blue boulders	9	0

	ft.	in.	ft.	in.
Measures	41	6		
COAL (Wigan 9 feet)	5	6	140	6
Measures	98	9		
COAL, inferior	1	0		
Measures	58	6		
Coal	1	3		
Warrant	0	8		
Coal	0	7	310	3
Measures	72	5		
KING COAL	3	11	386	7
Measures	58	1		
RAVEN COAL	1	6	446	2
Measures	122	9		
YARD COAL	2	5	571	4
Measures	133	4		
BONE COAL	1	10	706	6
Measures	82	3		
SMITH COAL (Bottoms)	2	3	781	9
Measures	41	3		
CANNEL COAL	0	2	830	2
Measures	150	3		
ARLEY MINE	4	6	984	11
Measures	15	9	1,000	8

Three large faults have been proved in these collieries, two running in a north-north-westerly direction, and having westerly downthrows, the other being a cross fault, ranging S.W. and N.E. from the Old Oak Tree, across Tan Yard Brook, until it is terminated by one of the faults mentioned above, which forms the eastern margin of this belt of Coal-measures, having a westerly downthrow at this point of not less than 320 yards, the Mountain Mine having been worked on the opposite side of the fault; further to the south, near Cowmoss, the throw of the fault is less, being 280 yards. From this point this fault it is believed lies in a S.S.E. direction towards North Hall, in 89 S.W., and even further. A boring made by Hic-bibi Brook, about 250 yards east of the railway, came on a fault, which is believed to be the same as that running across German Lane, and by Cowmoss, which I propose to call the Coppul fault. Its northerly extension in 89 S.W. is by Cowley Wood and Charnock Old Hall, until it is terminated by or coalesces with another fault, south of Bottom's Green, which has been described by Prof. Hull.

The downthrow of the fault forming the western margin of the Blainscough Hall tract is believed to be not less than 200 yards, but possibly its throw is much less. That of Tanyard Brook cross fault is 150 yards to the S.E., throwing in the Wigan 9 feet at the surface, at Spenmore Lane.

The construction of a line of railway between Chorley and Blackburn, has exposed many new sections, in one of them, between Brinscall Row and station, and the siding to the Abbey Mill, the crop of a coal overlying a bed of fire-clay may be seen; this coal and a thicker seam were formerly worked in the Withnell Collieries, and the fire-clay is largely wrought for fire-bricks, an adit level running into the hill just above the railway, where boulder clay with erratic pebbles overlies dark brown and mottled shales with ironstone nodules, dipping into the hill at 18° to the N.W., overlying a 16-inch coal resting on from 5 to 7 feet of fire-clay and 2½ to 3 feet of Gannister sandstone, overlying shale which has been proved by boring to be 14 feet in thickness, which rests on the 4-feet coal, one of the Mountain Mines.

These beds are cut off to the north by a N.E. fault, between Brinscall Row and Pike Lowe, which I propose calling the Brinscall Fault, which brings up the shales underlying the Rough Rock, which have been bored into to a depth of 120 yards, on the west side of the field south of Boardman's Heights, and at St. Paul's Church brings the Rough Rock to the surface, which is well seen in the adjacent quarry, between two beds of which is a bed of purple shale, from 3 to 7 inches thick, which has been found, by analysis, to contain a very large per-centage of metallic iron, in the form of per-oxide; in the grey micaceous beds here and at Pike Lowe Quarry, are many large purple patches with a yellow external shell, and many Calamites, and traces of other plant-remains lie in the planes of bedding, to the decomposition of which many long iron-stained markings may be due, which are particularly frequent in the finer grained portion of the grit, from which are made paving slabs 5½ feet by 5.

South of the church the fields are completely riddled with small shafts, in which the coals were formerly worked, the deepest being about 18 yards; the level in the fireclay below is about 48 yards long, and dips into the hill about 5°.

Another N.E. fault runs nearly parallel to the other, passing by the brick and tile kilns, throwing the coal and other beds down 15 yards to the S.E., preventing the thick from having much outcrop at the surface. At the time of my visit (May 1872), a shaft was being sunk south of this fault, near the gasometer of Withnell Mill, which had reached the thin coal at 8½ yards. About 150 yards south of this a large fault with a north-west downthrow occurs at the base of the escarpment of Kinder Scout Grit, forming the eastern side of the valley. At the base of this, a few yards further south, a small tract of Yoredale shales are thrown in by another fault ranging about N. 30 W., which is well seen in the large quarry in some fine-grained beds of Kinder Scout Grit, near Barnes. Higher up on the hill the grit is coarse, massive, and conglomeratic, containing large white quartz pebbles, the fine seams often exhibiting impressions of plants, some of them of a fine purple colour. The Yoredale shales, with fine grained interbedded bands of grit, are well seen by the side of the road in Edge Gate Lane, the dip being S.E., into the hill against that of the Millstone Grit higher up, which is N. 60 W. at 15°, or with the hill slope.

Flagstones (Second Grit), and black shales, are seen in Red Lee Wood, on the banks of the river Roddlesworth, the shales dipping N. 10 E. at 20°, on the east side, and N.N.W. at 5° on the other side, where a little higher is an old adit level which I am informed was driven to a 10-inch coal, but the quantity of water was too great to admit its being worked.

Hard massive Third Grit is seen above the engine house, and black shales, with small bullions of Gannister surrounded by a thin shell of coal, at the entrance to the tunnel of the waterworks, where the bank is above 80 feet in height. Between these points Rake Brook fall into the river, by the side of which a boring was made in 1849, half way between the road and the river, 240 feet in depth, a coal 1 ft. 3 in. being met with at a depth of 52½ feet; the lower beds accord well with the character of the basement bed of the Millstone Grit or Kinder Scout Grit.

High up on the Fell, at Bromley Pastures, a boring was made in the Kinder Scout Grit, 150 yards from the wall, in which a coal 1 ft. 8 in. was met with at 120 feet from the surface. In the old pit on Green Hill, Pimmes, 37 yards from the wall, the same coal occurs at 15 yards, and in an adjacent bore-hole, 9 yards from the wall, a 2-feet coal occurs at 20 yards, which was worked in the pit, distant 30 yards from the wall, which was 30 yards in depth.

East of these moors, a boring was made by Tilestone Quarry, Halliwell Fold Scar, 130 yards in depth, through alternating beds of grit, shale, flags, no coal being met with; the quarry was formerly largely worked; it is in the Third Millstone Grit.*—C. E. R.

CHAPTER V

COUNTRY EAST OF THE ANTICLINAL FAULT.

Lithological Description.

The lowest beds reached in this district are the shales below the Yoredale Grit; above these we have nearly the whole of the grit series up to the Rough Rock.

The following table gives a list of the measures, with their approximate thicknesses in feet:—

Millstone Grit.
- 17. Rough Rock - 100 to 225
- 16. Flags - 30 to 60 or 75
- 15. Shales - - 75 to 200 or 225
- 14. Grit A. - 30 to 180
- 13. Shales - 50 to 105
- 12. Grit B. - 25 to 80
- 11. Shales - 35 to 110
- 10. Grit C. - 55 to 200
- 9. Shales - 75 to upwards of 170
- 8. Grit D. - 30 to 200
- 7. Shales - 75 to 200
- 6. Kinder ⎫
- 5. Shales ⎬ - - 200 to 350
- 4. Kinder ⎭

Yoredale.
- 3. Shales.
- 2. Yoredale Grit.
- 1. Shales.

The bed No. 1, or Lower Yoredale shales, consists of blue, grey, and brown shales, with occasional limestone nodules and a little ironstone. There are, however, few good sections as the ground is much obscured by landslips.

This is succeeded by No. 2, which is a very irregular and troublesome formation. It consists sometimes of a thick, massive, coarse grit, precisely similar in character to the Kinder Scout Grit in its most typical form, and at other times of a few lenticular masses of sandstone, or of a series of shales and thin sandstones; and the changes are so abrupt from huge masses of grit to beds consisting almost entirely of shales, that we had often great difficulty in determining whether or no we had separate beds of grit and shale faulted against one another.

These changeable beds are succeeded by No. 3, a set of shales, sometimes black, containing fossils, nodules, and thin bands of limestone. A small area of these shales occurs between the anticlinal fault and the cliffs on the crest of Blackstone Edge; they may be seen in a brook near the fault and on the north side of the Roman Road. These are also, at least in their upper part, very changeable; there is sometimes a well-marked and tolerably sharp break between these and the overlying grit; while at other times the passage takes place by means of large, coarse, and massive lenticular beds of grit to No. 4, the lowest bed of Kinder. This is a coarse, massive grit; it is sometimes separated by a bed of shales, No. 5, with an occasional coal smut, from No. 6, another grit.

---

* For the detailed journal of this boring, and of others in the district, I am indebted to John Park, Esq., of Abbey Mill.—C. E. R.

The two beds, 4 and 6, together are generally of that coarse and massive character which, with few exceptions, the Kinder Grit has continually had throughout the Millstone Grit area to the south. It forms here too characteristic weathered stacks of queer shapes, full of pot holes formed by water or the weathering out of concretions, on the edges of escarpments or the brows of hills.—J. R. D.

Fig. 19.

Weathered Rocks on Blackstone Edge.

The accompanying figure of such weathered rocks on Blackstone Edge is from the pencil of Mr. Hull.

This rock not unfrequently makes a good building stone, and the fine church at Heptonstall is built of it.

It is overlaid by a series of shales and sandstone,* Nos. 7 to 15, inclusive.

These beds of sandstone vary much both in character and thickness, within sometimes but very short distances. Some of these beds are decidedly gritty, and some yield good flagstones, while others are of a hard calliardy nature and furnish capital road material. We have observed in the country south of our present district, say from Bradfield northward, that in ascending the Millstone Series we first come upon Gannister on or about the horizon of the Third Grit. These beds are coloured red as being on the general horizon of the Third Grit of Derbyshire.

The bed D. has for a distance of four miles, from Crimsworth to Luddenden, a coal, sometimes at its base, sometimes in the body of the rock, which has been worked. These seams of coal occasionally occur in connexion with some of the other beds. It is well known that all the grits and sandstones in the Millstone Grit Series are liable to have thin coals on them, but these are seldom workable.—J. R. D., J. C. W.

Over this series of shales and sandstones comes a bed of excellent flags, No. 16, generally largely quarried, overlaid by a well-marked and very persistent bed of coarse grit, No. 17. The flags are probably the equivalents of the Haslingden Flags of Lancashire ; the grit is the Rough Rock. Where these beds are separable we colour the flags green.—J. R. D.

---

* *See* Memoirs of the Geological Survey, Explanation of Quarter Sheet 88, S.E., p. 6.

Stratigraphical Description.

The Lower Yoredale Shales appear at Todmorden, and for about two miles east of that town, in the valley of the Calder.

They are overlaid by the Yoredale Grit, the base of which bed descends to the Calder near the foot of Stoodley Clough. This grit is overlaid on both sides of the Calder by the Upper Yoredale Shales, and these by the Kinder Grit. This latter rock is thrown down at Stoodley Pike to the level of the Yoredale Grit by a fault running down the Stoodley Clough, which we shall speak of as the Stoodley Clough fault.

South of this fault the Kinder Grit rises from the anticlinal fault, with a steep dip of 15° to 25°, till on the crest of the hill it becomes flat, and then falls away eastward, with a gentle dip slope, and passes regularly under the higher members of the series.

Immediately east of Langfield Edge the denudation has been great enough to expose the Yoredale Shales in a valley running down to St. John's. Here the country is much broken by faults, the details of which we reserve for future consideration. The Stoodley Clough fault is connected with these dislocations. North of that fault, and some branches of it, we find the Kinder crowning with a bold escarpment the sides of the Calder and Hebden valleys and the Colden and Haxbridge Cloughs, whose lower slopes of Yoredale Shales are for the most part well wooded.

On the west side of the Hebden the Kinder runs round in a narrow belt, enclosing outliers of Third Grit. On the east it is overlaid in regular succession by all the higher members of the Millstone Series. The structure of this part of the country will be best judged of by a glance at the sections, by Mr. J. C. Ward, Fig. 22, p. 114.

Both run east and west, No. 1, about a mile north of the Calder, No. 2, from two to two and a half miles north of the same river.

South of the Calder the Third Grit beds, with the exception of the lowest, run round as outliers to form the detached Soyland Moors included between the valley of St. John and Rippouden. The lowest bed D. runs down to the Calder at Sowerby Brig, whence with the overlying grits it rises up on the north side of that stream; and they all circle round the Luddenden Valley, enclosing in the bottom thereof an inlier of Kinder. In the N.E. the Rough Rock and its basement flags form the well-marked and lofty escarpment of Cold Edge.

"Anticlinal Fault."

It will be as well, before beginning the detailed description of the rocks, to give the evidence for the position of the anticlinal fault.

This fault is proved by the ending off of the escarpment of Kinder Scout Grit on Longfield Edge, and by the Yoredale Grit and shales abutting against beds of the Third Grit Series at Todmorden. On the north side of the Calder Valley, in Stanally Clough, on the line of direction of the fault, we see a fault which throws Kinder on the west against Yoredale Shales on the east; on the same line we have the grit escarpment of Hudson Moor broken by a fault which brings the Kinder Scout Grit on the east against a Third Grit on the west; there may be some doubt as to the exact spot where the fault crosses Paul Clough, but it cannot be far from where we have drawn it, as above that point we have, with the exception of one section in shale, coarse grit reaching right up to the Kinder Scout Grit moors of Hawkstones Common, and below it we have a regular succession of grits and shales in ascending order, which gives us all the grit beds above the Kinder Scout Grit, with the exception of the lowest Third Grit, D. It is possible that the lowest bed we get, consisting of sandstone, fine grit, shales, and tiles, may be the topmost part of D., but the main mass is nowhere got in Paul Clough. So the position of the fault is fixed very approximately. North of Paul Clough, Black Hambledon consists of Kinder bordered by shales of the Third Grit Series. In Black Clough vertical beds of sandstone and shale and a fault at the spot are seen. This will do very well for a point on the anticlinal fault. The Kinder escarpment of Gorplestones ends abruptly at Harestones: this gives another point.—J. R. D.

At the head of the Widdlop Valley the Kinder escarpments on either side end off abruptly, and the Yoredales Shales of the valley abut against one of the Lower Third Grit beds on the west. North of this the fault lessens its throw until joined in Heyslacks Clough by the fault that passes east of Greystone Hill, after which we again have the Yoredale Shales against the same Third Grit. Further north along the western flank of Boulsworth we find a mass of Kinder Grit ending off abruptly. This grit is seen in many places north of Round Hole Springs to be dipping N.N.W. at angles of from 40° to 60°, and striking at the uppermost Third Grit, which can be very well traced running parallel to the unmistakable Rough Rock of Willy Moor and Deerstone Moor. No other Third Grits are here seen, in fact there would not be room for them to crop out, and the run of the bed is inexplicable without a fault.

We will now give some evidence for the great east and west fault which passes between Boulsworth and Crow Hill.

The Kinder Scout Grit north of Round Hole Springs referred to above, and of Pot Brinks Moor, striking E. 35° N. and dipping at angles of from 40° to 60°, ends off rather abruptly, and in Saucer Hole Clough to the east nothing but shales and sandstones are seen, though the section is not a good one. At a point in the Clough close to the B in "Broad Head Moor" (1-inch map), two faults are seen, one of which throws grit against shale, and its direction apparently agrees with the line of fault inferred from the ending off of the grit of Pot Brinks Moor. Very little can be seen on Broad Head Moor, owing to the thick covering of peat, but the Kinder Scout Grit escarpment of the Great Saucer Stones appears to be cut off eastwards by a fault, and disturbances appear at Walshaw Dean Head and in Red Mires Clough, though the fault is not actually seen. It is well shown in Middle Moor Clough, a little further to the east. It is very evident a great fault must pass between the undoubted Kinder Grit of Walshaw Dean Head and Jackson's Ridge, and the grit of Crow Hill which there is good reason for believing is the Rough Rock. At Walshaw Dean Head the throw of the fault must be equal to the thickness of the whole grit series from the top of the Rough Rock to the top of the Kinder Scout Grit, probably upwards of 1,000 feet.—W. G.

Detailed Description.

We now proceed to describe the beds in detail, beginning with the lowest.

Yoredale Shales.

The Lower Yoredale Shales are only found in the immediate neighbourhood of Todmorden. They are to be seen in the clough, and New Broad Clough, immediately to the south of that town, where they contain limestone nodules, and goniatites, and pass rather gradually into the Yoredale Grit above. In the next clough to the eastward, viz., that descending from Lee, or Lumbutts Clough, these shales form the western side, where a very good junction is got between them and the overlying grits. In the next clough eastward, Shaw Clough, we have in the lower part rolling beds of these shales.

North of the Calder the railway cuttings give sections in these shales.

In the country north of Todmorden the Yoredale Grit seems to be absent, so that the Lower and Upper Yoredale shales form one unbroken series. We have the following sections:—In Stanally Clough, north of the anticlinal fault, we have beds of shale with the following dips, showing 25°, 10°, 5°, 30°, 20°, 7°, 15°, in an average west direction; these contain near the top of the clough beds of flaggy sandstone. Finally we have vertical beds of shale and sandstone; here we suppose a fault to cross in a N.W. direction, which cuts off the Kinder escarpment running down above Springs.

Below the Kinder escarpment we get universally sections in shales, rolling about a good deal, but as we approach the middle of the valley the ground is much obscured by huge landslips; these are doubtless due to the fact that the beds, which on the heights have an easterly dip, turn over and plunge towards the anticlinal fault, so that all the higher beds have entirely slipped away and been long since removed by denudation; while the lower beds, consisting of shales and occasional sandstones, and in places of the great series of interbedded massive grits and shales known as Yoredale Grit, strew the ground with their débris in the shape of huge slipped masses which have not yet been washed away.

At Cross Lea we have a quarry in a peculiar grey or white grit; this, if in place, which must be doubtful in a country so covered with landslips, may be Yoredale Sandstone.

East of Dungeon Top we have grey sandy shales with thin tiles, beds rolling much, a large number of dips being down hill, if they are true, then a mass of shales at high angles, then shales with beds of fine, hard, close-grained grit, interbedded in a very irregular fashion; this may possibly be the same bed as that at Cross Lea, which it is like. The clough running from near Winsley gives the following sections, as we descend the clough:—Shales and thin sandstones, then shales with the following dips, 6° E. by S., 8° N.E., 20° E., 8° S.E., 5° S.E., 12° S. by W., 14° S. by W., 15° S. by E., 30° S., 0°, 31° N. by W., 31° N.E., 27° N. by E., 90°, and then grit, and finally shales. The grit forms the Hough Stone between Wickenberry and Holebottom. We suppose it to be faulted on the S.W., as certain high dips and the sudden termination along a straight line of the massive Yoredale Grit of Cross Stone would seem to indicate. But it is uncertain whether the Hough Stone is in place, or whether it may not be a great slipped mass of Kinder from above. If it is in place, it would seem to be quite a casual bed, as we cannot trace it far eastward, and on the west it ends abruptly along the same line as the Yoredale Grit ends along. There is a similar detached mass of grit at Lawhill, which may be a slip, a faulted mass, or a casual detached or rather lenticular bed. This latter rock contains casts of shells, pronounced by Mr. Etheridge to be *Edmondia sulcata*.

Yoredale Grit and Shales.

East of Wickenberry Clough we have this series of beds:—
Kinder Scout Grit.
Shales.
Yoredale Grit.
Shales.

The Yoredale Grit at Cross Stones is a thick massive coarse grit and conglomerate. It ends off west of Cross Stone suddenly along a N.W. and S.E. line, which, combined with the disturbed state of the underlying shales, has induced us to put in a fault along this line; but the country is so covered with huge landslips of grit and shale immediately west of this line, and the Yoredale Grit is so apt to break up into lenticular masses of grit imbedded in shales, that it is quite possible, not to say probable, that the rock is not faulted along that line, but has merely slipped away; if so, it is probably because it here becomes split up into such lenticular masses, and so does not form a continuous solid stratum of stone; and then, the beds having a dip towards the valley near the anticlinal fault, these isolated masses of grit have slipped away from their place, leaving the solid mass of grit we find at Cross Stone standing up in a wall resembling the feature formed when a fault brings a hard against a soft bed.

In either case the Yoredale grit does not continue as a mapable bed to the north-west.

At the east end of Cross Stone a thick bed of shale of a singularly concretionary character, with thin veins of sandstone traversing it in a direction at right angles to the bedding, caused apparently by infiltration of sand, sets in quite suddenly, so suddenly as to look at a little distance exactly like a fault. The upper part of the grit is seen running on unbroken above this shale band. The lower part of the grit, too, is believed to run on under the shale, as there is a slight feature on the ground in that position; and in the next clough grit appears again about that horizon.

The upper part of the grit eastward becomes gradually a very thin bed of sandstone, till at the next clough it is only a few feet thick; there we see it over and under laid by shales. Immediately, however, on the east side of this clough grit sets in again very strong, being both coarse and massive, and upwards of 100 feet thick; but this probably includes a band of shale, which, however, can be but thin. In the next clough, Ingham Clough, all is again changed; instead of a great mass of grit and conglomerate, with perhaps a thin shale band, we have here a series of grits, sandstones, and shales, in which there is quite as much shale as sandstone, if not more. Here we get the following succession of beds.

At the bottom is a coarse grit, which is seen in the railway-cutting on the east of the clough dipping at angles of 25° to 50° to the east. In the clough what looks like the upper part of this bed consists of shales and sandstones, at the bottom of which we have good grit wedged in a queer way among the shales. The beds are very much disturbed, dipping at 50° or 60° to W.; it is possible there may be some fault. Above these beds comes a bed of pure shales, overlaid by a thin sandstone, which contains a shale band in it in places; and over all comes the main mass of Upper Yoredale Shales. The higher parts of these have thin sandstones interbedded. Some of the highest beds are blue shale, over which comes the Kinder.

If we trace out the beds of Yoredale Grit eastward, we find the lowest bed runs down to the Calder opposite Stoodley Clough. The highest bed seems to thicken eastward, becoming a massive sandstone with a little interbedded shale. Its top runs nearly to the level of the base of the Kinder above Eastwood station. This is owing to the beds being thrown down on the east by a fault, which, judging by the change in character of the ground, runs to Whitley Royd, up to which point the Kinder has a fair escarpment. This fault, after it reaches the valley, must turn sharply and run in a southerly direction along the valley, under the alluvium, as far as Stoodley Clough; here it again turns sharply and runs up that clough to where it breaks the Third Grit escarpment on Erringden.

The reason is apparent, for we find a grit on the east side of the Calder corresponding with the Kinder on the west. This grit runs from the foot of Parrock Clough as far as Stoodley Clough without being broken by the fault from Whitley Royd, and faces the Yoredale Grit on the opposite side of the valley, from which it must needs be parted by a fault in the valley; this fault in the valley is not seen, being concealed by alluvium and the great amount of detritus which forms the western bank of the valley. In Stoodley Clough the Kinder abuts against the Yoredale beds, and the fault is seen in the bank of the stream pointing for the end of the Third Grit escarpment. The Yoredale beds in Stoodley Clough are as follows: at the lower end of the clough we have shales dipping at 24° to E. This must be near the fault; these shales are overlaid by a bed of grit, this by a band of shales, and then by a sandstone, the topmost bed of the grit. These beds form no good escarpments southward of the clough; but they appear to be higher than the corresponding beds on the opposite side; there would thus seem to be a fault in the valley downthrowing on the west, but it is quite hidden by débris.

The next clough is Shaw Clough; here we have in the lower part shales rolling; these are succeeded upwards by grit dipping at 7° to 10° to E. and S.E., this by sandy shales and sandstones, these by a thin sandstone and shale. We do not get the very highest beds, as the Yoredale Grit is here faulted along its southern boundary. Westward the shale bands become feeble, and all along Raven Nest Wood and Doroad Scout we have a fine escarpment of coarse grit, which at the south end of the scout is thrown down into the valley, so as to form the east side of Lumbutts Clough; the western side is formed of shale, so there must be another fault along the clough. This is the fault which breaks the Kinder escarpment of Bald Scout; it throws down on the east.

The base of the Yoredale Grit runs up from the stream at Causeway Mill, and forms an escarpment along the west side of the clough. As we go westward the bed becomes very shaly, the rock at times seems to consist almost entirely of mere small detached masses of grit imbedded in shale. The country is much covered with landslips below the grit. Still further west, near Todmorden, the grit thickens again, consisting of massive coarse grit in irregular beds, in which are some large quarries. At Shewbroad the lowest bed consists of sandstone and shales dipping west at 16°. Near here the Yoredale beds abut against the anticlinal fault.

East of the Stoodley Clough fault we find the Yoredale shales underlying the Kinder Grit; sections are obtained in Parrock and Oak cloughs on the south side of the Calder. In Oak Clough the beds are dipping east at 20°, and consist of shales and sandstones. In Parrock Clough we have nearly continuous sections; first we have shales, above these, shales and sandstone, dipping at about 10° to the E. by N.; then shales, 10° to 17° to E.; the dip of 17° was taken near where we suppose a fault to cross.

The shales in Parrock Clough lie on the horizon of coarse Kinder Grit, which is found on both sides of the clough, forming the hill above Burnt Acres on the west, and on the east the massive escarpment below Cronstonstall; so that these shales are either, as we have mapped them, beds included between a pair of trough faults, which bring them against the Kinder Grit on either side; or else, which is a possible though perhaps not so probable an explanation, they are the equivalents of the Kinder Grit, which must then be supposed to change its character suddenly for a short distance and again recover it as suddenly. This is certainly possible, but as the Kinder Grit is on the whole very persistent in character, we think it more probable that the beds are faulted.

Beaumont Clough gives sections in Yoredale shales dipping at 6°, 10°, and 15° to S.E. The railway cutting gives sections in blue shales and interbedded sandstones, surrounded by a lot of grit rubbish. The beds are rolling.

In the country between the anticlinal and Lumbutts fault on the south side of Calder we have many sections in Upper Yoredale shales. These beds there consist of grey and brown shales, sandy shales, with beds of tiles and fine-grained grits interstratified. In the extreme west the beds have a dip of 5° a little to the E. of S. In Black Clough (which runs between High Lee and Wood Pasture) we have in ascending order dark grey shales, dip 2° to S.; grey sandy shales, dark grey sandy shales, dip 10° to S. by E.; grey shales, grey sandy shales, shales and tiles rolling, sandy shales, and fine-grained flaggy grits. In Heeley Clough, which runs nearly parallel with and close to Lumbutts fault on the north side of High Lee, there is a lot of disturbance, and the dips are much more easterly, of 10°, 32°, and 50°. On the east side of the above-mentioned fault we have not many sections; we have some in sandy shales near Mankinholes, and a fine section in dark blue shales immediately under the Kinder Scout Grit, just where the fault at Stoodley Pike crosses the escarpment.

In Withens Clough, which flows eastward from Langfield Edge to St. John's, we have shales interstratified with grits and sandstones, and also pure shales, dipping so as to pass under the Kinder Grit of the moor. These, then, are Yoredale shales. As we go down stream the beds become much disturbed. They consist of shales rolling about; we have dips to all points of the compass in the main stream or in its feeders, but on the whole we should say the dip is to S. by E.

On the north side of the Calder the ground is much obscured by landslips, but the great Calder Clough gives sections in Yoredale shales which in their upper part contain many intercalated masses of grit, which sometimes merge into the overlying Kinder Scout Grit. In Jumble Clough, which flows from Hippings, the grit beds on this horizon are so thick, massive, free from shale, that we have been forced to include them with the Kinder, from the normal base of which rock they are seen at the eastward to be parted merely by a thin shale band, while no shale at all was seen on that horizon in Jumble Clough, though it may exist there hidden by heaps of detritus.—J. R. D.

The Hebden Valley and Horsebridge Cloughs also give excellent sections in the Yoredale shales. In the upper part of the Hebden Valley, north of Walshaw, many lenticular sandstone masses, like those mentioned above, may be seen. These must not, however, be confounded with certain great slipped masses of Kinder Grit occurring on the eastern side of the valley, of which the line of crags called Hardcastle Crags is an example. The thin bands of limestone that occur in these beds are best seen in Horsebridge Clough. The following fossils, as determined by Mr. Etheridge, were found in this series:—

    Aviculopecten (Lanistes) rugosus, Phillips.
    Posidonomya Gibsoni, Brown.
      ,, vetusta (young of), Sowerby.
    Goniatites Listeri, Mart.
     ,, obtusus, Phillips.
     ,, crenistria, Phillips.
    Sigillaria, sp., Phillips.

The Yoredale shales of the Hebden Valley reach up stream as far as Blackden Bridge, where the stream bifurcates. In the Alcomden Water, which flows from the north, we have a natural base to the Kinder, but in Gorple Water, which flows from the west, we find ourselves suddenly on Kinder Grit, which therefore we conclude to be brought in by a fault.

In the head waters below Gorple we have a few sections in Yoredale shales containing interbedded grit and sandstone, dipping at 22° to the east and rolling about.—J. R. D.

Widdop.—This beautiful little valley of Yoredale shales is almost entirely surrounded by fine escarpments of Kinder Scout Grit, which at the Cludders, probably by the action of subaerial causes, have been made to assume the appearance of castellated ruins, while the slope below is strewn with masses of grit which have fallen from above. The view of the Cludders from a point on the road to the N.E. is very fine. Nearly close up to the rock sandy shale is seen in the paths which come into the valley from the southwards, and sections in shale and flaggy micaceous sandstone dipping S.E. at 27° are seen at Clough Head, near Ladies Walk. Between Ladies Walk and the Widdop Water shale is seen dipping E. at 30°, and east of Ladies Walk shale dipping E.S.E. at 10°. West of Wood Plumpton shales and sandstones are seen dipping westwards. At the head of the valley, nearly close to the anticlinal fault, shales and soft sandstones are seen dipping W.S.W. 20°. Further down the stream we get blue shales dipping N.W. 18°, and south of Pasture House two isolated sections in shales, one dipping 15° E.N.E., the other rolling. We see nothing further till we get south of Higher Houses. Then we get a continuous section for more than 300 yards in thick blue and grey shales. The beds dip 5° E.S.E., 7° E., 15° S.E., then flatten, dip 25° W.S.W., and we come to a fault ranging N.N.W. and throwing down on the west, referred to, p. 111. On the east of the fault the dip is E.N.E. 45°, then S.E. 15°, and soon after the section disappears. It will thus be seen that the valley is crossed by the great anticlinal, the beds, roughly speaking, dipping east and west from the centre of the valley. North and west of Higher Houses shales are seen, apparently horizontal. Here also are immense numbers of grit blocks strewn over the fields, "the Roughs," which may have fallen from The Scout when the valley was narrower; or they may be the remains of one of those lenticular beds of grit which in some places occur about this horizon. South-east of Lower Houses is a quarry in coarse reddish grit some distance below the Kinder escarpment, which is probably one of these occasional beds.

Hey Slacks Clough.—The stream which drains Hey Slacks gives good sections in the Yoredale shales. The beds are much disturbed, but the valley is evidently in an anticlinal, the direction of which, about N.E. and S.W., corresponds with the greatest length of the valley, the Kinder Grit on either side dipping away steeply to the N.W. and S.E. Near the anticlinal fault the beds dip W. 25°, but further up the clough many disturbances and several faults are seen. The little gulley that joins the clough from the southward called "Tom Groove" affords a good continuous section. In the lower part of Tom Groove *Aviculopectens* were observed in muddy shales. The beds dip pretty steadily to the southwards, so that we get an ascending section as we go up the gulley. Where a small east and west fault crosses dips of 70° S. and 15° S.S.W. were taken, and at the county boundary were shales and sandstones dipping S. at 20°. In the main stream a little above its junction with Tom Groove shales and thin earthy limestones are seen, and these are probably the lowest beds observed. Further up the clough the shales seen are rolling, but easterly dips prevail.

A small patch of Yoredale shales occurs north of the Great Saucer Stones on Boulsworth. The beds are nearly horizontal.—W. G.

Kinder Scout Grit.

The Kinder Scout Grit has a well-marked escarpment all along Langfield Edge from the anticlinal fault to that of Stoodley Pike. The rock is a massive coarse white grit with interstratified bands of grey shale near its base; these latter are particularly conspicuous in Jail Hole quarries north of Gadden Reservoir, where consequently are large landslips. Crossing Langfield Edge into the Withens country, we have on the north side of Withens Clough for some little distance a good escarpment of massive Kinder Grit; the escarpment gets weak both near the Stoodley Pike and the Lumbutts faults, and on the south side of the latter there is no escarpment at all, but we find grit on the moor, and in descending Withens Clough shales dipping to pass under the grit.

Round the head of Withens Clough the position of the base is thus very uncertain, as the rock forms no escarpment until we come to Buckstone Well, where it is faulted. Eastward from that point the position of the base is fairly fixed by sections in the streams, but the passage into the underlying shales is quite gradual. At Buckstones the rock is again faulted down so as to bring the base of a higher bed which forms the eastern end of Turley Holes Edge to the level of the base of the lower Kinder. Finally, the base forming an escarpment of massive grit runs down to a fault which crosses the vale of St. John by Priestley Ing, and which throws down the Kinder below the bed of the stream. This fault is the continuation of two long faults which we have traced, one from the coal-measure country south of Huddersfield by Lockwood. Beestonley, and across the Ryburn, a mile and a quarter below Ripponden, the other from the valley of the Colne near Slaithwaite by Clough House, West Hill, Ringstone Edge, and Rippondeu. This fault must in St. John's Valley throw out nearly the whole thickness of the Kinder, for it throws the shale above the Kinder against the base of that rock. The Kinder is soon brought to light again lower down the stream, partly by the dip and partly by a small fault.

Above the Priestly Ing fault we have the top of the Kinder running a little above and parallel to the road from Marshaw Bridge and Rochdale; the rock is exposed all along the brook and consists as usual of massive grit. It is also found all along Black Castle Clough in beds of massive grit and conglomerate with shale partings.

It forms the great stretch of moors which rise from the vale of St. John, with a dip slope to the height of 1,400 feet on the Lancashire border, whence the beds plunge westwards towards the anticlinal fault. The overlying beds, consisting of shales and sandstones, are seen in the upper part of Black Castle Clough, and on the high road near Blackcastle. The grit is also found in force in Booth Dean, where a great thickness of rock is displayed. The actual top is only seen in one or two places, and there we have grit at the very top; but generally it seems, from the numerous sections in streams descending into Booth Dean Clough from the south, that the upper beds consist of flags. These are succeeded by shales and fine and soft grits, flaggy grits, and sandstones, and then we reach the body of the bed, consisting of coarse grits and conglomerates more than 150 feet thick.

In the cloughs above Bar House the rock consists of a mass of grits, both coarse and flaggy, intermingled with shales and sandstones, all rolling about so much that no average dip can be got, though the slope

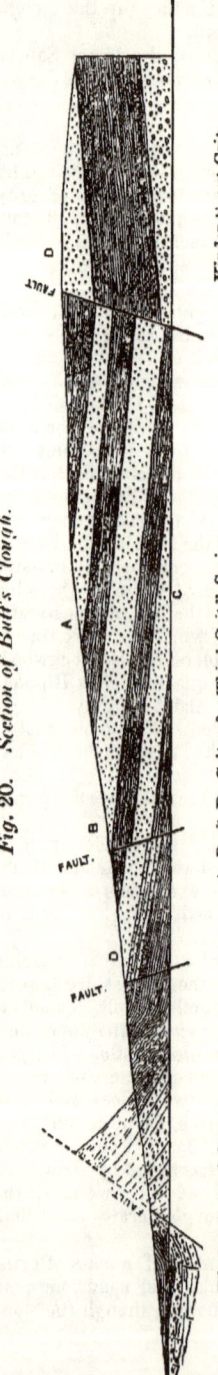

Fig. 20. Section of Butt's Clough.

of the ground indicates a general easterly dip. There is also much cross bedding. The main mass of grits and conglomerates continues in force all down Booth Dean Clough, forming below Spring Mill magnificent cliffs, 150 feet high, as far as Butts Green ; here the rock ends suddenly as if against a fault ranging along the south side of Butts Clough. West of this point, along the northern flank of Booth Dean, sections are scarce ; we have dotted the fault on to Castle Dean, where we saw signs of a fault, but further west it seems to die out.

Butts Clough gives the accompanying section.

On the north side of the fault through Butts Green we have the top beds of the Kinder in the stream. The overlying shales are seen in the lowest part of Butts Clough and above Godly on the west. The top line of the Kinder is here a bad one, except in the upper branches of the stream flowing through Nab End Wood. We have drawn it, guided partly by the slope of the ground, partly by occasional sections, so as to include the chief grits seen. The Kinder of this neighbourhood consists of a mass of grits and sandstones, with interbedded shales, bearing a general resemblance to the ordinary character of the rock, though somewhat more changeable and more shaly in parts than usual. This mass of grits and shales attains near Rishworth Hall a height of 225 feet above the river, becoming more shaly as we ascend. Below Swift Place the beds are disturbed at the weir ; a fault probably passes through this spot with a downthrow to the N.E., as below the weir we seem to have the top of the grit in the very bed of the stream overlaid by shales with thin irregular sandstones ; accordingly we have prolonged a fault seen at the waterfall in Butts Clough to this point.

This grit extends along the valley as far as Ripponden, where it is thrown down out of sight by a fault. It is as usual massive, and it has a thin coal on or near its top, which may be seen at Slitheroe Bridge.

Above the Kinder, on the south side of Ryburn, comes a mass of shales containing a few irregular sandstones ; three of these are seen in one part of the country, and we were able to trace two of them for a short way.

Over these comes the escarpment of Blackwood Edge.

If we follow up the Ryburn from the fault at Swift Place, we find beyond Nab End Wood a grit in the bed of the stream overlaid by shales and thin sandstones; this looks as if the top of the Kinder had been again thrown down into the stream; we have therefore taken a fault across by a break in the ground which is suggestive of one.

The dip of the beds soon brings up the mass of the Kinder again in Blackcastle Clough, which has already been mentioned.—J. R. D.

An inlier of Kinder Scout Grit occurs in the Calder valley just below the village of Sowerby. It also forms a long tongue running in the bottom of the Luddenden valley from the river Calder to Low Bridge. This inlier is thrown down out of sight by a fault a little to the west of Luddendenfoot Station.—J. C. W.

On the south side of the Calder we have in Beaumont Clough, immediately above the Yoredale shales, a thick bed of grit. After this we get no sections till we come to the bridge near Edge End. Here grit is seen in the path leading down to the bridge, and above the bridge we have beds of shale and sandy shale dipping at 12° to S.

East and west of Beaumont Clough the base of the Kinder is not well marked, though the upper part of the bed makes fine bold crags. We have here and there sections in coarse grit, but these seem to be in the body of the rock. Immediately south of Hebden Bridge we have in the railway cutting a junction between grit and underlying shales; eastwards of that spot the line is quite uncertain, but it must run up a bit, as is indicated by the slope of the ground. This change in the run of the base may be due to a roll or to a passage of grit into shale, a thing which often happens in this country. This would imply a change in its geological horizon. The base is above the river, where the Wadsworth fault crosses, because close to that point, at the edge of the alluvium, we have a section in shales dipping 28° to the east.

The top of the rock is very well marked by the shape of the ground.

On the north side of the Calder the Kinder grit forms generally a good escarpment. The lower bed is cut off above Lower Hartley by the anticlinal fault. It is thrown down on the east to Springs; there it consists of coarse massive grit and conglomerate in separate beds. Thence it forms a good escarpment by Dungeon Top, where it is thrown down by a trough fault, and Windy Harbour, Winsley, and Keelam, as far as Whitley Royd, where it is faulted down.

The Lower Kinder is found in the clough coming from Redmires Dam, called Hudson Clough. Its top is not well marked; but the country between it and the escarpment of Upper Kinder at Hawkstones is occupied by rolling beds of shale and flags; this may be seen in a lane leading to Hawkstones, and also in a brook on the west. In this brook a fault is seen which brings the said shales against beds of grit and tiles; this fault seems to bound the Upper Kinder till it runs into the anticlinal fault. Another fault breaks the escarpment west of Hawkstones. Near these faults the Kinder has a dip of from 15° to 40° towards S.E.; but its general dip on the west of these faults is 15° to S.W. The dip flattens as we rise the moor, and finally the beds turn over and dip eastward.

East of these faults the Upper Kinder makes a bold escarpment of castellated and tabular blocks of massive grit and conglomerate along Hawkstones; the dip is from 2° to 10° to E. by N. This escarpment is cut off by a fault, which is seen at Kebcote, throwing shales believed to be the overlying shales against the grit. From beneath these shales the grit rises in a dip slope, and forms a bold escarpment as far as Bridestones, where it is again thrown up to the east. Soon after this the escarpment dwindles away, probably from the band of shale between the two beds of grit either thinning out or becoming grit eastwards.

East of the fault at Whitley Royd we have, as stated above, included with the Kinder a lot of grit, which seems to lie really on a lower horizon. As we go eastward of Jumble Clough the passage through these beds, from the Kinder to the Yoredale shales, is very gradual, alternating beds of sandstone and shale being seen below what we have taken as the top of the Yoredales.

The Kinder Grit forms a well-marked and cliffy escarpment, capping the sides of the Hebden Valley and the Colden, and Horsebridge Cloughs.—J. R. D.

In the Hebden Valley, immediately above the junction of streams near Blackden Bridge, we reach the base of the Kinder in the Alcomden Water, which comes down from the north. About a mile up this water we come to the top of the grit, the overlying shales being exposed in a scar at the river side; but just there the beds are disturbed, and immediately after passing this point we find ourselves still on grit, which reaches some way above the stream on either side, at least 75 feet on the right bank; this appears to be the Kinder thrown up a little by a fault through the point of disturbance; the fault, however, is not actually seen. Soon after crossing the end of the stream coming from the Cascade we find ourselves rather suddenly on shales, whence we conclude that we have crossed another fault throwing down the overlying shales on the N.E. against the Kinder on the S.W.; indications of this fault are to be seen on the hills to the S.E., where we find a bed of massive grit discontinued; and to the N.W. at the Cascade, where we find the beds dipping at high angles to the E.N.E. We soon lose the shales in the streams and come upon grit which reaches about 300 yards up the valley to where Old Dike joins the main stream. This grit we suppose to be an inlier of one of the Third Grit beds, probably the lowest.

Thence as far as Grey Fosse Clough we are on the shales overlying this grit. These shales are rolling about a good deal and in places contorted. Above them we have on the east a series of sandstones and shales belonging to the Third Grit Series. Up Black Clough we have four beds of stone, and a fifth probably covers Withins Height End. The lowest of these beds thins out northwards, as also does the lowest but two, while southwards they get more important. The other beds are more persistent. These beds are more usually sandstones, though occasionally they deserve to be called grits.—J. R. D. and W. G.

Returning to the main stream we find that soon after passing Grey Fosse Clough we come suddenly on very coarse massive grit, which a little further up the valley is seen in one of the little side streams to reach to a height of at least 100 feet above the grit in the stream. This grit is the Kinder, and here we suppose the Greave Clough fault to cross Walshaw Dean. Close by in Grey Fosse Clough some disturbances are seen, but the exact place of the fault is not determinable. The fault must pass along Shoulder Nick between the two hills called Withins Height; the grit of Alcomden Stones on the northern Withins Height being almost certainly the same as that of Gablestone Edge on the western side of Walshaw Dean, which is continuous with that of Mere Stones, and is Lowest Third Grit; while the grit of the southern Withins Height as just mentioned has several beds of Third Grit below it, and one of these seen in Bakestone Clough must be thrown against the upper part of the Kinder Grit. Further to the east along this line of fault the beds near Withins are vertical, and there is much disturbance.

North of this fault in Walshaw Dean we have coarse massive grit in the bed of the stream till we come to Boft Hole, where several sections in shale and one in grit are seen. These shales are probably in the Kinder Grit, but it is possible they are the Yoredale shales below it. Beyond Boft Hole we get above these shales and on to the grit again, which is seen at intervals all the way up to Walshaw Dean Head, and over the county boundary into Lancashire. In very few places could any dips be taken, but the general dip is evidently between south and east. In Grey Fosse Clough dips S. 17° and S.E. 9° were observed, and at Walshaw Dean Head dips of S.S.E. 12° and N. 30°, but both these places are near faults.

Though many little streams run into the upper part of Walshaw Dean Water very few show anything more than sections in peat, and we nowhere see the junction of the Kinder Grit with the overlying shales.— W. G.

In ascending the Gorplewater from Blackdean Bridge we cross a fault, as was mentioned above (p. 106), on to the Kinder Grit. This bed is here of its usual marked character, a massive grit and conglomerate, with occasional partings of shale. In one of the side streams feeding Gorplewater we see at least 200 feet of solid grit. Rather more than two miles above Blackdean Bridge we reach the base of the Kinder, which descends sharply from the north at an angle of 30°, and more gently on the south, as the dip is on the whole south of east. The base runs up on each side and then the rock turns over, and dipping westward plunges towards the anticlinal fault, against which it abuts at Hare Stones.—J. R. D.

Crossing Gorple Stones and Black Moor into Widdop we have another splendid escarpment on each side of that valley. The base descends to the stream near the junction of the Widdop Water and Greave Clough, being thrown slightly down by a fault just before reaching that point. A short way up Greave Clough we come to an unmistakable fault, which brings up the Yoredale shales; while the base of the Kinder Grit forms a good escarpment a little way above the stream. The escarpment of Kinder along the northern side of Widdop is broken by a fault above Pasture House. It was a question with us whether the base on the east side of this fault did not abut against this fault; but we finally came to the conclusion that it was better to map it as we have done. This fault above Pasture House runs N.W. and S.E. along a face of rock, and on this line a fault was seen in Widdop water in shales, but as the escarpment on the south side of the valley is not broken, we suppose the fault to stop against a prolongation of the Greave Clough fault. The spread of Kinder Grit that forms Widdop Moor is interrupted by an oval patch of shales lying on the slope of the moor; these may either be the overlying shales, or merely some intercalated bed.—J. R. D. and W. G.

A little above where Pisser Clough joins Greave Clough the base of the Kinder comes into the stream, and then good and almost continuous sections are seen in coarse massive grit all the way up the Clough and its upper part, Hole Sike, to its head in the Peat Moss on the Lancashire side of the county boundary. On the Yorkshire side of the boundary Bullion Clough from the N.N.E. joins Hole Sike. A little way up this clough we pass off grits on to shales, which are rolling with a general dip to S.E. A little further on a fault is seen throwing thin sandstones against shales. The fault appears to run in a N.N.W. direction, and we suppose it to throw down on the west, and to continue northwards so far as to break the Kinder escarpment west of Great Saucer Stones. A few sections in shales and thin sandstones, some rolling, are seen further up the clough. At the Dove Stones, Ordnance Station, marked 1,573 on map, and at Great Saucer Stones, just by the *r* in "Boulsworth," fine escarpments of Kinder Scout Grit are to be seen. At the former place dips of S.S.E. 20° and S.E. were taken. Vast blocks which have probably slipped a little are seen below the escarpment. At Great Saucer Stones the beds dip gently to the S.S.E. We get sections in the beds near the top of the Kinder in two little streams running down the northern slope of Boulsworth. The general strike of the beds is N.E. and S.W. with steep dips to the N.W. and N.N.W. At Round Hole, formed by a slip, very coarse grit is seen dipping 40°. Further down are shales and sandstones vertical, then shales dipping 60°, flags 40°, and grit 50°. Going up the little stream to the east we get dips of 40°, 50°, 50° in grit, and then sandy shales vertical. Further up we get the only reverse dip, one to the S.E. There is good reason to believe that this last and the vertical shale beds are mere surface disturbances caused by the sliding of some heavy body down the slope. The agent was probably land ice. Many undoubted examples are given in other parts of the Memoir, see pp. 27, 31, 42, 134.

From the sudden ending off of two parallel ridges of grit which run steeply down from Lad Law towards Heyslacks Clough we suppose a fault to pass along here in an E.N.E. direction. The evidence for the N.N.W. fault which bounds the shales of Heyslacks on the east is very strong. On the one hand we have the ending off of the ridge of Kinder Grit from Dove Stones east of Warcock Hill, and on the other we have the steeply dipping Kinder Grit of Pot Brinks Moor striking against the nearly horizontal Yoredale Shales north of the Great Saucer Stones. It seems to be more a case of lateral shifting of beds than of vertical downthrow, the anticlinal being shifted a little northwards on the eastern side of the fault. At the Great Saucer Stones the vertical throw must be almost nothing.

Many groups of weathered grit blocks are scattered about on Boulsworth. Some of these are actually in place, others are but little removed from the places they originally occupied, having been isolated by the denudation of their surrounding portions of rock. Fine examples of pot-holes are here seen. One splendid pot-hole between Lad Law and the Great Saucer Stones is about 5 feet in diameter by 3½ feet deep, and has a worn slit in one side, descending to within a few inches of the bottom. Fine potholes are also seen on Crown Point, Grey Stone Hill. The Abbot Stone on Boulsworth is an interesting example of the action of rain. On a very steeply sloping surface of the stone about 8 or 9 feet long are seen a number of deep furrows. They all deepen towards the middle and shallow towards the bottom. The accompanying rough sketch will show their form, like rivers in miniature.

Fig. 21.

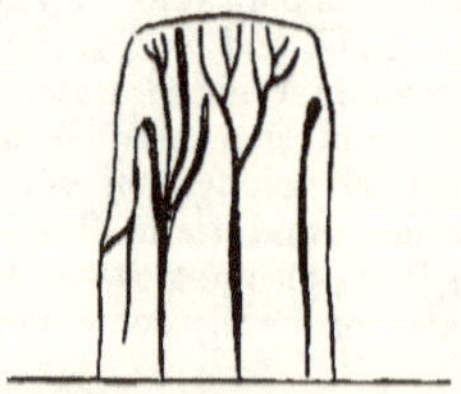

Abbot Stone on Boulsworth.

Third Grit Series.

On the east side of Greave Clough we have upwards of 175 feet of shales above the Kinder Grit before coming to the lowest Third Grit, viz., that forming Mere Stones. We suppose the Greave Clough fault to run close to and parallel with the S.E. face of this rock, which it throws down and repeats on the south. In fact the non-continuance of this apparently lower bed is one of the chief evidences for the direction of the said fault, and this is confirmed by our finding on the same general line of direction indications of a fault in Grey Fosse Clough.

A cross fault runs up alongside of the Cascade stream, which, as we mentioned above, p. 110, throws out the Kinder of Walshaw Dean. There is, however, some doubt about the correlation of the beds; it would seem that here our second bed of Third Grit splits into two, the lowest of which very soon dies out.—W. G.

South of Gorple Water we come upon the shales overlying the Kinder Grit near Reaps. The top of the grit, roughly speaking, runs up Red Carr Clough, and then across the moor along the head waters of various streams; thus, for instance, we found shales and occasional grits in the lower parts of the streams which flow into Noah Dale (No Dale on 1-inch map) dam from the N.W., while the upper portion of one of these cloughs consisted of coarse grit, in the cracks of which were thin veins of barytes. Again, we have shales and quantities of galena and barytes lying about in the clough called Lead Mine Clough, which joins Noah Dale from the S.W. It would seem that lead ore was once gotten here in past times.

Below Noah Dale dam we have shales with interbedded grits and sandstones; the like are also seen in Cross Clough. A short way below the foot of this last, we came upon Kinder which comes up with a sharp roll, or may possibly be brought up by a fault. Below this point we have Kinder all the way down the Calder Clough, to the base a mile below Jack Bridge stream; but for a mile above Jack Bridge, where the rock is thrown up on the east by a fault, it merely occupies the river flat.

The hills between Colden Clough and Gorple Water consist, besides shales, of two outliers of Third Grit; on that, forming Gainly Hill, we found Gannister lying about.

In the country south of Colden Clough we have generally a good top to the Kinder. The overlying shales are seen in the Hippings Clough, dipping at 10° to N.E. These shales contain beds of sandstone. As we ascend the hills on the north of this clough we have sandy shales with the same dip. At Moss Hall a bed of sandstone comes in, which makes a feature westward above Pole Hill, where it has been much quarried. The upper part of this bed, just below Mouse Nest, contains casts of shells. Above it we have shales, thin sandstone, shales, sandstones, shaly sandstone and shale, and Third Grit. This last is the bed which forms the outliers of Field Head Moor and Brown Hill. At Brown Hill it is a concretionary iron-stained gritty sandstone.—J. R. D.

On the east of the river Hebden the Third Grit beds occupy the summits of all the hills between the several valleys (see Section, Fig. 22, p. 114). The lowest bed (D.) occurs to the west of the Hebden Valley on the higher ground above the Kinder Grit escarpment, and again on the hill between the Hebden Valley and the Horsebridge Clough; in this last valley the bed is thrown down at one point between two parallel faults, which likewise shift the Kinder for a short distance into the stream. On the east side of the Horsebridge Clough and Hebden Valley this bed, D., sometimes clearly divided into two by a shale band, strikes from Grey Stones on the north southwards to Midgley, where it is thrown up by a fault to the east; it thence runs round and up the Luddenden

Fig. 22. (*Scale*, 1 *inch to the mile.*)

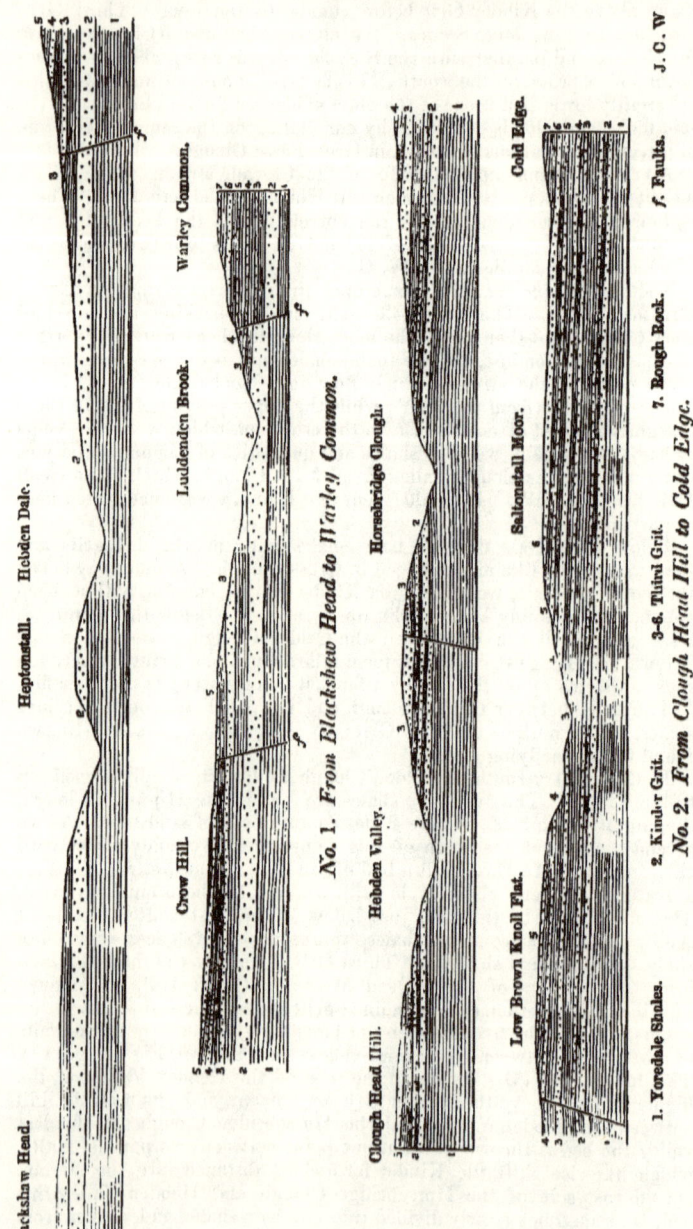

No. 1. *From Blackshaw Head to Warley Common.*

No. 2. *From Clough Head Hill to Cold Edge.*

1. Yoredale Shales. 2. Kinder Grit. 3–6. Third Grit. 7. Rough Rock. f. Faults.

MILLSTONE GRITS, EASTERN AREA. 115

Valley and back on the eastern side, some short way above the Kinder Grit.

A seam of coal occurs in connexion with this bed. The following section of a shaft and boring by the side of the footpath between Nook and Faugh, west of Crow Hill, will show the thickness of the coal and associated beds.

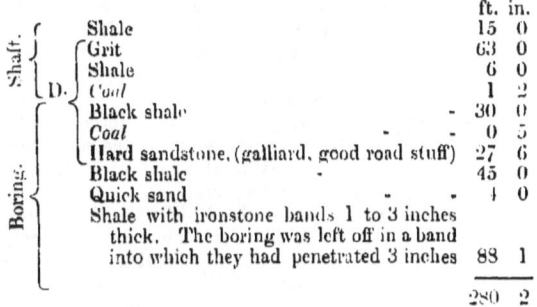

		ft.	in.
Shaft. {	Shale	15	0
	Grit	63	0
	Shale	6	0
D. {	Coal	1	2
Boring {	Black shale	30	0
	Coal	0	5
	Hard sandstone, (galliard, good road stuff)	27	6
	Black shale	45	0
	Quick sand	4	0
	Shale with ironstone bands 1 to 3 inches thick. The boring was left off in a band into which they had penetrated 3 inches	88	1
		280	2

Some of these beds may be well seen exposed in quarries along the outcrop. The upper of the two coals has been worked about Faugh to a depth of 45 yards; southwards, however, about Faugh Well, its thickness decreases to 5 inches, and workings cease, while northwards a shaft was sunk a little above Greystones to a depth of 15 yards, and then a boring made to the level of Dike without meeting with any coal; and the gritstone usually lying above it was found to be but a few yards thick. A little to the north both coal and gritstone occur between Lotham and Ballet on Moor End, between which places the coal has been worked to a depth of from 20 to 30 yards, and was found to be from 9 inches to 1 foot 9 inches in thickness. The overlying gritstone is here again some 20 yards thick. Farther to the north the coal can only be traced by an occasional smut found immediately beneath the rock. Southwards at Midgley the same coal occurs but 5 inches thick, and has been worked but very little. A shaft was sunk 50 to 60 yards in depth, hoping to find more coal, but without success. The ironstone was worked a little. In no other spot north of the Calder has this coal been met with.

The bed C. follows in strike the run of D.; but north of Midgley and east of Crow Hill it appears entirely to thin away, setting in again, however, still farther to the north and east. It furnishes in many parts good flagstones, especially along its western outcrop, from Han Royd northwards.

The bed B. exhibits the character of a more or less fine-grained sandstone. It occupies the greater part of the summit of the flat-topped hill between the Hebden and Luddenden valleys, from a little south of High Brown Knowl to just north of Midgley. On the eastern side of the Luddenden valley it conforms to the strike of the two previous beds.

The uppermost member, A., of the Third Grit series is generally a decided grit north of the Calder. It has a wider spread than any of the others, ranging from High Brown Knowl northwards to the Oxenhope Moors, and round the head of the Luddenden Valley, and southwards over Warley and Saltonstall Moors.

The country just described north of the Calder is very free from any large and persistent faults.

One of the most marked is that which throws the Third Grit bed (C) forming Shackleton Knowl (near the northern edge of sheet 88 N.W.) down to the south-west, and crossing Horsebridge Clough close to Outwood throws Kinder against bed (D) of Third Grit; farther to the east by Duck Hill its throw is very much lessened, though apparently again increasing across Low Brown Knowl Flat, on the hill top, where it throws down a small horse-shoe shaped mass of bed (A), while beyond the Luddenden Valley it seems to die away.

A fault meeting the northern end of this last, near Shackleton Knowl, and passing westwards close to Walshaw, and a little north of Widdop Gate, throws down to the north, though not to any great amount.

A short fault, roughly parallel to the first described, and a little to the north of Abel Cote, throws down north about 100 feet, or perhaps rather more, apparently, however, dying out westwards, and being met in Horsebridge Clough by a fault which just shifts the escarpments from Nase End to the stream, and passes southwards by Field Head to the Hebden Valley, beyond which it is not traceable.

An east and west fault, a little north of Miller's Grave, ranges from Dick Ing on the west to the Luddenden Valley on the east, passing just below Catherine House. At the hill top a short fault springs from this; both throw down to the north.

Another east and west fault farther down the Luddenden Valley throws up to the north, bringing together shale *in* the Kinder Grit and the *upper* mass of Kinder. Still farther down the same valley a fault crosses in a north-east and south-west direction; this throws a somewhat oval patch of the lowermost bed of Third Grit down southwards to a little west of Luddenden, and joins a north and south fault ranging through Midgley and throwing the Kinder Grit in the valley bottom down out of sight to the west.

All the faults above mentioned are more or less easily detected by their shifting of escarpments. A north and south fault just east of Hebden Bridge throws the Kinder down and out of sight to the east.

Two small faults were proved in the old coal-workings, one running north and south, east of Nook, having a throw of 7 feet down east, and another north of Old Town also throwing east to the amount of only 5 feet.
J. C. W.

On the south side of the Calder we have, as mentioned above, p. 109, immediately above the Kinder Scout Grit in Beaumont Clough a section in shales. On the west of this clough there rises a detached circular hill called Edge End Moor, composed of sandstone, which has a good escarpment facing N. and W., but, owing to a south-easterly dip, a poor one to the S. and E. This sandstone seems to be an outlier of an accidental bed of the Third Grit series. Below it we have, on the west side, shales dipping at 6° to S.E., and other shale sections on the south.

The position of the Wadsworth fault, which throws out of sight the Kinder a mile east of Hebden Bridge, is proved by our finding a series of sandy shales on the horizon of the Kinder Scout Grit in the brook next east of that fault.

These shales reach from the level of the alluvium upwards; they are succeeded by shales with their sandstones, at first lying flat, but as we ascend the hill the dip becomes steep, varying from 9° to 17° in a southerly direction.

Finally, a little below the road from Haven to Jumps we have sandstone in place on the horizon of the base of the Third Grit eastward. Above this sandstone we have a bed of shale overlaid by grit, which forms a small escarpment near the road. West of the fault we have, except near the fault, a very good escarpment running beneath Johnny Gap till it is broken off by the Stoodley Clough fault. The extension of this fault passes through Hill Top, where a grit escarpment is faulted. The base of the Third Grit is somewhat uncertain along the S.E. side; where the escarpment faces the Cragg Valley. Near Lumb the rock seems to be a mass of sandy shale and sandstone. There is a feature, like an escarpment, which we have taken to mark the base, with which it corresponds pretty well in position. This feature ends where the Stoodley Clough fault through Hill Top points to. Higher up we see sandstone in a lane; above that point the lane is occupied by sandy shales and sandstones, as high as the moor, which appears to be of grit.

North of Lumb the rock is a mass of grit, upwards of 300 feet thick, broken merely by an insignificant band of shale. The Broad Head Clough has cut through the rock into the underlying shales, which are crossed in many places in its course, and in the side streams that feed it. A large bay has thus been scooped out of the rock; and this amphitheatre is covered with great masses of coarse grit that have slipped from the surrounding wall of rock, which is here of so coarse and massive a character as to resemble the typical Kinder Scout Grit. This rock stretches across Erringden, and faces westward with a splendid escarpment of conglomerate, passing upwards into flagstone. Below it we have the upper part of Stoodley Clough occupied by shales rolling about, and much disturbed, particularly where we suppose the fault to cross that runs a little south of Rough Head. There is seen at a waterfall what may be the actual fault. Dips of 30°, 35°, 40°, and 50° are obtained hereabouts. Below these shales the top of the Kinder Scout Grit appears between the above mentioned and the Stoodley Clough faults.

The faulted bit of country between Stoodley Pike and the Stoodley Clough fault is very uncertain. There is no doubt about the Stoodley Clough fault itself; but there is some uncertainty about the correlation of the beds immediately south of it. On that side we have alternate beds of sandstone, shale, and grit, which in all probability belong to the Third Grit series. Close to the fault, and immediately west of the abrupt end of the Erringden escarpment, sandy shales are seen dipping at 30° to S. On the west of the hollow, where these shales occur, the grits of Stoodley Pike come on, reaching from top to bottom of the hill. This would appear to be the Erringden rock brought up again by a fault running S. by E. by Pasture, which breaks the Kinder Grit escarpment at Buck Stones.

At Stoodley Pike we have a good section, as follows:—

    Coarse massive grit, at the Pike.
    A space giving no section.
    Coal, 1 foot.
    Underclay.
    Grit.
    Massive flaggy grit.

That is a set of beds consisting of two main grits with a foot coal between them. As in the immediate neighbourhood the lowest Third Grits consist of two grits with a coal about a foot thick between them, though the lowest bed of grit does in places die out, we were led to consider the Stoodley Pike rock as the same bed. This grit is faulted against the Kinder Scout Grit immediately south of Stoodley Pike. On the general line of this fault we saw a fault in a clough at the *y* of the word "Stonyfield." This general direction corresponds too pretty well with that of the fault at Priestly Ing, mentioned above, p. 107.

The very broken ground between Withens Clough, and the Stoodley Clough fault must be uncertain. The alternations of grits and shales accords best with the supposition that these beds belong to the Third

Grit series, which is also consistent with the grit at Stoodley Pike being bed D.
J. R. D.

On the east side of St. John's Valley we have all the Third Grits forming an outlying mass of a somewhat oval shape. The highest point reached is 1,369 feet at Manshead End, and the next highest is the top of Crow Hill at a height of 1,250 feet. The rock forming the hill is the highest bed in this outlier, and is probably the Second Grit. It is a fine micaceous sandstone. From Crow Hill to Luddenden Foot all the members of the Third Grit series may be passed over to the Kinder in the valley of the Calder.

The lowermost bed of the Hathershelf Scout answers in character to the same bed at Midgley, consisting as it does of two grit beds parted by a shale band of some 20 feet in thickness, which contains a coal from 4 to 8 inches thick. Southwards, however, this coal is not traceable, though it reappears at Stoodley Pike. Vide supra, p. 117.

The following is the section at Hathershelf Scout Delf :—

	ft. in.
Grit	—
Coal	—
Shale	5 0
Grit	—
Shale	about 20 0
Coal	0 8 to 4 in.
Grit	—

the whole being nearly 200 feet thick.

Little Moor Quarry gives the following section :—

Grit changing into sandstone.
Black shale.
Coal, 2 inches.
Black shale, 10 feet.
Hard white fine-grained rock, 13 feet.
Shale, 1 foot.
Coal, 0 to 2 inches lying on a wavy surface of sandstone.

The upper members of the Third Grit series, being variable beds of sandstone, call for no particular notice. In many parts good flagstones are procurable from them. They are in some places rendered difficult to trace by the thinning away of some beds and thickening of others.

It may be as well to mention that the sandstone quarried low down on Marshaw Bank is a slipped mass.
J. C. W. and C. F. S.

The Kinder Scout Grit in Booth Dean is overlaid on the south side by about 300 feet of shales. Above these comes the escarpment of Third Grit along Mosleden Height and Way Stone Edge. This rock there consists of a basement bed of flags overlaid by coarse massive grit. The Way Stone is a block of this latter left subsided on the lower bed, on which it gradually settled down as the higher bed weathered away, whose former greater lateral extension it thus remains as a monument of. On the north side of Booth Dean we have an outlier of Third Grit forming a hill between Booth Dean and Black Castle Clough. On the north side, or that facing Black Castle Clough, this grit forms the escarpment of Blackwood Edge and Bench Holes; in the latter place are many landslips. This grit is coarse and massive. It is probably D. and C. combined as A. and C. are generally the important rocks, and B. and D. minor ones; moreover, we find that between Smith's Staff and White Hill the rock does split into two, of which the lowest is insignificant, while the uppermost forms the bold escarpment of Joiner Stones, and is the coarse grit of the moor; so, too, the basement beds under Blackwood Edge consist of fine flaggy grit, which is the representative of the lowest bed, where the rock is separated into two by a distinct band of shale.

East and north of Bench Holes and White Hill we have represented a faulted outlier of Third Grit. This rock, which forms a good escarpment, is not traceable beneath the grit of Bench Holes; we therefore think it is faulted, though the exact position of the fault is uncertain. We have taken the fault where a break in the ground suggests a possible fault. We have further evidence of such a fault in the fact that, while the valley above the fault at Swift Place is occupied by Kinder Grit as far as Nab End Wood, we have higher up the valley a grit in the stream overlaid by shales with thin sandstones; this looks as if the top of the Kinder had been thrown down into the stream.—J. R. D.

Agricultural Features.

Speaking generally, the northern, western, and southern borders of this district consist of lofty moorland, covered with peat and heather, the abode of mountain sheep, grouse, plover, and curlew. This is the character of the Oxenhope Moors, Boulsworth, Widdop, Black Moor, Black Hambledon, Heptonstall Moor, Warley Moor, the Kinder Grit slopes of Langfield and Blake moors, Blackstone Edge, and Rishworth, and the Third Grits of Erringden and Hassocks. The broad basin of the Calder beneath the escarpment of Kinder Scout Grit is dotted with houses and cut up into fields.

The actual valley of the Calder is in some places wooded, as are also the Hippings, Colden Clough, the valley of the Hebden, and Horsebridge Clough, which thus afford a striking contrast to the general treeless aspect of a millstone grit country.

On the east of the river Hebden the moory ground occupied by the uppermost member of the Third Grit, A., including Saltonstall and Warley Moors, and White Hill, is covered pretty uniformly with peat from 4 to 6 feet thick.

The hill top from High Brown Knowl southwards to a little north of Midgley presents for the most part the character of a heathery moor destitute of peat, and stony in many parts. The hill sides are, however, under cultivation, stone walls in plenty marking out the slopes. Much of the western side of the Luddenden Valley is covered with slips, as also is the left side of the Calder Valley from Luddenfoot to the anticlinal fault.

The greater part of the district forming the outlier of Third Grit between St. John's vale and the Ribourne, with the exception of a moory tract about Manshead, is under cultivation. There is little or no peat to be found here.

The alternation of sandstone and grit beds with shale gives a variety to the soil, though the fields on the hill slopes are very generally stony, owing to the constant washing of rocky material from above.

This country is a fine instance of the working of atmospheric denudation; the narrow but deep valley of the Hebden with its soft shaly sides and crowning line of crags speaking unmistakably of the power of weather, rain, and rivers. The constant slipping away of the sides, and consequent gradual widening of the valleys, is most clearly seen. A large slip of this nature occurred in the Hebden valley just opposite Lee Mill in Nov. 1866, which carried down into the river below a considerable length of the high road.

The effect of landslips in widening a valley is also well seen in the country immediately north of Todmorden.

The only part of this district in which drift has been observed is the northern flank of Boulsworth.

This will be more conveniently described in connexion with the great mass of drift to the north (see p. 137).

J. R. D.　　J. C. W.

CHAPTER VI.

THE PERMIAN SYSTEM.

The Permian Rocks, constituting, after the Carboniferous, the next great volume of the earth's history, are but poorly represented in this district. They occur in the Ribble below Waddow Hall near Clitheroe, in the Darwen above Roach Bridge, and probably in the Yarrow south of Euxton. In each case these detached masses lie near to great faults, and it seems probable that their low position, the two first lying in low valleys and the last on a great sea-side plain, is due not to their having been originally deposited at levels so low relatively to the neighbouring hills, but to their having been faulted down from the main mass, which must have been formed on a great plain now rudely represented by the hilly table land of the Carboniferous Rocks.

Roach Bridge Section.

Mr. Binney first called attention to this interesting section.* It commences about 300 yards above the bridge in the river Darwen. There on the left bank are black shales with ironstones, dipping south at 45° to 50°, and containing *Goniatites*, *Posidonomya*, coprolites, fish remains, and plants. The surface of these to the depth of two feet has changed its colour to pink and white, and these discoloured beds, not being well seen, might be easily mistaken for a part of the Permian.

My friend Mr. James Eccles and I, after grubbing a short time with our hammers, found that they were distinctly dipping south with the Carboniferous, and not W.N.W. with the Permian beds. The latter have an inclination of 10° to 15°.

The section was as follows:—

	ft.	in.
Red, yellow, and brown sandstones, about	220	0
Bright brick-red massive sandstones with ripple-marks	25	0
Red marl	0	4
Hard well-bedded fine red sandstones	2	6
Red sandy marls	0	5
Hard very fine red sandstone, containing lime	0	6
Red fine sandy marl	1	1
Red fine quartzose conglomerate, mottled with yellow and green, yellow at bottom	0	5
Red fine sandy marl	4	0
Pinkish white clayey shale, Carboniferous.		

In general terms this section may be described as consisting of red, yellow, and brown sandstones, with a marly base. The marls are no doubt derived from the waste of the Carboniferous shales on which they lie, and the little conglomerate has its pebbles from the Millstone Grits.

---

* Observations on the Permian and Triassic Strata of Lancashire, by E. W. Binney, F.R.S., F.G.S., Trans. Manchester Lit. and Phil. Soc., 2nd series, vol. xii. and xiv.

It will be seen that our estimate of the thickness of the sandstone differs from that of Mr. Binney. In the river itself there is nothing against his estimate of 400 feet, but reasoning from other sections we thought it probable that the fault between Permian and Trias came where we have placed it, and that therefore the whole of the space to where the Trias is first visible in the river, was not taken up by Permian Rocks. On the supposition that the dip continues at the same angle and that the fault exists as drawn, the greatest thickness that could come on would be 220 feet.

Waddow Hall Section.

From the weir below Waddow Hall near Clitheroe, to a point about 200 yards lower down, beds of dark Carboniferous Limestone, with alternations of shaley Carboniferous Limestone, are cut into by the Ribble. Then comes a fault ranging S. 20 E., and bringing on a sandstone which we suppose to be of Permian age. In or close to the fault itself and nearly vertical, is a band or bed of breccia composed, chiefly of angular fragments of Carboniferous Limestone in a red sandy matrix. Next to this is some red ferruginous clay, and then comes on the sandstone. It is difficult to say whether this breccia and clay are ordinary subaqueous deposits, or have been merely formed mechanically in the line of fracture. The breccia* is only a few inches, the clay several feet thick. Next one sees brown and yellow sandstones false-bedded with a general S.W. dip. They contain little nodular concretions, which consist of sand with some cementing material, probably lime. When broken they show a glistening fracture. A lead vein occurs near the fault, and almost parallel to it, in the sandstone. Lower down the Ribble the sandstones are red and appear to be nearly flat, but do not show their bedding very distinctly. Just above the iron footbridge by Low Moor Mills is a little vein of "Heavy-spar" running N.W.

The base of these beds may be well seen just above Stephen Bridge over Bashall Brook. There are limestones of the "Shales-with-Limestones" group, highly contorted, and resting on their truncated edges is a bright red sand dipping north-easterly. Just at its base occur some nodules of iron pyrites. These remind one forcibly of some in precisely the same position in the Permian Sandstone of Skillaw Clough. The rest of Bashall Brook to its junction with the Ribble, and that river for half a mile or more lower, runs over "Shales-with-Limestones," which have been coloured red by the Permian now removed, but in a westerly reach of the river, W. of Henthorn House some red sandstone with thin seams of marl may be seen under water. It appears to be dipping S.E. at 12°. Above Lower Henthorn, on the left bank, some red marls, mottled green and much disturbed at the surface, are visible. It is difficult to get a good dip, but N.W. seems the prevailing direction. The inclination is in some parts as much as 70° or 80°, but this I believe is due to surface disturbance. These marls probably overlie the sandstones before mentioned.

The sandstone at Stephen Bridge and about Low Moor Mills has been known to some local geologists, and taken by them for Old Red Sandstone. This is quite impossible from its position. I am not aware of anything having been written on this matter before I called Mr. Hull's attention to it. The special interest connected with it lies in this rock resting on beds so low in the Carboniferous System, and proving the

---

* I beg that geologists will treat tenderly this slender but interesting relic. I once had the pleasure of showing it to an enthusiastic friend, and the pain of seeing him, with a ponderous hammer, smash to pieces nearly all of it that was then visible.

removal of an enormous thickness of Carboniferous Rocks by denudation prior to the deposition of those of the Permian System next following.* And this occurred at the very locality above all others where the Carboniferous Rocks had their greatest development in the United Kingdom. Besides the Coal Measures there must have been removed of Millstone Grits 4,600, and of the Yoredale beds 3,000 feet.

Yarrow Section.

The little river Yarrow south of Euxton, near Preston, exhibits a section in Lower Coal Measures. The beds as you go down stream are seen to be dipping S.E. until you arrive at a small ruined mill called on the 6-inch map Chorley Holme Mill. Here they turn over rapidly to the west and a fault comes on. Beyond this under the right bank I found some red and green marls much disturbed. They seemed to me to bear a strong resemblance to the Permian Marls just mentioned near Lower Henthorn. They have therefore been coloured accordingly, but the section is an obscure one, and I do not feel certain about them. Their position here close to the great fault would only be analogous to the situation of the Skillaw Clough and Bentley Brook sections.—R. H. T.

CHAPTER VII.

TRIASSIC ROCKS.

Bashall Brook Section.—In Bashall Brook for 150 yards above Talbot Bridge are seen beds of purple grit, in places a conglomerate, with bands of sandy shale. They are dipping N. at about 35°. These beds are probably Upper Yoredale Grit. Above them is a little bed of sandy shale, and the next rock we see is some fine bright red sand with apparently the same dip, or if anything a little E. of N. It is much false-bedded, has white or yellow patches, and contains a few white quartz pebbles. It continues for about 60 yards along the stream, when it is concealed by stiff till. This is all that is seen of it. It is quite a different rock to that seen in the Ribble below Waddow, and is more probably a portion of the Pebble Beds of the Bunter Series.

If the fault which throws down the Permian at Waddow be produced on the map it will be seen that this sandstone lies on the downthrow, or west, side of it. That this same fault does run in this direction is rendered highly probable by the occurrence, in the same line, of the shift to the S. of the lower boundary of the Upper Yoredale Grit, as well as the great fault which interrupts the continuity of the limestone axis of the Slaidburn and Chipping anticlinal, and brings the grits down into the valley between Knowlmere Manor and Thornyholme. And this is a matter of no small importance, for it throws light upon the relations of the Permian and Trias to the Pennine Chain.

If these formations were deposited in their present position relatively to the neighbouring hill tops, then the waters of those periods must in this case merely have been narrow inlets of an archipelago district. Had this been the case we should be likely to find these patches in other places besides the neighbourhood of large faults, but none such exist. They would also be likely to have more the character of valley or shore deposits, and contain at any rate beds of coarser material. But in this district we do not find any portion of them so coarse as typical Millstone Grit, and *that* we know is not a shore deposit.—R. H. T.

---

* E. Hull, F.R.S., Quart. Geol. Journ., vol. xxiv. p. 327.

Pebble Beds around Preston.—This subdivision of the Bunter Series occupies a considerable area in quarter sheet 89 N.W. The chief tract extends from the boundary fault of the coal-measures at Cuerden Hall, near which place it dips under the Upper Mottled Sandstone; from thence the junction boundary of these two strata trends by Farrington and Charnock Moss to Drummock Hall, southward of which the western extension of the Bunter Series is cut off by the Scaresbrick Fault, which brings in the Keuper Marls; but neither the fault nor the actual junction of the Upper Mottled Sandstone with the Pebble Beds is seen, owing to the thick covering of glacial and post-glacial drifts enveloping the country. But the position of the fault is inferred from the outcrop of Pebble Beds in the Ribble dipping easterly, and from that of Keuper Marls in the Douglas dipping westerly.

None of the deep brook valleys of the district are as yet cut sufficiently deep to expose any section of the Pebble Beds, except in Bezza Brook, east of Samelsbury Hall, in the immediate neighbourhood of the Ribble plain, which, as well as that of its tributary the Darwen, is cut down as low as high-water mark near the junction of the two rivers. The rock forms a flat uniform surface, extending under the drift plains as well as under the alluvial flat, as shown in the following diagram.

Fig. 23.

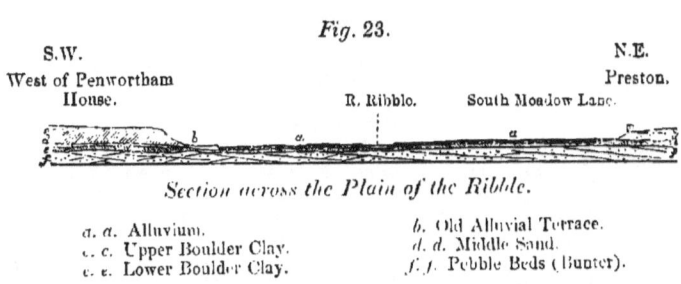

S.W.   West of Penwortham House.   R. Ribble.   South Meadow Lane.   N.E. Preston.

Section across the Plain of the Ribble.

a. a. Alluvium.
c. c. Upper Boulder Clay.
e. e. Lower Boulder Clay.
b. Old Alluvial Terrace.
d. d. Middle Sand.
f. f. Pebble Beds (Bunter).

§ From the above it will be seen that the river Ribble at Preston has merely excavated its valley in the drift deposits, a channel being cut in the rock to a slight depth, only at a few points, where the original surface of the rock rose above the level of the inclined plane, representing the lowest possible "gradient of fall" attainable by the river, so long as the sea retains its present level, a gradient, which it has already reached, its denuding powers being confined to lateral denudation of the banks bounding the river, and to the wearing back at points of the bluffs forming the limits of the valley. The rock-surface slopes from the land towards the sea, but the gradient of its fall being rather steeper than that of the river bed, causes the banks of the latter to be more and more composed of rock in advancing from Preston inland.

Advancing seawards and westwards the river bed is nearly level, and tidal, while the slope of the rock surface continues, causing the base of the glacial drifts to be never seen in the coast lines of the area comprised in map 89 N.W., the rock-surface lying 50 or 60 feet beneath high-water mark.

The first indication of the Pebble Beds of the Trias is seen under the alluvium on the south bank of the Ribble, opposite Old Roan Wood, where deep brick-red coloured sandstone was found even west of this point, on the deepening of the river by the Ribble Navigation Company many years ago; the excavated material may still be seen on the banks and built into the river wall; it is rather a fine-grained sandstone, with occasional seams of red shaly marl, and with pockets filled with the same material of the deepest purple colour. At a short distance east of Hangman's Rack Wood, in a low cliff, at the base of the bluff of Boulder Clay, a rounded knoll of sandstone, covered by terrace alluvium, rises to a height of eight or ten feet above the level of the ordinary alluvial plain; it presents the following section, the true dip of the Pebble Beds being about E.S.E., inclining to the horizon at an angle of four degrees.

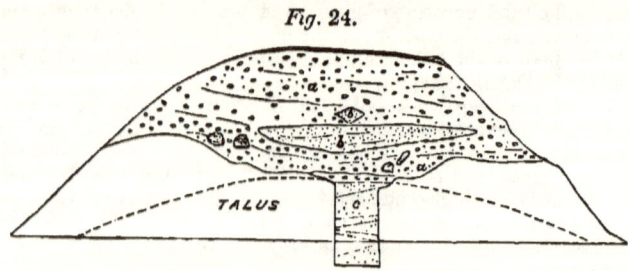

Fig. 24.

Section at Hangman's Rack Wood.

  a. Old River Gravel.   *b.* Masses of Red Sandstone.
     c. Rather soft Red and Brown Sandstone.

Further east, in the bed of the river opposite Marsh End, and also near the tramroad bridge, the Pebble Beds were formerly quarried for building purposes; the stone being thoroughly saturated with moisture, was naturally found to be wanting in durability, and a church built of it at Preston had eventually to be pulled down. These quarries have been filled up during the last 20 years, and the river flows over the site. The rock is, however, well seen in the river bed at the latter locality, where it was first described by Mr. Binney, F.R.S.;* it is a rather hard brownish red sandstone, with a few pebbles, dipping, on the west of the bridge, to the S.S.E. at 25 degrees, and on the east, S. 10 E. at 12 degrees; between this point and Walton Bridge, it is seen at several points under the alluvial bank on the south side of the river, at two points; hard beds of red sandstones, with greyish seams, strike entirely across the river, dipping up stream.

Hard greyish red beds are also seen in the river on the south side of Walton Bridge, dipping from south-east to S.S.E. It has been penetrated to a short distance in a shallow well at the adjoining cotton factory, where I was informed by Mr. Calvert the rock was found to be soft and spongy. The next section is that at the base of the river cliff, below the parish church of Walton-le-dale, above the village; about five feet of hard grey sandstone with current bedding, and numerous small quartz pebbles, dip to W. 20 S., at from five to six degrees. From this point eastwards, both in the Ribble and Darwen sections, the Pebble Beds dip in a westerly direction, and there is no doubt, between Samelsbury and Preston, the beds lie in a synclinal curve, running from west to east; the trough of which appears to pass through Walton-le-dale in a N.N.E. and S.S.W. direction, being prolonged southwards under the Upper Mottled Sandstone of Croston, Horscar Moss, Bickerstaffe, and the east of Ormskirk, to Kirkby, around which

---

* Mem. Lit. Phil. Soc. Man.

latter place the tract of Pebble Beds, overlying the Lower Mottled Sandstones of Croxteth, described by Mr. Hull, F.R.S., in the memoir on quarter sheet 89 S.W. of the Geological Survey Map, occurs, which tract was evidently continuously connected with that around Preston before the two sides of this centroclinal trough was cut off by the coal-measure boundary fault to the east and the Scarisbrick fault to the west.

A portion of the eastern extension of this basin of Pebble Beds, occurs at Knowles Wood and Eccleston Green, where hard red and brown sandstone with a few scattered pebbles is seen dipping under the soft red basement bed of the Upper Mottled Sandstone already described. The dip of the Pebble Beds near the boundary fault of the coal-measures is about 26°, becoming less in advancing westwards, and slightly rolling in the brook section of Knowles Wood, where pebbles are tolerably numerous; the dip of the beds, where they disappear under Upper Mottled Sandstone, is not more than four to five degrees. The southern boundary of this small tract is probably formed by a small fault, but it is not visible owing to the country south of the brook being obscured by drift.

Returning to the sections on the Ribble, the next section to that beneath Walton-le-Dale occurs about a mile further up the river at Mete House at the base of the alluvial bank in the middle of the plain. The dip of the sandstone, which is of a deep purple colour, is to the W.S.W. at a low angle. No other section occurs on this side (north) of the river until that south of Alston Hall is reached, but the south bank of the river, and indeed the bed of the river itself, is composed of the sandstones of the Pebble Beds from a little N.E. of Brockholes, Samelsbury, to Seed House, the beds being capped by terrace alluvium. The sandstones are much current-bedded and contain a few coloured quartz pebbles and seams of grey or greenish sandstone of similar lithological aspect to the surrounding mass. The dip is generally at low angles to the W.N.W. Rather soft red sandstone is seen in the bed of Bezza Brook a few yards west of the bridge over it at Cunliffe's, and also in a pit at the base of the hill, where the red shaly partings are used for ruddle, as mentioned by Mr. Binney.

At Lower Hall as at Samelsbury it forms the base of a low cliff capped with alluvium, but from a point a little north of this, up to the boundary fault, it forms the base of a cliff composed of the thick glacial drifts. It is here of a dark purple colour, contains numerous pebbles and is much current bedded, dipping from 6 to 8 degrees W. 10 N. The rock is much like the pebble beds around Birkenhead.

Returning to Walton-le-Dale the Pebble Beds are seen in the river Darwen; at the eastern end of Holland Wood north of Coopers Fold the sandstone is very hard and of a greyish yellow colour, dipping 23° W. 8 S. Its outcrop across the river has been used for the base of a weir in connexion with the sluice at Walton Mill. From this point the rock does not appear until Carver Bridge above Bannister Hall is reached, above which the river falls over about 30 feet of soft red sandstone containing a few scattered quartz pebbles, pockets of red marl, and seams of sandy grey marl.

Below the waterfall the river runs through a narrow gorge excavated in the rock, the level of the surface of which gradually diminishes towards Carver Bridge, below which 12 feet of river gravel cemented by iron overlies 2 feet of rock above the ordinary river level; this is about 30 feet above ordnance datum line, while the surface of the rock above the fall is 64 feet, giving 30 feet fall between the two points, which are 750 feet apart or 1 in 25, while that from Carver Bridge to the Holland Wood exposure is much smaller, being only 7 feet in 8,200, or 1 in 1,171, or nearly horizontal, being the eastern margin of the great "plain of marine denudation" which underlies the whole of the drift deposits of Western Lancashire from the mouth of the river Mersey to that of the Lune, the greatest width of which occurs from the sea coast to the sections under consideration. The rock surface from these sections to the Permian section on the Darwen at Roach Bridge rises 70 feet in 4,000, or 1 in 57. It is therefore clear that the slope is more steep between Bannister Hall Print Works and the waterfall than that either above or below, and it is not impossible that it marks the actual sea margin of the plane of denudation to the west of it, which is described in the portion of this Memoir treating on the Drift Deposits.

The fine section at the waterfall was first observed by Mr. Binney, F.R.S., who considered the beds to belong to the Pebble Beds of the Bunter, which view further investigation has corroborated, and he was also the first to point out that the sandstones of the Ribble and Darwen lie in a synclinal curve.

After carefully examining the dips of the sandstones in these rivers, I think it probable that the Pebble Beds are not less than 800 feet in thickness in this area.

Upper Mottled Sandstone.—It is well seen at Black Moss between Rufford and Mawdesley, where a section occurs in the pit north of the lane of soft red, and light coloured, mottled sandstone dipping N. 50 W. at five degrees, the sandstones being much current-bedded to the N.W.; in the pit to the south, current-bedded, bright red coloured sandstone, streaked with greyish yellow, is seen, the dip being to the south-west at three or four degrees, that of the current-bedding to the west.*

Current-bedded, rather soft red sandstones are seen in the brook running between Hurst Green to the banks under Mawdesley Hall, the dip being to the W.N.W. In the bank above the road about 20 feet of sandstone is exposed traversed by thin seams of shaly marl, and by joints, the surface being here and there marked with greyish white patches or seams running for a few inches with the bedding, which dips from four to six degrees to the W.N.W. Towards the Black Bull the bank gradually lessens in height and is capped by boulder clay, but the rock is seen in the roadway as well as in the bank, between the road and the brook, beneath as well as in the brook running down from Ambrose's House to Mawdesley, where it dips to the north-west at 7 degrees.

Another rock exposure occurs at the meeting of the three lanes at the Boarded Barn, where rather hard red sandstone dips north-west at 5 degrees ; a few yards further it is also seen in a garden attached to the above cottage, of which it forms the western wall ; the space occupied by the garden has been quarried out ; it also occurs in a hole by the road side a little beyond, where it dips 4 degrees in the same direction ; the road at this point is composed of the rock which has some traces of having been glaciated from the north-west, and a little further west it is overlaid by particularly yellow-coloured Upper Boulder Clay. The water from the spring called Robin Hood Well is evidently supplied by these sandstones, as is also that to the south-east near Salt Pit House, in which the water is salt, containing several grains per gallon.

In the country between Liverpool and Southport the Upper Mottled Sandstone was found to consist of three divisions, the upper being a hard yellowish sandstone with softer beds, the whole often containing

---

* See Lancashire 6-inch Geological Survey Map, 84.

cavities or empty pockets, small pebbles of white quartz being not entirely unknown.

The middle bed is of a bright red colour with greenish-grey seams or mottlings associated with extremely soft pinkish-red sandstones, with an almost equal amount of yellow mottlings; to this division, the beds of Mawdesley described above belong, the upper hard beds being absent in quarter sheet 89 N.W. The lowest division is of a bright deep red colour, extremely fine grained, seldom current bedded, in Cheshire; it is well seen at the mouth of Bromborough Pool, as described by Prof. Hull, F.R.S. ;* and in the present map (89 N.W.) it is seen resting on the Pebble Beds at Syd Brook at Eccleston Green; it dips at Mill Lane 4 degrees to the north-west.

Fig. 25.

N.E.   S.W.

Upper Mottled Sandstone.

Lower   Pebble Beds (Bunter).
Coal-measures.

Section at Syd Brook, Eccleston Green.

Soft red sandstones running nearly level are met with beneath the alluvium in the wells at Croston; they probably belong to the middle beds and are continuous with those of Mawdesley.

Keuper Marls.—Are the newest strata found in West Lancashire; they are generally of a pale greenish grey colour, especially in the upper portion, but red seams sometimes occur. At the Palace Hotel, Birkdale Park, Southport, they were bored into to a depth of 567 feet without the bottom being reached; between this boring and Bescar Lane, in 90 S.E. and 90 N.E., several sections occur; the strike is generally north-east and south-west, the dip being small and subject to small rolls or undulations towards the north-west and south-east, entering the south-western margin of the Preston quarter sheet. The red and grey marls occur beneath the Boulder Clay knolls W.S.W. of Bescar Lane Station, and under the peat, in the fields north of the sluice, to the east, where masses of indurated grey shales with pseudomorphous crystals of salt are turned up plentifully in digging the deep peat drains; they also occur, associated with fibrous gypsum, at the bottom of the sluice, at the engine-house, at Wigans Bridge, north of Berry House, where about three feet of Shirdley Hill Sand, six of Boulder Clay, and 15 of Red Marls, with gypsum, were passed through.

Indurated shales also occur at the bottom of the deserted canal cutting from Merscar Brook, near Dochyles, just off the margin of the map.

The Marls are also found underlying the Boulder Clay in the bed of the Douglas, at Placks, near Tarleton, and at the mouth of the river, between Hesketh Bank and Freckleton Point. They also occur under the cliff of Boulder Clay, at the latter locality, which is sometimes called the Naze, or Neb, of Freckleton; here, after the severe storm of February 1868, I observed the marls, consisting of red clay with partings of grey shales, covered with pseudomorphous crystals of salt, and associated with much fibrous gypsum, the sand and shingle generally covering the section having been swept away; some fragments of marl and gypsum occur in the base of the boulder clay, and can be seen at any time; the marls beneath appeared to be nearly horizontal; further to the west,

---

* " Explanation of Horizontal Section, sheet 68."

at Warton, I am informed by Mr. Gregson, of Lytham, they were bored into some years ago in a fruitless search for coal. Shaly marls have also been brought up from the bottom of clay wells at Kirkham, and further to the west, off the margin of the map, a boring was made in 1837, at Poulton-le-Fylde, to search for coal; the bore, after penetrating through the beds of the Glacial Drift, passed through the Keuper Marls (as mentioned by Mr. Binney, F.R.S.) to a considerable depth without reaching their base, the total depth from the surface being 179 yards; this, however, was reached in an unsuccessful well made for the War Office at Fleetwood, a bed of grit being reached at a depth of 170 feet. This grit may possibly be Lower Keuper Sandstone, but in this case, as well as that of the deep well at Scarisbrick, which penetrated the marls, there appears to be a slight probability of the underlying beds being of Yoredale age in the former case, and of Lower Millstone Grit age in the latter; the grounds for this possibility are stated in the account of the faults of the district.

The basement beds of the Keuper Marls are not seen anywhere within the map, and the Lower Keuper Sandstone is absent, the next strata in descending order being the Upper Mottled Sandstone of the Bunter series.—C. E. R.

CHAPTER VIII.

GLACIAL AND POST-GLACIAL DRIFT DEPOSITS.

Observations in this district go to confirm the generally recognised fact that all the main physical features of hill and dale had been sculptured out before the commencement of the Glacial Epoch. We find the deposits of boulder clay, sand, and gravel referable to this period occupying not only the plains, but partially filling in the lateral valleys, and resting on the denuded edges of the Carboniferous Rocks. As regards the vertical limit which erratic pebbles attain in this district, there is some uncertainty, because it is quite possible they may occur in hollows concealed by peat at elevations higher than where exposed to view; but I can state from my own observation that erratic pebbles may be traced to an elevation of 1,300 feet, both on Black Hambledon and Boulsworth.* Above these elevations wandering stones and ice-borne boulders are not found, though both on Pendle, Boulsworth, and other hills are strewn large blocks of Millstone Grit, relics of rock masses which once overlay the beds now forming the surface of these hills.†
Thus, on the western escarpment of Pendle Hill, may be seen large masses of grit and conglomerate, differing in coarseness of grain and general appearance from the beds of finer grit of the Yoredale series, which forms the summit of the hill. These coarser blocks are clearly referable to the Kinder Scout Grit of the Millstone series, which once overspread the Yoredale beds, and has left behind these memorials of its former

---

* This statement accords with that of the late Mr. Whitaker of Burnley, who stated that he could trace the drift to a height of 1,300 or 1,400 feet, but no higher. He likewise mentions the occurrence on Pendle of the semi-arctic plant, *Rubus chamæmorus*, which does not grow on the opposite side of the Burnley Valley, attributing its presence on Pendle to the view of its having been left unsubmerged during the glacial period. Trans. Geol. Soc. Manchester, vol. 4, p. 115.

† Mr. T. T. Wilkinson, F.R.A.S., considers that some of these blocks containing basin-shaped hollows have been used for purposes of Druidical worship. See Trans. Hist. Soc. Lanc., vol. 5, N. S. Mr. Gunn, of the Geological Survey, states that all the blocks of stone on Boulsworth are millstone grit and not erratic boulders.

presence. In a similar way has Mr. Prestwich accounted for the "Sarsen stones" of Wiltshire and Dorset; which, although belonging to the early Tertiary series, are found scattered over districts formed of Cretaceous and Oolitic formations.

Although the subdivisions of the Drift series, consisting of a lower and upper till, divided by beds of sand and gravel, which occur in South Lancashire and Cheshire, have not been as yet traced out on the maps over the Burnley and Blackburn districts, I have little doubt they will be found to maintain their relative characters and position in this more northern and hilly region. These three members of the Drift are displayed in at least one fine section, and confirm me in the view I have stated. The section is laid open in the left bank of the Ribble, near Balderston Hall, below Ribchester, of which a sketch is annexed. Here the river takes a semicircular sweep of 750 yards in diameter, bounded by a cliff wholly composed of these glacial deposits rising 170 feet above the bed of the stream, as shown in the sketch below.

Fig. 26.

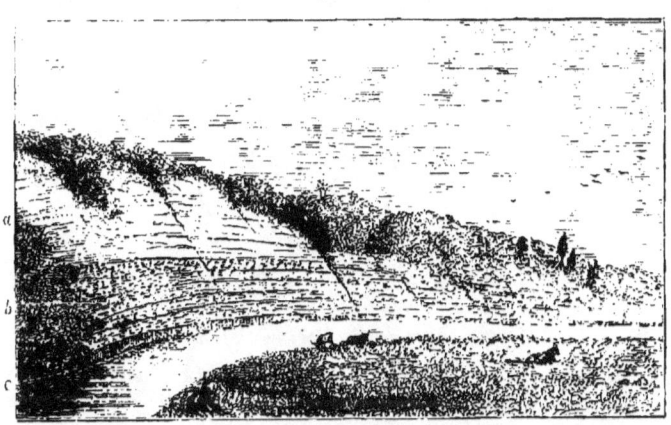

Drift Deposits in the Banks of the Ribble.

The following is the description of the beds as seen in the section :—

a, Upper Till or Boulder Clay. { Red clay, apparently laminated, resting on the underlying sand in an even well-defined line of demarcation, 55 to 60 feet.

b, Middle Sand. { Fine reddish sand with beds of gravel of rounded waterworn pebbles, and with a gravel bed at the base, 50 feet.

c, Lower Till. { Dark brown and bluish stiff clay, apparently not laminated, and with angular pebbles and boulders, more than 10 feet.

These divisions, which correspond exactly with those in the neighbourhood of Manchester, may be frequently traced along some of the deep dells which intersect the country north and west of Blackburn. Thus in the lower part of Old Park Wood, near Oxendale Hall, the Middle Sand is found resting on the Lower Till, and in the upper, the Upper Till is found resting on the Middle Sand. A similar series of deposits may be again observed in Merserfield Wood and Bezzabrook. At Blackburn, and downwards along the valley of the river Darwen as far as Witton Park, we find the Upper Till (much used in brick-making) resting on the Middle Sand. In the neighbourhood of Pleasington the sand reappears from beneath the Upper Till, and on the right bank of the river Darwen, above Hoghton Bottoms, forms rounded banks and mounds, such as are characteristic of these beds when they occur in force. On the left bank of the valley, however, the Upper Till again overlies the sand and forms a flattish country.

In the neighbourhood of Manchester, Oldham, and Stockport I have shown that the several subdivisions of the Drift have a tendency to rise and fall to higher and lower levels as they approach the hills or recede from them, thus accommodating themselves to the original form of the ground.* Similar phenomena may be observed in the district of the Pendle Range, and on the flanks of the hills east of the Burnley Basin. As an illustration in the former district, I here give a sketch showing the position of the Lower Till underlying the Middle Sand at an elevation of 600 feet near Blackburn, and which is also interesting as showing the form of the shelving bank of flagstone rocks over which the more recent beds were deposited.

Fig. 27.

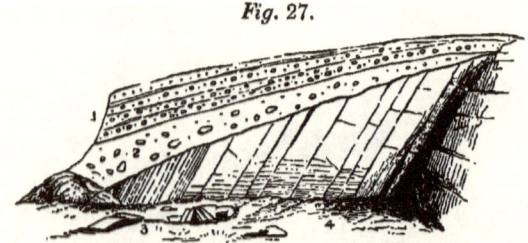

Sketch in a Quarry above Little Harwood Hall, near Blackburn.

		feet.
1. Gravel, principally local rocks	-	2 to 8
2. Dark blue Till	-	0 to 9
3. Sandy shales and grits -	}	Dip 55°.
4. Fine-grained grits and flagstones		

In the valley of the Calder above Whalley we find the Middle Sand and gravel surmounting the Lower Till, and rising gradually from the plain by the old Roman camp to an elevation of 500 feet at Clerk Hill.

On the southern slopes of Padiham Heights several brooks descend along the bottom of deep cloughs to empty themselves into the Calder. These banks are often composed of the Middle Sand overlying the Lower Till which forms the bed of the brook. The line of boundary between the sand and the Till can be very plainly traced by the outburst of springs and marshy ground ; and the beds have a slope towards the plain nearly corresponding to that of the surface of the ground.

Amongst the deep cloughs or valleys which stretch far into the hills

---

* "Geology of the Country around Oldham, &c.," Mem. Geol. Survey. Also, "On the Drift Deposits in the neighbourhood of Manchester," Mem. Lit. and Phil. Soc. of Manchester, vol. 2, 3rd series, p. 456–8.

along the eastern side of the Burnley coal-basin we find the Drift deposits assuming very peculiar appearances, and offering several points of interest. These valleys are of older date than the Drift itself, and have subsequently been filled in, sometimes to depths of 150 or 200 feet, with masses of gravel resting on boulder clay.* The upper surface of those deposits slopes upwards with the valleys, and upon the re-elevation of the land the brooks, with their innumerable tributary rills, have cut down channels through this gravel, so as sometimes almost to obliterate the original surface. Besides this, the gravel itself has been worn into conical mounds or hummocks by rain and brook action. These have somewhat the appearance of old glacial moraines, such as may be seen in the lake district of Cumberland and Westmoreland; but when we come to examine them closely, we can trace the original surface of the gravel bed by joining the tops of the mounds with an imaginary surface, and observing the successive stages of formation, from the mound shape to the original solid gravel-bed.

Fig. 28.

Hummock-shaped Mounds of Drift Gravel near Worsthorn.

The valley of Hurstwood Brook near Worsthorn (Fig. 28) affords a good example. Here the north bank of shale and flags has formed the original sea cliff against which the gravel has been deposited, and from this it has been spread over the moorland for a considerable distance upwards. The gravel is now worn into great numbers of mounds, but the original sloping surface can be very well observed from several positions at a little distance. Similar appearances may be observed in Thursden Clough near the new bridge, in Swains Plat Clough, along the higher part of Catlow Brook, and in the upper part of Cant Clough Beck. We find similar phenomena presented by the drift gravel in a locality at some distance from the above, but not less marked, and evidently referable to similar causes. From the foot of the high escarpment of Hapton Scout there extends a tract of moorland sloping downwards towards the north, and over which drift deposits are often deeply spread, as shown in Thorny Bank Clough. These deposits form an old sea beach, extending to the base of the cliff, and spreading over the tract now known as the Lower Park. On approaching the ravine formed by a brook course which is often dry, the surface of the gravel-bed begins to be intersected by little rills, which gradually deepen their channels, and being connected by cross gutters at length spread a network over the whole ground. As the deepening of the rills and gutters proceeds the hummock-shaped appearance of the enclosed masses of gravel becomes more developed, till at length all traces of the original surface have disappeared, and we find in its stead a space of ground dotted over with conical mounds, like a colony of gigantic anthills.

I have stated above that these brook courses on the western slopes of

---

* Or sometimes one of these without the other.

Boulsworth and Black Hambledon afford evidence of having been scooped out before the Drift period. The figure showing the section of the Hurstwood valley is of course an illustration of this; another good example is afforded by the valley of Swinden Water below Extwistle Mill.

Here you walk along the bed of the brook in a valley which is lined on one side by banks of boulder clay, and on the other by nearly horizontal beds of flagstone and shale. The cliff of flagstone is clearly the old cliff against which the Drift has been deposited, but it has been laid open here again by the brook which has cut down its channel along the old boundary wall (Fig. 29).

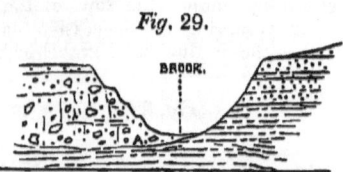

Fig. 29.

Section across the Valley of Swinden Brook, showing the Margin of an old Valley filled in with Boulder Clay.

That the glacial deposits have been spread in great masses over this district, and have subsequently suffered considerable denudation, will be apparent on observing the valleys of Catlow, Thursden, and Swinden Brooks. That they have also been transported from the northward or north-westward will be apparent from a comparison of the rock fragments they contain. The majority of the stones are Carboniferous limestone and grits, besides which are Silurian grits, traps, quartzites, granites, and ironstone; but very surprising is the abundance of limestone pebbles which occur, and which were, before the introduction of good roads, canals, and railways, largely used for lime. Along the valleys above named we frequently come upon the ruins of little limekilns, in which the limestone boulders were calcined, and sometimes we find little heaps of the stones themselves collected, in which are often mixed pieces of trap or other strange rocks, which the limeburners probably thought would make lime if only burnt enough. In Cant Clough Beck, at an elevation of 1,000 feet above the sea, there appears to have been a little colony of limeburners, for the ruins of their kilns, huts, and roadways may be traced very clearly along the base of Hazel Edge.

As might be inferred the stones and boulders of the Till generally show glaciated surfaces. This is particularly the case with the limestone blocks when freshly disinterred; their close and uniform texture apparently well fitting them to receive and retain the marks of the polishing and grooving to which they have been subjected.

An observer cannot fail to be struck with the contrast between the western and eastern sides of the Penine chain as regards the glacial deposits. All along the western flanks of the chain and extending into the plains these deposits are spread in masses often attaining a thickness of 150 or 200 feet, but on the Yorkshire side these deposits are absent, the strata everywhere appearing at the surface or only covered by soil. It is clear that during the glacial epoch, and during the submergence of the land to a depth of 1,800 feet, the sea on the western side of the Penine Hills was surcharged with mud and rock-laden ice, while on the others there was clear water. And if we assume that the earliest stage of the drift was a sheet of land-ice, we must infer a state of things still more strange, for in the same country and divided only by a narrow band of high land we should have to suppose the plains on one side to be overspread with an ice-sheet stretching all the way southward from the Cumberland mountains, and on the other similar plains without the ice-covering. A plausible reason might be assigned for this in the fact that it is only on the west side that the Cumberland mountains are found, and that it is on this side that the greatest amount of rain (or snow) fell, as at the present day. The subject, however, is still involved in obscurity, notwithstanding all that has been written upon it, and I merely throw out these hints in order to show some of the difficulties which have to be grappled with.

Glaciated Rock Surfaces (*Roches Moutonnées*).—That the land was originally covered by an ice-sheet as far south as the Pendle range in the Burnley district seems highly probable from the ice-worn surfaces of the rocks in several places. The most remarkable example is that of the large *roche moutonnée* in the valley of the Calder about half a mile above Whalley, discovered by Mr. Tiddeman and myself during an excursion we made together in the spring of 1865. Here a large boss of Millstone Grit rises above the alluvial flat of the river on the north side of Nab Wood. The rock has a rounded outline and is partially covered by vegetation, but where exposed presents, more or less clearly examples of parallel groovings, unmistakably of glacial origin.* The direction of the grooves is N. 15 W. or S. 15 E. The opposite side of the valley is filled up with thick masses of glacial deposits.

It would appear therefore that this gap in the Pendle Range through which the Calder flows had been scooped out before the glacial epoch. It was originally about half a mile in diameter, bounded on the north by the millstone rocks of Clerk Hill and on the south by those of Nab Wood and Whalley Banks. Through this gap an offshoot from the great ice-sheet seems to have protruded, debouching on the plain of the Calder, and bringing some portion of those thick masses of Till which fill the valley of the Calder in this neighbourhood.

The surface of the Mountain Limestone at Chatburn when freshly uncovered is seen to be scored with ice-groves and scratches, and my colleague Mr. Tiddeman tells me that these appearances are not unfrequent over the region to the northward; but as he himself will in a future page describe these examples I will not further allude to them myself. There are some rock surfaces in Colne Edge formed of Millstone Grit, which from their moulded outline present every appearance of ice-worn surfaces, but my colleague and myself have not been able to detect any ice-marks. This, however, may be easily accounted for, as the surfaces have apparently been long exposed to the weather.

In connexion with these examples of land-ice action, we cannot forget the far more perfect cases which were observed and described by Mr. G. H. Morton in 1862† on both banks of the Mersey at Liverpool, where the surface of the New Red Sandstone, recently stripped of its covering of Till, was found to be sculptured over with parallel grooves and

---

* Professor Ramsay has since (1867) visited this boss of rock, and fully concurs in the opinion of the glacial origin of these markings.
† Journ. Geol. Soc., London, vol. xviii, p. 377.

flutings pointing in a general direction west of north, and extending over square yards in four several localities. These markings were subsequently visited and described by myself in 1863,* and are represented on the six-inch Geological Maps of Liverpool (Sheets 106, 113).

Ice-markings on Horwich Moor.—The ice-markings above described occur at comparatively low elevations, and may be attributed to an ice sheet which overspread the plains and lower eminences; but the ice marks which I am about to describe are found at a comparatively high elevation, and are to be referred in all probability to the stranding or scraping of an iceberg during the later glacial period of submersion of the land, rather than to land ice.

The rock surface to which I now refer occurs a short distance to the south of the margin of the Blackburn one-inch map, in the adjoining map 89 N.E., and has not hitherto been described in the Memoirs of the Geological Survey. The rock consists of Upper Millstone Grit, and runs along the crest of an escarpment called Burnt Edge, overlooking the valley of Dean Brook. The site is near to the side of the road which crosses the hill from the southward, and the surface is exposed over an area of several square yards. On this surface numerous perfectly distinct and parallel groovings were observed, ranging in a direction W. 10° N. and E. 10° S. This direction, it will be observed, is different from that of the groovings at Liverpool and Whalley, and taken together with the difference of level are reasons why I am disposed to attribute them respectively to different causes.

I may add that on first finding these ice-marks in so unlooked-for a position, I felt some diffidence regarding their nature and origin; but on a subsequent visit with Professor Ramsay, whose experience in questions of this kind is second to none, I was quite confirmed by him in the view regarding their glacial origin.†

Reverting to the question of land ice and the supposed ice sheet,‡ I may draw attention to a remarkable appearance exhibited by the rocks of the Pendle Range along many parts of the lower ridges, for I do not think they extend to elevations above 1,000 feet. In our survey of the millstone beds of the range above Blackburn, Harwood, and Padiham, Mr. Tiddeman and myself have often been struck with the fact that the ends of the beds, as they reach the surface with a high southerly declination, are generally bent over so as to seem to dip in an opposite direction. This is a matter of so common occurrence that it is hardly necessary to specify instances; and my colleague has observed cases not only on the southern slopes but on the northern slopes, where the fall of the ground is against the dip. In pondering over the matter, and endeavouring to determine some cause for a phenomenon extending over several miles of hilly ground, it seemed clear to us that some general cause ought to be assigned for so general an effect. Nor can we suppose that any but a very powerful influence has been sufficient to bend over strata weighing many thousands of tons from their normal position.

Recollecting the evidences which this district affords (some of which have been just described) of the extensive development of land ice, it has occurred to us that the overturning of the beds may have been one of the

---

* Trans. Geol. Soc. Manchester, vol. 4, p. 288.
† The directions of these ice-marks are now engraved on the 6-inch Ordnance and Geological Map No. 86. Those at Liverpool are also engraved on the 6-inch maps of that town, 106 and 113.
‡ A generalization of Professor Agassiz, many years ago.

effects of the ice-sheet which may have been forced over the lower portions of the barrier offered to its progress southward by the Pendle Range of hills.

Another explanation, however, offers itself. These overturnings may, in some cases, be attributed to floating ice, which impelled by north-west winds and currents during the period of submersion, and coming in contact with the sunken reefs of the Pendle Range, would tend to force the beds over towards the southward in the manner observed.*—E. H.

Drifts; Ribble Valley.

So much space has been given to the Drifts of the neighbourhood by my colleagues Prof. Hull and Mr. De Rance, that I need not repeat their chief characteristics or succession. For my views as to the glaciation of this district by an Ice-sheet, the arrangement of the Till as to colour and materials, and the period at which it occupied the country, I must refer the reader to what I have written elsewhere.† I will merely give a few sections from my note-book.

At Euxton Hall, near Preston, a well sunk to the depth of 60 feet, with the exception of a few feet of clay at top, passed entirely through sand and gravel (Middle Drift). Ditto, a well at the priest's house, near the Roman Catholic Church, Euxton, to the depth of 45 feet.

At Dobridden Wood, near Sunderland Hall, on the left bank of the Ribble, occurs the following interesting section:

	feet.
Upper Boulder Clay, of the usual type, about	20
Buff sand, with thin occasional beds of gravel	10
Hard sandy stiff Brown Boulder Clay, with scratched stones, and a laminated sandy bed, two feet from the top	5
Buff sand, as above, containing a well-scratched small boulder of Limestone	10
Lower Boulder Clay, or Till, with large scratched stones	25 seen.

At Copy Wheel, on the Calder, near its junction with the Ribble, beneath 10 feet of alluvial mud, was the following descending succession, River Gravel, Alluvial Mud, Boulder Clay, Gravel, Boulder Clay. As there were several landslips close by, this cannot be firmly trusted.

The terraces of Middle Drift, on both sides of the Calder above Whalley, are very interesting. They consist of sand and gravel tolerably well bedded. On the north side of a sand pit a little above the turnpike coarse sand of buff colour overlies a bed of sandy gravel. On the south side is sand (about 6 feet shown) containing obscure frag-

---

* If the cause was floating ice, the overturning of the beds ought only to be found when the rock was likely to oppose an obstacle to the berg, such as the top of a ridge or knowl.

While these pages are passing through the press, Mr. Tiddeman has published a valuable paper "On the Evidence for the Ice-sheet in North Lancashire," &c., accompanied by a map in which the glacial markings are laid down (Quart. Journ. Geol. Soc., vol. 28, p. 471). The description of the phenomena on the Pendle Ridge above given having been written about six years ago, my colleague will observe that some of his views were anticipated. (Nov. 1872.)

† "On the Evidence for the Ice-sheet in North Lancashire," &c., Quart. Jonrn. Geol. Soc., vol. xxviii., pp. 471-491, 1872. "The Older Deposits in the Victoria Cave, Settle;" Geological Magazine, vol. x., p. 11. "The Age of the North of England Ice-sheet," Ibid., p. 140. "The Relation of Man to the Ice-sheet in the North of England," Nature, vol. ix., No. 210, p. 14.

ments of marine shells and pebbles of coal; beneath this a bed of red marly alluvial clay, roughly laminated, 9 inches thick, resting on sand. The entire thickness of the Middle Drift is not exposed, but there must be about 40 feet of it. Boulder Clay or Till is seen in the bank of the river below. It consists of a very stiff grey sandy clay, full of angular, subangular, and rounded fragments of Carboniferous Rocks, chiefly Limestone. The boulders are all local, except two or three pebbles of Silurian Grit, which may have come down Ribblesdale. The junction of the Middle Drift and Till is well seen below the sandpit.

The woodcut, Fig. 30, is intended to represent a drift section in the bank of the Ribble near Waddow Hall.

Its chief interest lies in a patch of Boulder Clay deposited in an eroded hollow in the sand and gravel A., and evenly covered by a bed of sharp sand B.*

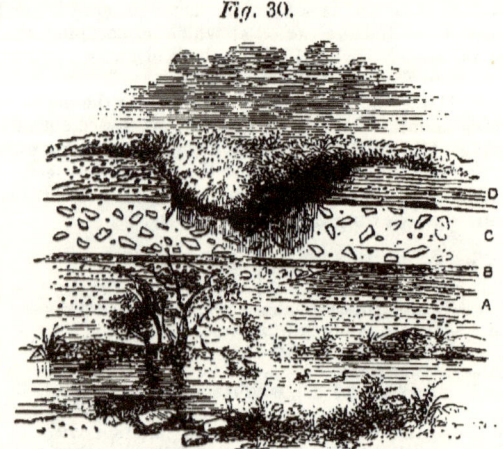

Fig. 30.

Section on the Ribble, near *Waddow Hall.*

A. Sand and gravel, with Boulder-clay resting in an eroded hollow at the top of it.
B. Seam of fine sand.
C. Boulder-clay with well-scratched stones.
D. Sand and gravel.

On the left bank of Smithies Beck, a little way above the road from Chatburn to Sawley, is the following section:

    Fine and coarse gravel and sand   10 ft.
    A bed of very hard stiff mud, with scratched stones at bottom, but tolerably free from them above, about - 3 ft.
    Boulder Clay, containing large and small scratched boulders.

This stiff mud occurs in several places. In Bashall Brook there are several beds of it intercalated with gravel.

A curious ridge of gravel runs from near West Bradford towards Stoneyhurst. In some places it is very coarse, but fine beds occur in it. The

---

* The engraver by inserting some unusually large ducks, which were not in my original sketch, has rather dwarfed both the river and the cliff.—R. H. T.

pebbles of which it is composed seem to be derived from the Boulder Clay, and are local: they have only faint scratches, but have been subjected to a good deal of pounding. It probably represents an old beach or shoal. In elevation the deposit lies between the 250 and 300 foot levels above Ordnance datum.—R. H. T.

Drifts near Colne.

In the country north and east of Colne it was not possible to recognise and trace the three divisions of the Drift before mentioned. Stiff bluish clay studded with numerous blocks of Carboniferous grit and limestone is the prevailing drift; the limestones are almost always well scratched. In places, *e.g.* along the northern flanks of Boulsworth and in parts of the low limestone country north of Barnoldswick, the clay is not so stiff and the stones are more rounded, so that it may be called a gravelly clay; but this is full of scratched stones. Mounds of gravel composed of well-rounded waterworn stones, and also clean running sand, are seen here and there. Some of these will be noticed further on. The highest point to which I have traced undoubted glacial drift is a little over 1,200 feet above the sea. This is about three-quarters of a mile west of Lad Law, the highest point of Boulsworth Hill, and some distance N.N.E. of Red Spa, and it runs up to heights varying from 1,000 to 1,200 feet all along the northern flank of Boulsworth. North of Combe Hill it goes up to about 1,150 feet at Round Holes and near Cowloughton Dam. In the high ground S.W. of Carlton I have not traced it higher than about 1,050 feet at Park Head quarry, but it may go higher. The grit of White Moor and of the Carlton Synclinal seems almost entirely free from drift, and yet very small portions of these areas rise above the 1,200 feet contour line, and a great part of the Carlton Synclinal which is bare of drift is between 600 and 1,000 feet. Even in the low ground there are many considerable areas where there is little or no drift. It is thickest in the valleys, as a rule, thinner on the hill slopes, and thinnest on the grit and limestones ridges. I have long thought that its present rather capricious distribution is due in a great measure to its irregular original deposition, though it has of course been denuded more in some places than in others; and I cannot help thinking that the low ground which is now bare of drift had little or no drift on it at the close of the glacial period, and that the drift never was so deep in valleys as its position on their sides would lead one to suppose.

There appears to be no glacial drift in the valleys east of Black Hambledon, Black Moor, and Boulsworth. The lowest points on the watershed about here are at Widdop Cross 1,286 and at Harestones about 1,240 feet respectively. These heights appear to have been too great for the drift coming from the west to get over either into the Widdop Valley or into that of Gorple Water. But when we get further north to where the watershed descends to about 1,120 above the sea, between Combe Hill and Crow Hill, we find that the glacial drift has gone over the pass. For the drift is very thick at Smithy Clough and can be seen up to about 1,050 feet, and though none is seen at the watershed near Barn Hill, nor any for some distance down the Worth on the other side, drift is found in the Worth Valley about a mile and a half east of the watershed, and one does not see how it could get there unless it came over this pass.

The Boulder Clay is very thick about Ickornshaw as shown in the banks of Ickornshaw Beck and Summerhouse Clough. Immediately N.E. of the village there must be from 75 to 100 feet of it at least. It is also thick about Laneshaw Bridge.

The following is the detail of a boring made east of the latter village so far as it related to drift:—

	ft.	in.
Stony clay	34	6
Smooth clay	15	5
Gravel -	4	6
Stony clay	20	0
	74	5

Another boring near showed 88 ft. 9 in. of drift made up of blue marl, gravel, and sand.

In several places along the northern side of Boulsworth are remains of old limekilns built to burn limestone boulders got out of the drift, which appears to have been a good deal worked for limestone at Beaver, Stack Hill, and Smithy Clough Scars. The moundy ground near the two last-mentioned scars is no doubt in a great measure the result of these old workings. Old limekilns can also be seen down Ickornshaw and Gill Becks, near gravel pits at Cowloughton and on Emmott Moor, and they are very numerous near the Laneshaw Brook at Higher Scars, about three-quarters of a mile S.E. of Laneshaw House. Here I found some quartz boulders, the only place in this district where I have noticed any rocks in the drift other than Carboniferous limestones, grits, and cherts.

There appear to be two distinct sets of gravel drifts in this district, one occurring at heights between 900 and 1,150 feet above the sea, the other not being found much above 500 feet above the sea. I cannot say there is any essential difference in character between the two sets.

To the former of these belongs the gravel seen in pits on Emmott Moor about a mile E. of Emmott Hall, by the side of Laneshaw Beck on Ickornshaw Moor where the word "quarry" is seen on the 1-inch map, and near Cowloughton. These gravels seem generally to have the form of mounds. At all these places the gravel is consolidated, and near Cowloughton this drift conglomerate assumes very fantastic forms, due no doubt in a great measure to the quarrying it has undergone.

The other gravels I shall refer to belong to the lower set. There is a long mound of gravel W. of Colne between Haverholt and the Cotton factory, but not much of a section in it, and there is a sand pit at Sand Hole, near the smallest canal reservoir.

East of Foulridge station there is a section in brown sandy earth with large stones; further on gravel occurs, and still further on, on the same level, a kind of brick earth free from stones, but what relation these bear to one another is not clear. On the opposite side of the railway is a large excavation in drift gravel made for ballast. The stones here are mostly rounded, consisting of Carboniferous limestones, rotten stones, grits, and gannister. At the N. end the gravel is converted into a conglomerate. But the finest section by far in these drift sands and gravels was exposed in the making of the Barnoldswick railway in a cutting near Salterforth, between Cross Lane and Salterforth Lane. The section showed beautifully stratified clean running sands and fine gravels, exceedingly current bedded, and several small shifts in the beds were apparent.* The dip of the beds was to the south-east, conformable to the slope of the surface. On the western side of Cross Lane, boulder clay of the ordinary

---

* For much of this description of the Salterforth sand and gravel I am indebted to my colleague, Mr. Dalton, who kindly visited the cutting for me while the railway was being made.

type was seen reposing on a steeply dipping surface of Lower Yoredale Grit; and in one place close to the lane there was a very thin coating of gravel on the boulder clay, which might be the base of the sands, the dip of which corresponded pretty nearly with the slope of the top of the clay. Covering up the clay, and filling up a hollow between the sand and the solid rock, occurred a mass of loamy clayey sand, like rainwash, nearly free from stones except near the surface, where there were seen a few scattered pebbles of sandstone and glaciated limestone. It is possible that this may be the equivalent of the Upper Boulder Clay of South Lancashire, and that the sands are the Middle Sands. The rainwash-like material evidently lies in a denudation hollow of the sands. No trace of a fossil has been found in these drifts after very careful search.

The drift between Blacko and Foulridge seems locally derived. It is very sandy, and contains apparently little besides large blocks of coarse grit.

On the north side of Black Clough Head (Hey Slacks Clough), between Tom Groove and the sheepfold, and at a distance of about 100 yards from the stream, is a flat-topped bank which looks like the remains of a terrace. The side affords very poor sections, but the bank seems to consist of rather small angular gravel. If one looks from here towards the Dove Stones a terrace-like flat is seen running nearly all round the clough at about the same level, which is 1,400 feet above the sea. This is covered with peat; but at about the same height, halfway up Tom Groove, toward the county boundary, well-rounded gravel can be seen high up at the top of the shale bank. This looks very much as if the clough had formerly been here the site of an old lake, possibly dammed by a moraine, of which the bank first mentioned may be the remains. No far-travelled boulders were seen here, nothing but locally derived stones.

On the west side of Laneshaw Brook, about 300 yards south of Laushaw House, a mass of coarse grit was seen apparently dipping south at a considerable angle. I cannot account for the position of this unless it is a boulder.

Perhaps some reference should be made here to the peculiar moundy shape of the low ground between Gisburn and Skipton, which is in some measure due to the drift, though I do not think that many of the numerous hills in this district are composed entirely of drift clay, which is the prevailing drift here. I have no doubt if the drift were removed we should see an undulating surface, the main features of which would be much the same as the present, only that the contours would be sharper, and the limestone would stand out in serrated ridges here and there, as it does at Marton Scar. No doubt the peculiar smoothed rounded outline of the hills is principally due to the coating of drift. The surface of the limestone rock in a quarry close to the canal, west of Gill church is glaciated, the scratches trending S. 10 W.   W. G.

River Terraces and Alluvium.—Several of the principal streams in this district offer examples of old river terraces higher by several feet than the more modern alluvial flats. These gravels show the position which the rivers formerly occupied. In some of these gravels in the ancient Forest of Rossendale antlers and bones of red deer have been found by Captain Aitken and Mr. J. H. Ashworth,[*] and a doubtful specimen of a flint arrowhead by the late Mr. Whitaker, of Burnley, from Barrowford.—E. H.

---

[*] Trans. Geol. Soc. Manchester, vol. 4, p. 333.

GLACIAL DRIFT DEPOSITS OF THE DISTRICT AROUND PRESTON AND CHORLEY.

As in the adjoining district of Blackburn, the classification of the Glacial Drifts, established by Professor Hull for the Manchester neighbourhood, is found to hold good in the country around Preston, Kirkham, and Chorley. The whole of the great plain already described as extended from the foot of the hills of the Pendle range consists of three drifts, the Upper Boulder generally forming the surface, and the rock-surface nearly everywhere deeply concealed by the thick covering of drift which near Standish reaches a great thickness. In the western portion of the area between Coppull and Preston the rock-surface is often completely beneath high-water mark, so that were the Boulder drifts absent, the whole of the western edge of Lancashire, from the river Alt north of Liverpool, to the Lune south of Lancaster, would be beneath the sea, up to a line ranging by Bay Horse, Garstang, the Ribble at Alston Hall, Roach Bridge, Euxton, Rufford, Ormskirk, and Maghull.

Lower Boulder Clay.—This deposit is well seen in a cutting near the Preston Waggon Works at Marsh End, where it underlies the Middle Sand and Upper Boulder Clay. It is of reddish brown colour, and contains numerous stones and boulders, chiefly derived from the lake district of Cumberland and Westmoreland; one boulder of greyish white granite, with large crystals of black mica, is 5 × 4 × 3 feet in size; it, as well as most of the smaller erratics, are scratched in more than one direction. The lower and upper clays precisely resemble each other in this section, the lower perhaps having the greater number of stones per cubic yard, and occasionally containing large boulders of altered volcanic breccias from the Lake District, more than a yard in length.

On the 1-inch map a brook is seen to bifurcate near Laurel Bank, north-west of Preston, one branch running south of Fulwood Barracks towards Sion Hill, the other to run north of them, and to again divide at Clayton Villa, the northern branch running N.E. to Gerard's Hall. All these brooks east of Withy Trees flow through flat alluvial plain, bounded by cliffs, in some instances smooth and rounded, resembling those bounding the Ribble plain, which I have hereafter described as bluffs. These cliffs or bluffs exhibit fine sections of the Middle Drift and Upper Boulder Clay, and the former, on the south side of the brook from Gerard's Hall to Clayton Villa, is seen resting on the Lower Boulder Clay, which is here a dull red-coloured clay, well packed with stones. It is also obscurely seen in the north cliff of the brook from the latter villa to the "Hall," also in the brook south of the Barracks, both on the north side below the word "Manor House" and on the south side at Holmes Slack.

It occurs at the base of the long winding bluff forming the northern limit of the Ribble plain from above Walton-le-dale Bridge to Alston Hall, and beyond the margin of the map, at Red Scar, this bluff has been eaten by the river into a cliff, at the base of which the Lower Boulder Clay is well seen as a rather stiff clay, with a great number of scratched pebbles, the clay being of the usual red colour (Fig. 35).

Till.—In a railway cutting made in 1868, at Brinscall, east of Chorley, the Upper Boulder Clay and Middle Drift I found to rest upon a boss of tough hard lead-coloured clay or Till. It contained many sub-angular blocks of Black Carboniferous Limestone and other rocks derived from the western slopes of the Penine chain to the north, but no Lake District erratics, nor, as far as I could see, any marine shells or traces of oblique lamination. About 13 feet of the clay was exposed in the cutting, the cost of making which was considerably increased by the intense hardness of the Till, which resisted the action of blasting. This clay would appear to have been formed by a small in-shut flowing from the N.N.E. over the watershed south of the Roddlesworth valley. Its surface at the highest portion of the boss is 437 feet above O. D. L. It is there covered by 17 feet of Middle Drift and 5 feet of ordinary red Upper Boulder Clay, with erratic pebbles, gradually increasing in thickness in every direction.

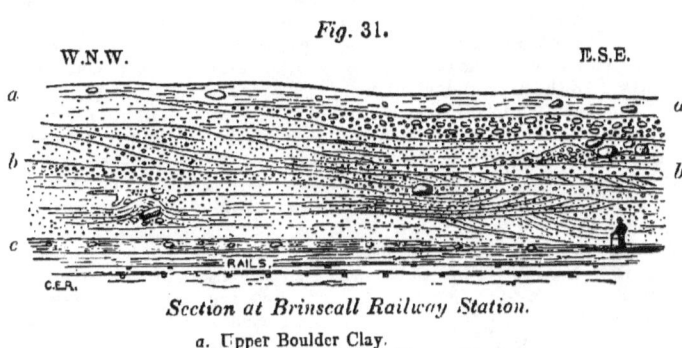

Fig. 31.

Section at Brinscall Railway Station.

 a. Upper Boulder Clay.
 b. Middle drift, sand, and large shingle.
 c. Till.

Traces of this Till are met with at several other points, but in none are they so clearly seen as in the above section. This clay would appear to have been formed partly before and partly contemporaneously with the red Lower Boulder Clay of the lowlands.

Middle Drift.—The surface of the sand and gravel appears to have been everywhere deposited in a series of mounds resembling the sand banks of the modern sea coast, causing a very unequal thickness of Upper Boulder Clay to rest upon it. Thus at the village of Leyland, which is built upon the crest or summit of one of these mounds, the Middle Drift has hardly any covering of clay; while in a well at the India Rubber Works, Golden Hill, only half a mile distant, 75 feet of Upper Boulder Clay was passed through before the Middle Drift was reached; and as the surface of the ground at the latter place is 10 feet below the former, it follows that the surface of the Middle Drift has sunk 85 feet in a space of about 2,000 feet in length. The actual fall is probably much more abrupt, for south of Leyland, in Shaw Brook, the junction of the Middle Drift and the Upper Boulder Clay only attains an elevation of 95 feet, while immediately to the north in the park it reaches 146 feet, and immediately to the south it rises in numerous knolls to an elevation of 156 feet.

Many examples of these knolls are found both north and south of Chorley, between Leyland and Farrington, and a small one on the high road between Blackburn and Preston. In this case, as at Leyland and Euxton, the elevation of the Middle Drift in the knoll is higher than that exposed in the banks of the adjoining valley, proving apparently that the present brooks and modern drainage have a tendency to flow along the lines of natural depression in the surface of the Middle Drift; but this may be partly due to the fact that rain falling on sand would have a tendency to sink rather than to flow away as a stream, and that therefore the thicker the clay the less the power of soakage of rain, and the more likelihood of its forming a stream along the line of thickest clay, or in other words, along the line of *lowest* Middle Drift.

In addition to the small knolls above described, there are four large patches, or small sheets, of Middle Drift, each of which has been chosen for the site of a town by the Northmen when they first colonized the district, namely, Preston, Kirkham, Chorley, and Leyland, the four largest towns and villages in this district. The cause of this may possibly be due to the occurrence of beds of loam in the sand, which, supporting small sheets of water, cause it to issue as springs in these tracts, which would make them extremely valuable in a springless country, like the Upper Boulder region surrounding these towns.

In the north-western corner of the map the Middle Drift is seen in the churchyard of the Roman Catholic chapel of the Willows. It consists of fine yellow sand, and is connected with that found at Kirkham by a strip of sand cropping to the surface across Wrong Way Lane, running to Church Street and there expanding and forming the subsoil of the greater part of the town. In a pit dug behind Preston Street 24 feet of false-bedded white and yellow sand was seen, each bed being divided by a thin hard seam of sand, cemented together by the action of carbonated water on the lime derived from the included shells on horizons of the drift, where water is supported and held in sheets by impermeable seams of loam. The level of the surface of the sand in the pit is 102 feet above O. D. L. The town of Kirkham is built partly on the continuation of the Boulder Clay plain described in the Preston district, running from the foot of the Yoredale Grit Fells, at an elevation of 500 feet, by Preston (175 feet), and Kirkham (100 feet), to Blackpool, where it is abruptly terminated by the sea, forming a cliff 70 feet high, which is being worn back into the country, at an average rate of nearly a yard a year, by the combined action of frost, rain, and wind, which, removing vast quantities of sand, cause large masses of the Upper Boulder Clay to fall to the base of the cliff, from which they are removed by the spring tides.

This uniformity of the level of the slope of the country, though subject to undulations below the plane, is constant in the district, stray hills never rising above. These undulations, though unimportant in the greater part of the Preston sheet, increase in number and depth in the country to the north and west, where they form "swamp hollows," generally filled in at the bottom with peat, occasionally underlain by lacustrine clay. The hollows are entirely excavated in the glacial deposits, and the rock-surface is invariably at a considerable depth below the bottom of the hollow, trending, like the surface of the drift, in an inclined plane from the grit fells at Chipping, at 500 feet, towards the sea; but, unlike the Drift, its surface is at least 20 feet below the Ordnance datum line, instead of nearly 100 feet above it. It is clear, therefore, that though the rock and the Drift surface start from the same point, the fall of the latter is less steep than the former by nearly 120 feet, and that the whole of the space intervening between the two planes must be filled in with drift of that thickness; and thus it is that in the present quarter sheet the rock in the western portions is only found at the bottom of the Ribble valley and near the mouths of its immediate tributaries, and that in all the deep brook-valleys north of the river, even in that of Tun Brook, near Redscar, nearly 80 feet in depth, that rock is never found, and that it is only on approaching the grit hills that it crops to the surface, first appearing in the neighbouring country in map 91 S.E., under the drift at the bottom of the deep brook gorges.

It may be well here to state that in the centre of the area occupied by that sheet, Grit hills from 200 to 600 feet in height occupy a zone of country intervening between the drift plain and the high fells, rising to more than 1,700 feet, while in the country to the south these Fells rise at once abruptly from the rock-plane below, which occupies even a lower level than represented above. This intermediate range of low flanking hills, from thus occupying the space where the Boulder Drift should have been, causes the Glacial Deposits to be sparsely represented in the district, the hills being barely covered, and the drift plain to terminate at 180 feet above the sea, instead of at nearly 600. This is more and more the case in advancing to the north, the hills gradually approaching the existing coast line, until at Hest Bank the plain commences and ends within a mile, and at Warton Crag, beyond Carnforth, it has disappeared altogether, and Glacial Drift, to the north and west, nearly surrounds the hills and fills up the valleys, the tops of the hills and mountains being bare of drift, as in the country around Ormskirk and Wigan to the *south* of the plain, while, in the great drift plain extending from north of Ormskirk to south of Lancaster, the crests of the hills are the exact points where the drift is thickest.

It is therefore probable that the drift only exhibits the phenomenon of a smooth inclined plain, dipping towards the sea, where it rests upon a similar plane of rock beneath, and that where the original "rock-surface" was undulating the drift has merely filled in the hollows, most of which have been in great measure re-excavated out by modern stream action, examples of which are seen in the spur of the Penine chain, called by Professor Hull the Pendle range, part of which runs along the south-eastern edge of the present sheet (89 N.W.). It would therefore appear that the inclined plain of Glacial Drift owes its origin to being deposited during subsidence upon an old rock plain of marine denudation, the eastern limit of which forms a bay running from near Rufford, by Eccleston, Euxton, Bamber Bridge, Samelsbury, and the Ribble valley in the direction of Ribchester, and thence westward by Broughton, Garstang, and Cockerham to the sea, westward of which line the rock surface is either little above, at, and often below, the existing sea level, in which, from the superposition of Glacial Drifts, the surface of the country is often 170 feet above it.

If it be admitted that the surface of the Glacial Drift plain owed its inclination to deposition, and not to denudation, then it is easy to understand that the surface would be subject to a certain amount of undulation, and hollows would be formed, the level of the entrances to which would be often above that of the central part of the depression, resembling in this respect the "Rock-basin" of Prof. Ramsay, though due to so entirely different a cause. And it would also explain the fact that these swamp-hollows, when numerous, are connected with each other by "cols," and are entirely unconnected with the drainage of the country.

At Kirkham one of these "swamp-hollows" breaks the continuity of the plain from north to south; it is covered with a thin bed of alluvium overlying peat, brought down partly since the hollow has been drained by a sluice carrying its waters into Freckleton (or Dow) Brook, below Kirkham Windmill, and near the point where the Roman Road crossed the moss, on the surface of which can still be traced the stones which they piled on it to give it solidity; a little to the west, at the foot of the bluff or escarpment forming the margin of the hollow, the site of a Roman bath occurs, supplied by what was once called the "New England Spring," and near which the splendid Roman shield dedicated to Minerva was discovered in 1793. These facts are of importance, owing to the level of the ground here being only 30 feet above the Ordnance datum line, and therefore only about eight feet above highest spring-tide mark, proving that the country cannot have risen many feet, if at all, since the Roman epoch, and also that the growth of peat appears even then to have ceased in this district.

The bluff or escarpment, mentioned above as forming the limit of the swamp hollow, abruptly terminates the plain on which Kirkham is built, Carr Lane, Church Street, Mill Street, and Wrong Way Lane, being built on the bluff itself; this between Church and Mill Street is very steep, forming a cliff facing north, 50 feet high, composed of Middle Drift, the base concealed by peat; but its upper surface is seen underlying the Upper Boulder Clay the pit mentioned above, and also in Poulton Street; the Middle Sand here is therefore at least 60 feet in thickness. To the west, just off the margin of the sheet, it rises to the surface in several knolls towards Great Plumpton, near which it is well seen in the railway cutting (of the Blackpool and Preston line), but in the Lytham branch, near Westby Bridge, the Upper Boulder Clay reaches a thickness of 31 feet without the base being exposed. Eastward, near the canal, east of Salwick Hall, the Upper Boulder Clay thin to four feet, and the Middle Sand is seen in pits (at an elevation of the surface of 84 feet). It is also seen on the north bluff of the Ribble, near New Lea Hall, where the following section is observable:—

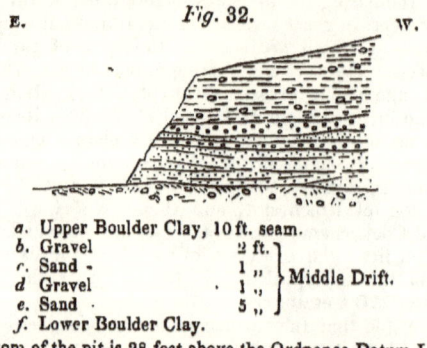

E. *Fig. 32.* W.

a. Upper Boulder Clay, 10 ft. seam.
b. Gravel 2 ft.
c. Sand - 1 "
d Gravel 1 " } Middle Drift.
e. Sand 5 "
f. Lower Boulder Clay.

(The bottom of the pit is 28 feet above the Ordnance Datum Level.)

Section in the **Ribble** *near New Lea Hall.*

In a clay-pit near here large masses of consolidated Middle Drift Shingle occur, three or four yards in length, resembling in general appearance, variation of size of pebble seams, &c., the Kinder Scout division of the Millstone Grit. The spring called St. Catherine's well, near New Lea Hall, appears to issue from the junction of the Middle Drift with the Upper Boulder Clay.

In the Preston 6-inch map (61) the Middle Drift forms both sides of a valley running from the Wheat Sheaf Inn (below Tulketh Hall, by Ox Heys Mill, near the L. N. W. R.), to Deepdale, but before reaching this point it expands, occupying the surface of the plain from Gallows Hill, southward, in almost one continuous sheet to Avenham, dipping beneath the Upper Boulder Clay to the west and to the east, the western boundary running by Maudlands Bridge Lane, between Friar Gate and Lune Street, across the middle of Fishergate, the north side of Winckley Square, top of Avenham, to Bank Parade; thence it runs to the north-east to Church Street, whence it turns to the north, running to St. Thomas' Church, to the north of which it joins that portion which runs up by Deepdale. The following buildings in Preston are built upon the Middle Drift, the Town Hall, Avenham Institution, Dr. Shepherd's library, &c., St. John's, St. Peter's, St. Thomas' churches, and the Roman Catholic cathedral, with a thin bed of Upper Boulder Clay intervening. The Middle Drift dips suddenly westwards on the west side of the tract described above, reappearing near the foot of the bluff forming the margin of the Ribble valley, running from Tulketh Hall, by Stanley Terrace, Avenham Park, to Fishwick, where the bluff has been cut back into a remarkable semicircular cliff by the action of the Ribble; between this outcrop and the Middle Drift at the top of the plain and the top and edge of the bluff there is generally an overlap of Upper Till, in fact at Avenham Park it descends nearly to the alluvium below. Near the Wheat Sheaf Inn the following artificial section of the bluff has been exposed, Fig. 31, and the sand has been recently cut into in the cliff below Stanley Terrace.

Fig. 33.

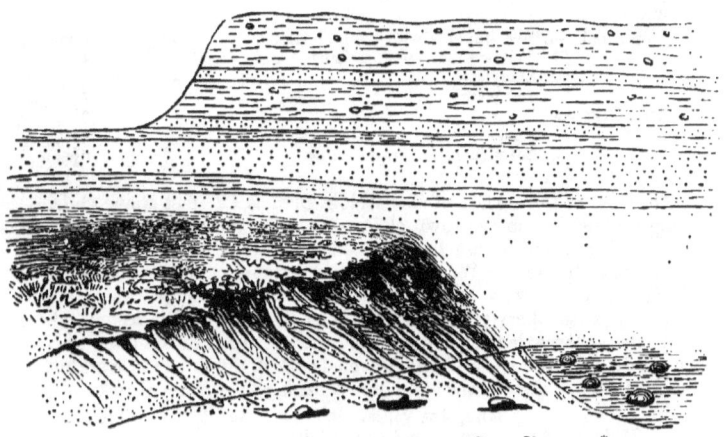

Section near *Wheat Sheaf Inn*, Preston.*

		ft.			ft.	
	a. Boulder Clay	8	*d.* Sand	2	
			e. Loam	2	
Upper Till	*b.* Sand	1	*f.* Sand	3	} Middle Drift.
			g. Clay	1	
	c. Boulder Clay	3	*h.* Sand	13	
	i. Lower Boulder Clay, 12 ft.				

The Middle Drift, capped by Upper Boulder Clay, is well seen in the sand pit on the south side of the canal, near Tulketh factory, the clay being from 6 to 22 feet thick, resting on at least 30 feet of sand, which is slightly current-bedded in a S.S.E. direction, and traversed by hard seams of consolidated sand, running in sheets about four inches in thickness; on the opposite side of the canal, near Brookhouse Mill, the sand crops to the surface; a few yards to the south-east, 10 feet of sand is visible, capped by

---

* The bank of sand in the foreground has been made too dark by the engraver. At the time the section was taken (1870), it had not been cut back as far as the artificial cliff behind it, which joins it. C. E. R.

four feet of clay ; still further south the latter has thickened to 10 feet, and an additional 10 feet of sand is seen in a well. On the opposite side of the Fylde Road, near the site of St. Mary Magdalen's church, the following section occurs :—

Fig. 34.

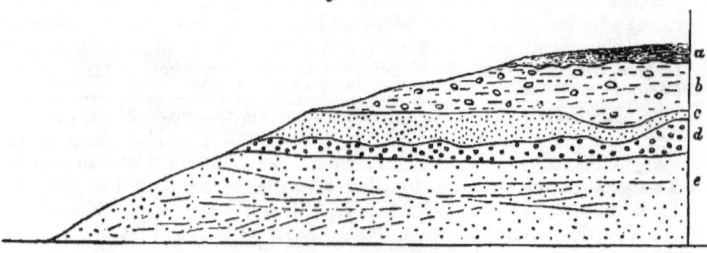

Section in *Fylde Road, Preston.*

a. Vegetable mould.
b. Upper Boulder Clay.
e. Yellow Sand - ⎱
d. Gravel -   ⎬ Middle Drift.
e. Sand   ⎰

At Deepdale, in a pit to the N.E. of the station, the following section occurs :—

Upper Till.—*a.* Red stiff clay with erratic boulders - - - 6 feet.
Middle Drift. {
b. Yellow sand (the upper surface undulating) - 4 „
c. Shingle - - - 3 „
d. Soft yellow sand - - 27 „
e. Loam - 0½ „
f. Sand - - 6½ „
g. Loam (used for brass casting) 0¾ „
h. Sand - - (+)
}

The level of the surface of this Middle Drift is 124 feet above O. D. L., being higher than in the town of Preston or in the country to the west of it. The sand is much false-bedded in a S.S.E. direction, and occasionally to the S.S.W. the shingle bed contains fragments of shells of *Turritella terebra* and *Cardium edule*. No decided shingle bed occurs anywhere between this section and the coast at Blackpool (where they attain a large developement), with the exception of the seam at New Lea Hall. To the west of the Deepdale Road, north of the workhouse, the following is seen in the steep north bank of Deepdale Brook :—

Middle Drift {
Sand, fine yellow - - 30 feet.
Loam, with *Ostrea edulis* - - 6 „
Sand - - 5 „
}

Further west, also on the north bank of the same brook (here called Moor Brook), the following is seen at a point opposite Brookfield Mill :—

Middle Drift {
Sand - - - - - 1 foot.
Clay, thickening to the west - - 3 to 4 feet.
Laminated fine yellow sand, with fragments of shells 24 feet visible.
}

A few yards to the north, towards Moor Park, the sand is capped with 6 or 8 feet of Upper Till which is worked for bricks.

It may be here mentioned as a singular fact that the steep (and therefore section side) of nearly all the brooks which flow from east to west is the north side, and it is a saying with the country people that "sand " always faces the 12 o'clock sun." The southern sides of the brook valleys are almost invariably rounded bluffs covered with rainwash, concealing the Middle Drift beneath.

The following is the section exposed in digging a gasometer bed in 1869 at Ribbleton Lane, for the new gasworks:—

Upper Till {	Red Boulder clay, with erratic pebbles	12 feet.
	Wet loam	4 ,,
	Red clay, in part brown, with few stones	9 ,,
Middle Drift	Sand, wet running	5 ,,

At the foot of the long winding bluff forming the northern margin of the Ribble valley, extending from Fishwick to Alston Hall, and beyond the margin of the sheet, the Middle Drift is invariably found; it generally occurs as sand, but seams of shingle with shells of *Turritella terebra*, *Cardium edule*, *Tellina Balthica*, occur at Throsbock Wood and at Red Scar, where, as already mentioned, the Lower Till is so well seen, as shown in the following section:—

Fig. 35.

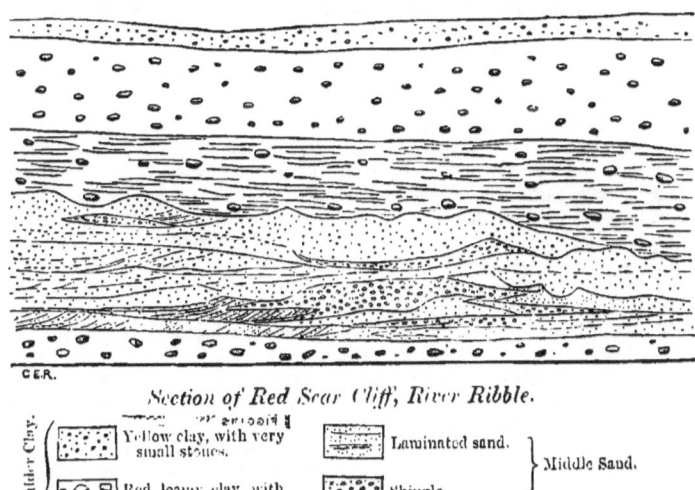

Section of Red Scar Cliff, River Ribble.

A similar section occurs on the south bank of the river at Spring Wood north of Lower Hall. The sand here is equally thick, but the pebble seams are thinner. The clay beneath is, however, rather changed, being a deep blue in colour and of a much stiffer consistence than the Red Lower Boulder Clay of Red Scar. The level of the latter is 27 feet, that of the former 50 feet, above Ordnance datum line, resting on the Pebble Beds of the Bunter. On this bank of the Ribble, a little further east, is the Balderstone section described by Professor Hull, F.R.S., at p. 129 of this Memoir. The sand from Spring Wood runs round by Bezza Farm to Bezza Brook, where it forms the chief part of each side of the valley, the Upper Till forming a plateau along which a road is carried; the sand from the south bank of Bezza valley runs by the Seed Park, the level of its surface gradually decreasing to the brook north of Samelsbury Church, where it disappears beneath the Boulder Clay, but it crops to the surface through the Upper Boulder Clay at the top of the hill, east of South Bank, of the 6-inch map 61, and east of the "6" and

K 2

"5," by the turnpike road of the 1-inch map. About 26 feet of sand, but no gravel, is here seen in the north bank of a brook which runs out at Ribble Bank. In Tun Brook on the north side of the river and in all the other brook valleys near, the Middle Drift is well seen underlying the Upper Boulder Clay, running to the north beyond the margin to that point where the level of the bottom of the brook is coincident with that of the surface of the Middle Drift ; this in Elston Hall brook is 164 feet, descending to 70 feet at Elston Wood ; is 146 feet at Tun Brook east of Grimsargh Hall, descending to 130 feet at Red Scar Wood. North of the Ribble in the brook valley south of the Barracks it is at 130 feet, in that north of the Barracks it is at 130 feet, cropping to the surface above the valley south of Fulwood Row at 180 feet ; in this patch there is a pit, where 5 feet of stiff chocolate-coloured Upper Boulder Clay is seen resting on 25 feet of current-bedded sand with seams of very coarse sand and small shingle dipping to the S.E. In the tributary brook north of this the Middle Sand is last seen at a point in the bed of the brook beyond the margin of the sheet, leaving the sheet at an elevation of 150 feet.

Returning to the brooks with short courses running immediately into the Ribble, between Tun Brook and Boilton, the first to the west is a very small one, rising at an elevation of 150 feet, and only flowing eastwards 450 feet before reaching the Ribble plain at a level of 60 feet, falling therefore 90 feet in 450, or one foot in five, the gorge thus excavated being V-shaped in plan and section. In plan the V being longer and longer in these valleys, in proportion as there is a greater and a greater distance between the source and the outfall ; as the springs gradually work backward so this lengthens, and the angle of the fall of the stream becomes less and less steep. On examination of the brook valleys, with long courses, flowing west from Grimsargh and Fulwood to the Ribble plain east of Clifton, I found the gradients of their beds to be one uniform slope from the source to the plain, in other words, these brooks have cut their channels as low as it is possible, so long as the level of the sea (and therefore of the plain on which they debouch) and that of their sources remains at their existing level.

In most streams denudation appears to be threefold, *vertical*, cutting the channels downwards, always acting from below upwards, or from the outfall, towards the source ; *longitudinal*, cutting away the banks on both sides when the streams flow straight, and the outer side only when it flows in a curve ; and *horizontal* where the stream or river through excessive curving, in addition to wearing the edge of its channel, wears it completely backwards, causing the water to be thrown diagonally at the hill or bluff forming the limit of the valley, which thus receives the whole horizontal width of the stream.

In streams supplied by springs flowing out of the hill sides, it follows that if the strata dip inwards into the hill, that the further back the stream is cut the lower will be the level of the springs (which must necessarily issue at the junction of permeable with impermeable beds) ; in these instances the cutting back of the stream by denudation will have the effect of lowering the level of the source. If, on the contrary, the strata have an outward dip, the source will be raised until it reach the line along which the water-bearing stratum crops to the surface, on the top of the hill, the slope corresponding to the dip of the bed, and not to the gradient of the stream below. In the Glacial Drift beds the water-bearing seams occur at the surface of the Lower Boulder clay, and on thin beds of clayey loam which occur in the Middle Drift as well as above the seams of consolidated sand and gravel found in that formation. The boss or mound-shaped arrangement so often found in these beds, an envelope of upper boulder clay, covering a central mass of Middle Drift, or a thick bed of the latter covering an internal core of Lower Boulder Clay, resembles to a certain extent the outward dip of strata ; and the sources of nearly all the small brooks intersecting the boulder-clay plain running to the Ribble occur in beds with this arrangement, and have no tendency to cut down their channels at their source, but rather in cutting back to ascend the hill. The ground above a spring is generally steep, and from 10 to 30 feet in height, up to the top of the plain, or to the undulating hollow, which forms the upper portion of most of the brook valleys running into the Ribble ; it is particularly well seen in the valley behind Fishwick Hall, of which Fig. 36 is a diagram.

Fig. 36.

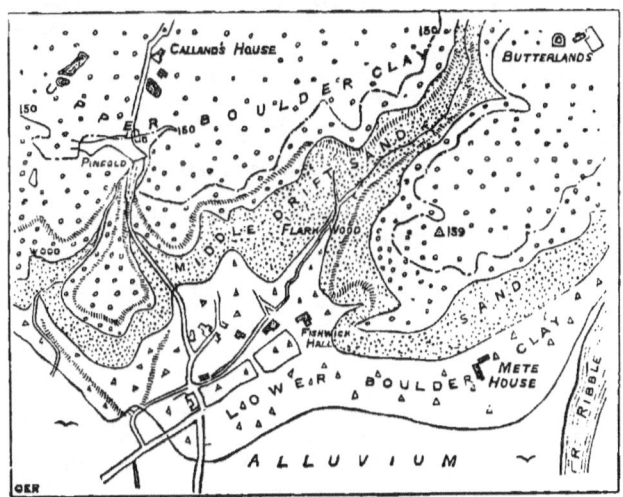

Sketch Map of Brooks, near Fishwick Hall.
(Scale, 6 inches = 1 mile.)

In the little brook valley of Boilton Wood the Middle Drift does not rise to a level higher than that which it occupies on the face of the escarpment. In the next valley breaching it, the stream rises in a spring at an elevation of 170 feet, a few feet from a col, in the watershed, which runs a little to the north of the Ribble, separated the water slopes drained by the long-course brooks flowing west, from the short-course flowing south, direct to the Ribble. From a spring (west of Higher Boilton) it flows 500 feet across a field with a very slight fall in the centre of an undulation of the ground ; it then falls from 164 feet to 75 feet at the Ribble plain, in 900 feet, or one in 10. The Middle Drift, in this valley, is last seen at an elevation of 170 feet, or that in which it occurs in the bluff or escarpment, for in these short valleys the surface of the Middle Drift is not seen for a sufficient length of space for undulations to any extent to occur.

The next valley, that of Brockholes Wood, is longer, the stream running 2,500 feet, from the springs to the Ribble plain, falling 119 feet in its course, the Middle Drift disappearing under the boulder clay at 144 feet, being 120 at Eyes Wood in the bluff, to the south of the Blackburn and Preston turnpike road, where the sand is very fine-grained and rather dark coloured. Further west another very short brook runs down from Butterlands to Mote House Brook, at the bottom of which Middle Drift sand is seen, its surface at its junction with the Upper Boulder Clay being 140 feet, its lower surface at its junction with the Lower Till 100 feet, the brook falling 125 feet in 800 feet, from its source to the Ribble.

To the west, near Fishwick Hall, several brook valleys meet, as shown in Fig. 36. The chief of these is that immediately behind the Hall; the brook takes its source from a spring by the side of the high road mentioned above, opposite the Preston cemetery, in which a large knoll of Middle Drift sand crops to the plain through the Upper Boulder Clay, the highest point of junction being 185 feet, west of Farrington Hall. Between the edge of this patch and the spring, 10 feet of boulder clay was seen resting on the Middle Sand; further south it is no doubt thicker, the sand not being seen in the brook until the 125 feet contour is reached, the Lower Boulder Clay underlying its base at the 75 feet, the sand being therefore 50 feet in thickness. This brook runs 3,000 feet and falls 117 (167-50) feet. Immediately to the west two brooks take their source, to the south of Pinfold, the edges of the two valleys being at the upper ends within 40 feet of each other, as shown in Fig. 36.

The south bluff of the Ribble has been so seldom denuded that very few sections occur, and no Middle Drift is seen westward of Samelsbury Church, until reaching the cliff below Walton-le-dale Church, where the following section occurs :—

    a. Upper Boulder Clay, 81 feet seen.
    b. Middle Drift. Rather loamy sand, densely packed with pebbles, some of large size, all apparently derived from the Lower Till, some retaining the scratches,* 18 feet.
    c. Lower Boulder Clay. Red loamy clay, densely packed with erratic pebbles. 7 feet under water.
    d. Rock. (Pebble-beds of Bunter.)

About a mile to the south-east the Middle Drift is also seen in the south bank of the river Darwen at Mosney Wood, where it is a pure reddish-coloured sand, the surface being about 75 feet above O. D. L. It will be seen by reference to the appended table, p. 107, the surface of the Middle Drift is 225 feet lower than at Pleasington, five miles east on the same river, giving a fall 45 feet per mile; the trend of the highest Middle Drift runs from N.E. to S.W., following the western slopes of the Pendle range; thus at Brindle it is seen in a deep pit, at an elevation of 450 feet, while at Bezza Brook to the due north it is only 118 feet, proving that this deposit invariably conforms to the level of the rocky floor beneath, as might be expected from the shallow-water conditions observable in it, successive lines of Middle Drift sand and shingle being formed as the country gradually subsided, offering higher and still higher coast-lines to the denuding action of its breakers; thus in the lowlands the included stones are invariably such as might be found in the Lower Boulder Clay, while higher up pebbles of Coal, Millstone, and Yoredale Grits first make their appearance, becoming more and more local as greater heights are reached.

This is especially seen to be the case in the curious gravel mounds occurring between Chorley and Henpey, one of which, known as Pickering Castle, rises to a height of 75 feet above the average level of the surrounding ground. No Upper Boulder Clay is found on any of these mounds, though it occurs at their feet; it is therefore just

---

* Erratic boulders, now being washed out of the boulder clay by the sea, on the shores of Morecambe Bay, and deposited in recent sand, still retain their scratches. The same fact has been observed by Professor Ramsay, F.R.S., on the coast of Anglesea.

possible that they may be *Esker Mounds* resting upon the Upper Boulder Clay.

At Chorley, in a sand pit on the north side of the town, current-bedded sand and shingle occurs, the latter consisting of 86 per cent. of erratic, and 14 per cent. of local (carboniferous) pebbles. They are associated with shells of *Turritella*, &c.

A great number of shells occur in a gravel pit* between Whittle-le-Woods and Leyland, where I also obtained some bones, pronounced by Mr. Boyd Dawkins, F.R.S. (who kindly examined them), to have belonged to a herbivorous animal.

Between Chorley and the villages of Euxton and Leyland a great number of brook valleys occur, the brooks being tributary to the rivers Yarrow and Lostock; in nearly all of these the Middle Drift Sand is seen beneath the Upper Boulder Clay, shingle never occurs. West of Euxton its surface dips down entirely beneath the Upper Boulder Clay, with the exception of one section at Eccleston, which is the most western exposure of Middle Drift (south of the Ribble), in map 89 N.W., it is shown in the following section :—

Fig. 37.

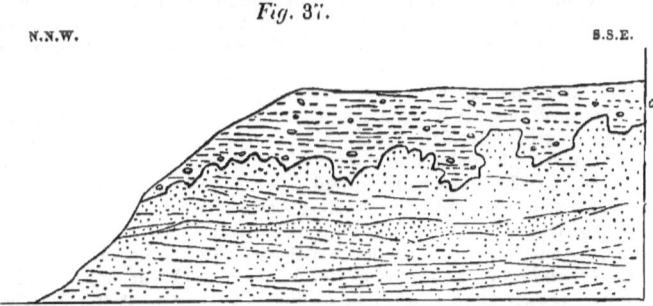

N.N.W.   S.S.E.

Section in Sand Pit, south of St. Mary's Church, Eccleston.

a. Upper Boulder Clay - - - 4 feet
b. False-bedded sand (Middle Drift) 12 „

Upper Boulder Clay.—This deposit consists, near its junction with the Middle Drift beneath, of a deep chocolate coloured stiff clay with a few subangular stones, 4 to 7 inches in length, all more or less scratched. This bed, which is nearly 20 feet thick at Red Scar, on the banks of the Ribble (on the opposite bank to the Balderstone Hall section described by Mr. Hull), is there succeeded by a thick bed of reddish clay with a considerable number of stones of all sizes, but generally 2 or 3 inches in length, but sometimes from 3 to 4 feet; the smaller stones are often more or less perfectly rounded and not invariably scratched; this clay may be considered to be the typical Upper Boulder Clay of the district, the uppermost bed being generally denuded away by the wash of rain over the surface of the land; it consists of bright yellow or pure white moderately stiff clay, with bands of very small stones, slightly stratified. It is well seen in the fields on both sides of the railway, west of Grimsargh Station. The Boulder Clay is also rudely laminated and stratified in a pit by the side of the lane, running between the letters *t* and *h*, in the word "*Penwortham*" (in the 1-inch map), the section occurring between them and the word "*Crooknoss*," also in the lane

---

* From the superficial deposit overlying the gravel of this pit I obtained portions of two early British rough-baked urns ornamented with the string pattern, which I have deposited in the Museum of Practical Geology, Jermyn Street. The urns were broken by the workmen in digging them out a few minutes before I visited the pit in May 1869.—C. E. R.

west of the word "New Gate," south-west of Charnock Moss in Penwortham.

The following villages and hamlets, in this quarter sheet, are situated on the Upper Boulder Clay :—Wren Green, Warton, Freckleton, Treales, Newton, Scales, Lund, Clifton, Ashton, Fullwood, Ribbleton, Samlesbury, Penwortham, Howick, Hutton, Longton, Farrington, Bamber Bridge, Moon's Mill, Houghton, Hesketh Bank, Hesketh with Becconsall, Much Hoole, Tarleton, Bretherton, Eccleston, Euxton, Brindle, Whittle-le-Woods, Heapey, Duxbury, Coppull, Mawdesley, and Rufford.

South of the Ribble and west of the river Lostock, from Farrington to Croston, no section of the Middle Drift is seen, and as far as I am aware it has never been bored into; but there appears strong reason to believe that the Boulder Clay of the Hoole district belongs to the upper division, because it has every appearance of being continuously connected, without a break, with the undoubted Upper Boulder Clay of Charnock Moss to the north and Eccleston to the south; but at the same time it is certain that the Hook Clay rests directly upon rock (Keuper Marls) at several points in the River Douglas, and the Middle Drift and Lower Clay must therefore be absent; this is, however, often locally the case both in the Euxton and other districts, for in a quarry near Euxton undoubted Upper Boulder Clay is seen resting on the Lower Coal-measure Sandstone, which is glacially striated in a direction (first observed by my colleague Mr. Tiddeman).

West of the River Douglas there is a tract of Boulder Clay, extending from Hesketh Bank to Sollom, being about three and a half miles in length, and a mile and a half in breadth at the former place and at Tarleton. There can be little doubt that this clay is a portion of the same bed as that occurring on the eastern bank of the river, and if the latter is true Upper Boulder Clay this bed must be so also. This view is the more strengthened by the fact that the true Upper Boulder Clay seen at Freckleton Point, on the north side of the river Ribble, opposite Hesketh Bank, rests on red marl without the intervention of the Middle and Lower Drifts; but in these low tracts of country, including that between Southport and Liverpool, it is difficult to know whether a clay belongs to the lower or upper division, because when they both happen to be present divided by the Middle Drift, they are found to be precisely similar in character, the Lower Clay only becoming distinct when higher elevations are reached.

The surface of the country, composed of the Upper Boulder Clay, from the stiffness of the clay and the level nature of the ground is extremely wet, requiring much drainage, and when uncultivated covered with rushes and rank grass and patches of marsh.

In Tun Brook, a little north of the margin of the sheet, a bed of sand occurs precisely similar to that in the Middle Drift, 10 feet in thickness, capped by 10 feet of boulder clay, 28 feet of which occurs between the sand and the Middle Drift beneath. Small seams of sand also sometimes occurs near the base of the Upper Boulder Clay, and when excavated under they are generally found to communicate with the surface of the Middle Drift below, apparently proving the still-water conditions under which this clay was formed ; at the same time the occasional occurrence of perfect shells (*Turritella terebra, Cardium edule, C. echinatum, Tellina Balthica*), and the constant presence of fragments, as well as of oblique lamination caused by currents, that the sea in which it was deposited was of no very great depth. The surface of Middle Drift, on which the clay rests though often minutely eroded, as seen in Fig. 38, is sometimes a perfectly smooth and level surface as far as regards a small space, though invariably dipping westward as a whole. Whether the Middle

UPPER BOULDER CLAY.

Drift was elevated above the sea and again submerged in the Upper Boulder Clay period, it is difficult to say, but it appears more probable that the surface of the sand became here and there eroded, when through great precipitation of sand the crest of the bank at certain points rose above the level of the sea, causing it to experience sub-aërial denudation, which furrowed and minutely eroded its surface, which had sufficiently hardened and consolidated to resist the action of the waves when again submerged, and through alteration of the direction of the currents bearing sand, it was never again covered with this deposit. That some such sequence of events took place is partly proved by the fact that in the midst of the Middle Drift when two beds are seen resting on each other, it is not uncommon to find the surface of the lower bed eroded, and having the appearance of being sub-aërially denuded, and also that in modern sand-banks, exposed at low tide, the sand is often worn into hollow concave channels by water, which flows into little streamlets from the crest of the ridge down to low-water mark, which hollows occasionally harden, and in a few instances become filled up with sand, often of a different sized grain from the higher part of the bank ; the surface of this is smoothed by the returning tide, which on retreating, deposits a thin pellicle of fresh sand over the whole, after which fresh streams of water again denude the surface into hollows.

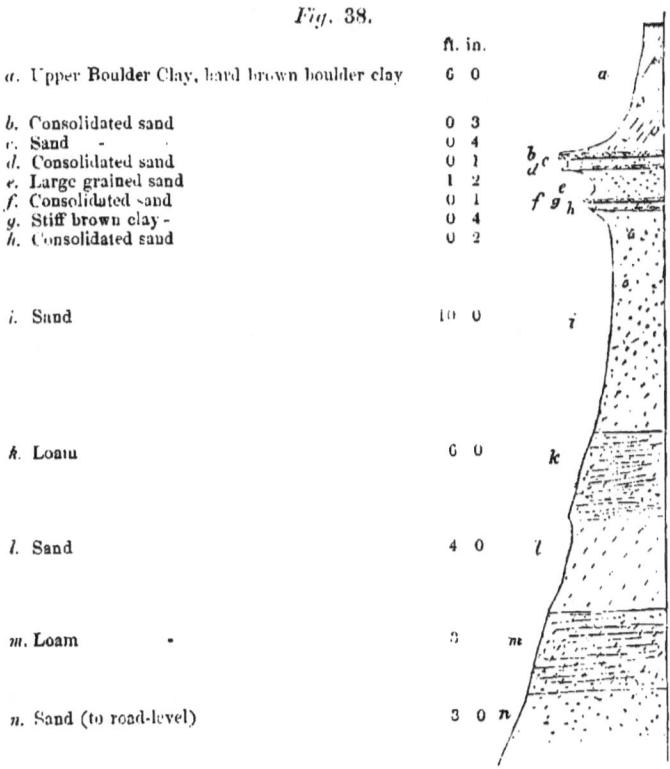

Fig. 38.

		ft.	in.
a.	Upper Boulder Clay, hard brown boulder clay	6	0
b.	Consolidated sand	0	3
c.	Sand	0	4
d.	Consolidated sand	0	1
e.	Large grained sand	1	2
f.	Consolidated sand	0	1
g.	Stiff brown clay	0	4
h.	Consolidated sand	0	2
i.	Sand	10	0
k.	Loam	6	0
l.	Sand	4	0
m.	Loam	3	
n.	Sand (to road-level)	3	0

Section in a Sand Pit at Town's Brow, Clayton-le-Woods
(scale 8 feet = 1 inch).

Sequence of Glacial Deposits.—Two types of Clay with Boulders have been shown to occur below the Middle Drift in this district; the one resembling the Upper Boulder Clay in general consistence, colour, and character of included fragments: the other occurring at elevations of 300 or 400 feet, entirely differing from it and from the low-level, Lower Boulder Clay. The low-level clay was deposited on an old plain of marine denudation, formed in the rocks in early or pre-glacial times, much of it being still 50 and even 70 feet below the present sea-level. The deeper valleys above the old sea-margin were occupied by large glaciers, which deposited the high-level Till, but which as the country subsided and the climate ameliorated became, like the low-level clay, covered with Middle Drift.

Level of the Upper Surface of the Middle Drift at its Junction with the Upper Boulder Clay. From East to West.	Brook.	River Bluff or Cliff.	Plain or Knoll.	Level of the Upper Surface of the Middle Drift at its Junction with the Upper Boulder Clay. From East to West.	Brook.	River Bluff or Cliff.	Plain or Knoll.
IN 89 N.E. LANCASHIRE: 6-INCH MAP, 62.				6-INCH MAP, 69.			
Little Harwood, east of Blackburn	—	430	—	Oram House, Brindle	240	—	240
Duke's Brow, Blackburn	—	512	—	St. James' Church, Brindle	—	—	400
North of Nova Scotia	—	340	—	Sandy Lane, Brindle	—	—	450
				Car Brook, Clayton Green	300	—	—
IN 6-INCH MAP, 70.				West of Whittle-le-Woods	—	—	250
Cherry Tree Station	—	370	—	River Lostock	—	—	240
Livesey Hall	—	330	—	" " Clayton	—	—	250
River Darwen, Pleasington	—	360	—	Gravel Hole Wood	—	—	234
" " east of Hoghton Towers	—	330	—	River Lostock, Cuerden Park	—	200	—
				" " Bamber Bridge	—	150	—
IN 89 N.W. 6-INCH MAP, 61.				" " Darwen, Mosney Wood	—	75	—
River Ribble, Elston Hall	—	70	—	Slony Lane, Cuerden	—	—	150
" " Red Scar	—	130	—	Farington, N.E. of	—	—	160
Tun Brook	146	—	—	" " N.W. of	—	—	120
Boilton Brook	—	—	—	Leyland, east of	—	—	170
" " No. 2	170	—	—	" " west of	—	—	127
Brockholes Brook	140	—	—				
River Ribble, Eye Wood	120	—	—	6-INCH SHEET, 77.			
Butterland Brook	116	—	—	Whittle Springs, north of	—	360	—
Fishwick Brook	125	—	—	Pickering Castle	—	—	407
Pinfold "	120	—	—	Dalton Pits, west of	—	—	300
Throstle "	115	—	—	Brinscall, west of	—	—	—
Preston Cemetery	—	—	185	Heapy Bleach Works	400	—	—
Holmes Slack Brook	110	—	—	Botany Bay	—	—	358
Fulwood Wood	127	—	—	Chorley	—	—	292
" " Upper	130	—	180	Lambrick	—	375	—
River Ribble, Fishwick	—	130	—	Adlington Station, north of	—	—	330
Gasworks	—	—	126	River Yarrow, east of Park	—	300	—
Deepdale	121	—	—	" " Duxbury Wood	—	240	—
Moor Brook	105	—	—	Higham House, east of Chorley	—	—	250
Preston (several)	—	—	110	River Eber, Astley Park	—	210	—
Winckley Square	—	—	96	Ackhurst, west of	—	—	240
Tulketh Hall	—	—	78	Worden Brook	—	180	—
				Worden Hall, north of	—	—	146
6-INCH MAP, 60.				Shaw Brook	—	125	—
New Lea Hall	—	37	—	" " south of	—	—	180
Salwick	—	—	89	Alicar Brook	—	—	160
Kirkham	—	—	102	" " south of	75	—	110
Westby Mill	—	—	95	Ransnap Brook	130	—	—
				Bank Lane Woods	90	—	—
SOUTH BANK OF THE RIBBLE, 6-INCH MAP, 61.				Chorley Cemetery	—	—	225
				Wheat Sheaf, east of Coppul Sta.	—	—	250
Hozza Brook	—	—	—	River Yarrow, Birkacre Print Works	—	—	175
River Ribble, Walton-le-Dale	118	—	—	River Yarrow, Parkers Wood	160	—	—
Penwortham	—	70	—	" " Charnock Richard	150	—	—
	—	50	—	Park Hall, Charnock Richard	—	—	200
				River Yarrow, Eccleston	—	50	—

C. E. D.

CHAPTER IX.
POST-GLACIAL DEPOSITS OF THE WESTERN PLAIN.

High Level Alluvium.

THE alluvial plain of the Ribble above Preston is fringed on either side by an inclined plane or terrace, with the exception of those points where the curves of the horse-shoe bends of the river have worn away the terrace and denuded the old cliff behind them, as at Red Scar and under Penwortham Church, and a little north of Mete House, in Fishwick; in the northern end of this cliff, in the direction of Brockholes Bridge, the river is now denuding one of these inclined terraces into a cliff, which exhibits the following section:—

I.
- *a.* Wet alluvial yellow *sand* — 10 feet.
- *b. Gravel*, with small and large pebbles — 15 „
- *c.* Red marl, stiff " low " boulder clay (water level) — 4 „

A little further north, where a small brook crosses the wood, the following sections occur:—

II.
	ft.
a. Sand	8
b. Gravel	13
d. Peat and clay	2
f. Boulder clay (water)	6
	29

III.
	ft.
b. Gravel	0
c. Clay	6
d. Peat	2
e. Gravel	1
f. Boulder clay (water)	2
	10 (+)

Still further north the old bluff of Boulder Clay is not only faced with the deposits given in Section III., but modern alluvium masks the secondary cliff, forming the face of the old alluvium, as shown in the following diagram, Fig. 39, and the alluvial sand is in its turn cliffed, and masked by still more recent alluvium.

Fig. 39.

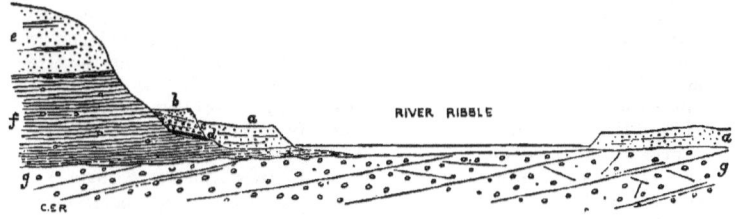

Diagram of Banks of the Ribble, in Mete House Wood.

- *a.* Recent Alluvium.
- *b.* Sand.
- *c.* Gravel.
- *d.* Peaty Clay.
- *e.* Middle Drift Sand.
- *f.* Lower Boulder Clay.
- *g.* Pebble Beds of Bunter Sandstone.

Inclined terraces, similar to those of the Ribble, also occur, at the edge of the alluvium of the river Darwen; between its junction with the Ribble and Roach Bridge the terrace rises to height of at least 30 feet above the present water level, above Bannister Hall, at which point the river falls about 10 feet over the rocks of the Pebble-Beds, flowing for 210 yards through a rocky gorge, formed by the retreat of the waterfall; from the end of this rock gorge the river flows through one excavated in ordinary alluvium, the banks being about 20 feet in height, but diminishing in height towards Walton, where another, though slighter, rock barrier exists; from this point to Walton Hall the river has denuded away all the alluvium on the south bank, flowing against the Boulder Clay cliff at Mosney Wood, but evinces a tendency to retreat from it, which it does at Walton Hall, flowing across the alluvial plain to its junction with Ribble above Preston.

High-level alluvium also occurs at Lostock above the river Lostock; from Woodcock Hall to Pinfold House it is a sandy gravel, the pebbles of which have apparently been derived from the Boulder clay; a little east of Lostock Hall a small island of the High-level emerges from the ordinary alluvium.

Terraces of High-level Alluvium also occur in patches in some of the brook valleys north of Preston, and a patch of sand at Golden Hill, near Leyland, is possibly referable to this age.

Shirdley Hill Sand.

In the "Description of the Country between Liverpool and Southport,"* the Triassic hills westward of Ormskirk and Halsall were described as being terminated by an escarpment, more or less covered with Glacial Drift, forming the margin of the great peat-moss plain, which plain occupies the greater portion of the country south of the Ribble, in quarter sheet 90 N.E., and extends from it nearly to the river Douglas (in the Preston quarter sheet), the patch of Boulder clay on the left bank extending from Hesketh Bank to Sollom, already described, further south, the clay ceasing with the island, so to speak, at Rufford Park; it dips under the alluvium of the river Douglas, and reappears in Croston Moss, on the opposite bank of the river, its eastward extension being terminated by an escarpment of thin Upper Boulder Clay lying on the red sandstone of the Pebble Beds of the Bunter, running from Croston through Moss Side and Back House to the Hall west of Mawdesley. This short escarpment may be considered as a prolongation of that described above as limiting the great peat-moss plain at Halsall; in fact, with the exception of the break caused by the valley of the Douglas (the rock bottom of which is below low-water mark), it is continuously connected with it. the escarpment trending round from Scarisbrick, by New Lane Railway Station (where the clay is worked for bricks), St. John's to Newburgh (all in 89 S.W.), where it approaches that coming down from the Hall, Mawdesley, to Lancaster House, Parbold, where the two escarpments are only about half a mile apart; at this point the valley of the Douglas may be said to commence, running to the south-east from this inland to Wigan, for westward it runs almost without fall over plain to the sea, cutting its channel through the peat, which it has covered with alluvium in the neighbourhood of its banks during floods.

In the "Description of quarter sheet 90 S.E.," already alluded to, a line of ancient sand dunes is described as occurring at the base of the escarpment, dividing the lower from the upper plain in the former, as forming the substratum of the Cyclas Clay and Peat, being the actual bottom of the sea, upon the shores of which were blown up the sand dunes.

The Boulder clay and Triassic rocks forming the country between Ormskirk and Liverpool were also described as being covered more

---

* Memoirs of the Geological Survey of England and Wales, Explanation of Quarter-sheet 90 S.E.

or less with a coating of sand blown over the country from the line of ancient sand dunes, the most typical of which is Shirdley Hill, after which the sand is named. In the Preston quarter sheet this sand is only found in the south-west corner, the chief patch being nearly two miles in length, extending from Holmes, in Tarleton, to Mere Side and Mere Windmill, in Rufford; in the centre of this tract it rises so as to form a ridge or sandy cliff running in a south-easterly direction, probably facing Boulder clay, which appears at the surface at the hamlet of Holmes.

On the northern portion of this ridge is Holmes Wood westward towards Martin Mere, the sand occupies the plain, sloping gradually from the ridge, until a little north of the main sluice it dips under the peat, which occupies the site of the old lake Martin Mere, to which the cliff or ridge of Holmes Wood must have at one time formed the northern limit.

A little north of the Boulder clay of Holmes, and separated from it by a strip of peat, is another Boulder clay knoll, rising from under the peat, similarly to that of Rufford Park, extending from the farm south of "Moss Side" to the farm called Wignalls. Both this knoll, that of Holmes, and Hesketh Park, as well as that at Clay Brow and Tarlscough, are covered on their western sides by thick deposits of the Shirdley Hill Sand, which is also found at Black Moor, west of Mawdesley, where the following section was observable in April 1869 :—

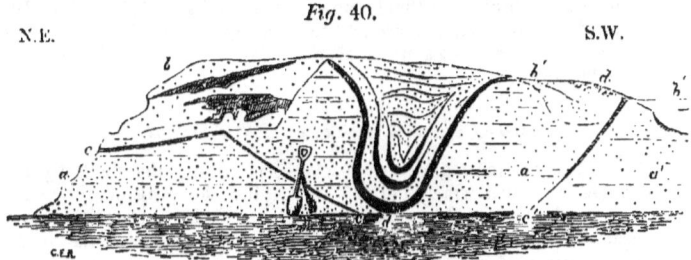

Fig. 40.

N.E.   S.W.

Section in a Sand Pit east of Rufford.

a. Fine-grained yellow sand.
a'. Coarse   „   „   „
b. Ashy-grey sand.
b'. Broken mixed sand.
c. Seam of oxide of iron, one inch thick.
c'. Seam of oxide of iron, four inches thick.
d. Peat in masses, and in synclinal folds.

Following the escarpment of the Boulder Clay to Croston three small patches of sand are seen, the first a little west of Back House, the second on the north side of Moss Farm, the third in the middle of Croston Moss. In quarter-sheet 89 S.W., a continuation of the peat-moss plain runs up the trumpet-shaped estuary of the Douglas to the entrance of the true river-valley. Between the Boulder Clay of New Lane, St. John's, and Burscough, and the peat, a tract of Shirdley Hill Sand intervenes, extending from Tarlscough to St. John's, occupying an area of nearly two square miles, and eastward in Horscar Moss (north of the Southport and Manchester Railway) four patches of the sand rise to the sand from the peat below. No sand crops to the surface in a southerly direction, on the east bank of the Douglas, after the patch at Black Moss, Mawdesley, the sand having been removed as far as

Parbold by the denudation of that river, which here and there in Wilbrahams and Low Meadows has formed a true fluviatile alluvium, that overlying the peat being chiefly estuarine or marine formed by tidal action, the spring tides at the present time still travelling further than the southern margin of map 89 N.W., to Wanes Blade Bridge in 89 S.W.

No bed of sand or other deposit, corresponding to that of Shirdley Hill, occurs north of the Ribble either in the Preston or the Lytham districts, but the great peat-moss plain occurring in the Fylde district between Fleetwood and Lancaster I find to be often margined by a bed of shingle, with a little sand occurring at the edge of the Boulder Clay forming the boundary of the peat, and dipping in a seaward direction under the peat and the Grey Clay beneath. This shingle is particularly well seen at Presall (in map 91 S.W.), where it is dug in a pit and used for mending the roads. In this pit I found fragments of hematite iron-ore, pebbles of Permian Breccia, and the following shells :—

| *Cardium edule.* | *Natica monilifera.* |
| ,, *echinatum.* | *Turritella terebra.* |

The first and last of the above mollusca also occur in the Shirdley Hill Sand; and considering that both hold the same position relative to the great plain, since covered with peat, extending throughout the whole of western Lancashire, both being slightly above the sea level, and at an identical level in regard to each other, and both being older than the peat, I think that there can be little doubt that Shirdley Hill Sand and the Presall shingle are of the same age; a similar deposit to the latter I have also examined at Rampside,* on the north side of Morecambe Bay.

Lower Cyclas *and* Scrobicularia *Clays.*

The grey lacustrine and estuarine clays containing the shells *Cyclas cornea* and *Scrobicularia piperata,* underlying the peat of the plains between Southport and Liverpool (90 N.E. and S.E.) also occurs under the peat which covers the area occupied by the whole of the south-western corner of the map, with the exception of the knolls of Boulder clay, which rise above the peat plain, as far east as the Douglas, described above. In both tracts, or rather in the whole tract, the grey clays only occur on the western portion, thinning out to the east, where the peat rests directly on the Shirdley Sand. Their eastern limits in the present quarter sheet extends from the end of the lane N.W. of Midge Hall to near the point where Long Ditch joins the great Martin Mere Sluice, thence turning to the north-west it leaves the sheet, where the first lane, north of the Sluice, enters map 90 N.E., in which sheet it disappears under the estuary of the Ribble at Crossens, reappearing on the opposite shore in Lytham Moss, the western part of which rests on Cyclas Clay, the eastern on the Boulder Clay; the termination of the moss northwards is made by undulating hills of Boulder Clay, which terminate abruptly at Blackpool in a line of cliffs, below which peat or Lacustro-Estuarine Clays are never seen, but on the beach adjoining the peat tracts of Lytham and South Shore to the south, and Fleetwood to the north these deposits are found between tide-marks, patches extending to the lowest low-water mark of spring tides. At South Shore the bluff of Boulder Clay terminating the peat plain runs into

---

* From this deposit several species of shells have been recorded by Miss Hodgson in the North Lonsdale Magazine.

the sea at the Royal Hotel. On the shore immediately below, to the south of this bluff, in March 1869, I observed about six feet of peat resting on four feet of grey clay with many vegetable fibres of water plants and rushes, showing the limit of the Cyclas Clay in this part of Lancashire to be coincident with the limits of the peat plain itself.

Ancient Low-level Fluviatile Alluvium.

At the base of the ordinary low-level alluvium, both of the Ribble and its tributary brooks, there is a deposit of peat, with fallen trees and twigs, and occasionally traces of leaves and nut shell of *Corylus avellana*. This peaty seam invariably rests, both in the low-level Ribble alluvium and in that of the brooks, on a bed of grey freshwater clay, similar to the "Lower Cyclas Clay" underlying the peat in the plains, the clay itself generally resting on a seam of river gravel, with large stones, and occasionally on the Boulder Clay, which is always beneath the latter.

A good section of the beds is exposed in the Ribble near Mete House, of which Fig. 41 is a section.

Fig. 41.

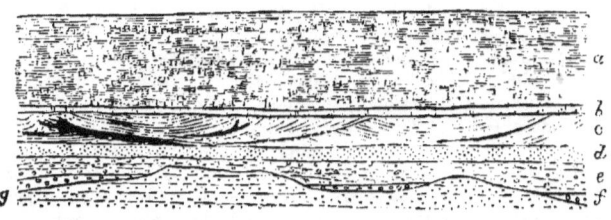

a. Reddish loamy clay	12 feet.
b. White earthy clay	1 foot.
c. Grey clay, with much vegetable matter, and young trees	4 feet.
d. Yellow sand	1 foot.
e. Red ochreous gravel	2 to 4 feet.
f. Lower boulder clay	0 to 3 feet.
g. Sandstone (Pebble Bunter)	1 foot.

(Both *f.* and *g.* are generally below the water-level.)

As these deposits could not well be separated from the more modern alluvium resting upon them, I have described them with the recent alluvial deposits.

Peat.

The great tract of peat lying in the south corner of the sheet on either side of the river Douglas has already been incidentally described in the account of the underlying "Lower Cyclas and Scrobicularia Clays," and Shirdley Hill Sand, the Douglas as having cut its channel through the peat, occasionally exposing the boulder clay beneath, the plain in which it occurs being part of that described as girdling Lancashire in the "Description of 90 S.E."

If a line is drawn from Bonney Barn to Wignals (in Tarleton), all the peat north of it and south of the knoll of boulder clay (running from Hesketh Bank to Sollom) will rest upon the latter clay, the surface of which commences to rise from the line towards the knoll. Thus the peat forming Hesketh and Tarleton mosses is a "slope peat," rising from 14·4 feet above the Ordnance datum line, at Sugar Stubs, to 46 feet at Boundary Meanygate, at a point halfway between its junction with Bottom's and Johnson's Meanygates (lanes). Still further to the

N.E. towards Becconsall Windmill, the level of the surface of the peat diminishes to 35 feet, where it thins out on the Boulder clay.

The peat in this tract, and in fact in all the "slope peats" of the district, though reaching a thickness of 12 and even 16 feet, appears to have been nearly entirely formed of heath and small brushwood laid down in successive thin layers, which can be separated from each other like leaves of paper. The peat is much drier than that formed on the flat surfaces, and when perfectly free from moisture is a bright yellowish brown, while the "swamp peat" is invariably a very dark brown colour, and possesses from three to seven times the specific gravity of the heath or slope peat.

Another considerable tract of "slope peat" occurs around Midge Hall Station. It includes the tracts known as Farrington Moss, Longton Moss, Little Hoole Moss, Leyland Moss, and Great Moss, in all about five square miles, the boundary of which was traced by my former colleague, Mr. Shelswell, who reports the average thickness as being 12 feet, and also the presence of oaks and other trees lying at the bottom of the moss. At one point north of White Houses (the second house from the east in the lane running north of the word "Londonderry" on the 1-inch map) I observed the base of a large oak tree *in situ*, with its roots in the Boulder clay beneath, the colour of the surface of which here is changed from the usual red to a pale blue, which phenomena is generally found to be the case elsewhere, when the peat rests on red clay. Here and there, however, I found under this moss patches of pale grey (Great glacial) lacustrine clay, formed in basin-shaped hollows in the Boulder clay. The average elevation of this moss is about 65 feet, but rises to a level of more than 100 feet at Farington Moss, where the peat is 18 feet thick, forming a dome-shaped mass, which is being gradually reduced by the peat being cut down to a certain level, and occasionally the peat is set on fire, so as to reduce the height and prepare its surface for cultivation, the heathy peat above being too porous to admit it, the peat becoming denser at the bottom, but still containing a far greater proportion of heathy matter than would be found in a swamp pit.

North of these mosses the greater part of the parish of Penwortham is a very slightly undulating plain of Upper Boulder clay, with an average elevation of 75 feet above O. D. L. The peat mosses just described may be considered as resting upon and covering up the southward extension of this plain, and indeed they appear formerly to have extended northwards nearly to the margin of the Ribble valley, for nearly everywhere this plain is covered with a thin coating of peat earth and plant growths hardly carbonized enough to be called peat, resting on the clay, in the hollows of which here and there, in one part of a field and not in another, seams of true peat, from two inches to four feet, are found to occur. The peat in this tract has been probably removed by cutting, by cultivation, and by natural denudation at a comparatively late date, as the local names testify to its presence, Charnock *Moss*, Hutton *Moor*, and Howick *Turn Moor*, &c.

It is evident that after the deposition of the Shirdley Sand an elevation took place, and that from the flatness of the ground and probable distance from the sea (for peat has been met with in boring at a depth of 60 feet below the present sea level), the deposition of lacustrine silt (Lower Cyclas Clay) took place up to a certain limit west of the margin of the plain, on the surface of which as well as on that of the sand further east the growth of an immense forest commenced, and that after the trees had obtained a considerable size an obstruction of drainage ensued, which eventually caused the trunks to be covered with water to a depth of rather more than two feet, making them rotten at that level, and thus causing them during a heavy gale or hurricane from the W.N.W. to be all blown down in one direction with their heads E.N.E. and N.E., and to lie side by side with their broken stocks. The fallen trees still more obstructing the drainage, the growth of peat ensued, reaching in some instances a thickness of nearly 30 feet ; during this growth the whole of these great plains would be covered by water, remnants of which have remained down to late historical periods in Martin Mere (in this sheet), and in Barton Mere, and the White and the Black Otter in 90 S.E.

Martin Mere.—The bank of Shirdley Hill sand at Holmes Wood has been described as formerly constituting the northern limit of Martin Mere, which old lake was described by Leland as "the biggest meare in " Lancashire, iiii miles in length and ii in breadth," and in 1692 it had an area, before it was drained, of 3,000 acres ; in that year T. Fleetwood, Esq., of Bank Hall, began the great work of reclaiming it by cutting a sluice 24 feet in width from the sea through the salt marsh bordering the coast near Crossens (which at that early period was already embanked) through a thick deposit of peat for a mile and a half up to the margin of the lake ; the water gradually flowed away into the sea which before only escaped during floods into the river Douglas.

This flood channel appears to have run between the Boulder Clay patches already described at Tarlscough and Rufford, following the line of "Boundary Lane" of the 1-in. map, joining the Douglas at the point where a sort of arm of alluvium juts out in the direction of Lathom Windmill. In a small portion of the area occupied by the peat in the south-west corner of the sheet it is found to rest on the grey lacustrine "Cyclas Clay," as in 90 S.E., but all around this tract (between the " M " of Martin Mere and the " L " of Longditch) the peat rests directly on the Shirdley Hill Sand, lying directly on the Boulder Clay plain.

To prevent the sea entering the sluice at spring tides Mr. Fleetwood built flood gates which in 1755, five years after his lease expired, were thrown down by a high tide ; the sea rushing in these were again rebuilt, but the sluice was neglected, and in the winter of that year the site of the lake was again covered with water; but in 1781 Mr. T. Eccleston procured the assistance of Mr. Gilbert of Worsley, who had worked for the Duke of Bridgewater, who commenced operations in 1783 by making sea-gates * at the coast, "stop-gates" half a mile from it, and " flushing-gates " at the junction of the sluice with the lake. In the following year these were so far successful that several acres were sown with corn, and Mr. Eccleston was voted a gold medal by the Society for the Encouragement of Arts, Manufactures, and Commerce, and when he died in 1809 an elaborate Latin inscription recorded how he with immense pains had drained the great Martinesian lake ; but unfortunately in 1813 the sea destroyed the two outer gates, which were replaced by cast-iron cylinders with valves instead of flood-gates, which though often choked have answered ever since ; but immense damage has since been done on several occasions through the sea breaking down the banks and flooding the country, most of which is ten feet below spring-tide high-water mark.

When the water was first drained off the lake eight canoes were found, one of which is now in the British Museum, cut out of a single tree. Some of them are stated by Leigh to have contained plates of iron. He also mentions the discovery of an elk's head 4 yards within marl, under the moss between Martin Mere and Meales, now North Meales,

---

* Trans. Geol. Soc., Vol. VII.

the brow antlers being thicker than a man's arm, their beam two yards, and from tip to tip two yards. This head probably belonged to the *Cervus megaceros*. He also says that it has been stated * that trunks of birch, ash, oak, and pine trees were found in separate plots on drawing off the water of the Mere, as if they had been planted by man.

Dr. Harrison, chaplain to Lord Cobham, in describing the course of the Lancashire rivers, nearly 300 years ago, said, "the Duglesse meeteth also on the same side with Merton Meere water, in which meere is an island called Netherholme, besides others," the former no doubt the present Holmes Wood Hill, which would denote that the water covered the peat to the north of the hill as well as that to the south, and the channel from the Mere to the Douglas ran from here between the Boulder Clay knoll at Rufford Park and that of Sollom. This view is strengthened by the fact that a deposit of sand occurs on the peat in the tract extending from "Mere side" to the plantation west of the word "*Mere Sands*," but possibly there may have been a channel by the "Boundary Lane" route also.†

Recent Alluvium.

Fluviatile.—A deposit of peaty clay has been described as occurring at the base of the ordinary alluvium in the Ribble near Mete House; it is also seen on the south bank of the river from Walton Bridge or Penwortham Bridge, a little ; it is also occasionally seen on the north bank of the river between Mete House and Walton Bridge ; it is also seen in nearly all the sections in the south bank from that bridge to Penwortham, a little east of the tramroad. The following is the sequence of beds :—

(1.) Loam — 6 feet.
    Peat - - 1 foot.
    Grey clay - 3 feet.
    Gravel - 4 „
    Red sandstone - 1 foot (above water).

Further to the east towards the Darwen :—

(2.) Laminated sandy clay — 4 feet.
    Peaty clay - 2 „
    Gravel - 6 „

West of Number 1 :—

(3.) Laminated sand — 3 feet.
    Yellow clay - 4 „
    Grey clay, with pebbles - 2 „
    Peaty beds, with hazel nuts 1 „
    Gravel - 3 „

At Penwortham Bridge four of laminated clay rest on six of sandy gravel, resting on two of pale grey clay, the base of which is not seen. Following the river to the north, the section near Mete House, Fig. 41, is seen, while a little further north three terraces occur, the highest consisting of the high level alluvium, the lowest of sand, which thins in the direction of Brockholes Bridge, on the south of which three feet of

---

* History of Lancashire. Oxford, 1700.

† In 1741 great quantities of bones of men and horses were found near the river Douglas, at Wigan, which are believed by some to have resulted from five battles fought by King Arthur, which the old historian (*Historia Britonum, auctore* Nennio, lxv, lxvi,) described as having taken place, "super aliud flumen, quod vocatur Duglas, quod est in regione *Linius*," which points to the whole district being named after "the lake," Martin Mere, on the borders of which, according to tradition, lived Sir Launcelot of the Lake, Knight of the Round Table.

sand on three of Lower Boulder Clay. On the north it thickens to 18 feet, below Lower Brockholes, still further the sand is covered with sandy clay, containing many large rounded stones resting on sand. Opposite Seed Houses the river bank consists of 16 feet of yellowish alluvial clay. Still further north, near Higher Brockholes, 15 feet of clay is seen resting on five of Lower Boulder Clay, which has a shaly appearance. Between Red Scar and Elston Hall the clay is replaced by sand, about 15 feet of which is seen in the river bank for a couple of miles. Near the Hall about three feet of alluvial sand rests on 12 feet of gravel. In one place a basin-shaped mass of grey clay occurs, the surface of which is level with that of the gravel. On the opposite side of the river the banks of Bezza Brook near the foot bridge consists of four feet of alluvial sand on two of gravel, resting on two of blue sandy loam lying on the red sandstone of the Middle Bunter. The south bank of the Ribble from Seed Houses to Charnleys consists of a bank 20 feet high, the lower portion of which consists of red sandstone in place, dipping to the W.S.W., capped by gravel and loam, which may be said to belong to the first terrace of high-level alluvium rather than to ordinary alluvium, as it is out of reach of the present flood level of the river.

The alluvium of the Darwen generally consists of about 16 feet of sand, above which rises one and sometimes two terraces of high-level deposits. At Carver Bridge, Bannister Hall, six or seven different sands and gravels are seen wedging one into the other, the long axes pointing down the river. Near the base is a bed of thin stones much impregnated with iron, having at a little distance a resemblance to old copper money. At the base a seam of large stones occurs, from 10 inches to 18.

At Long Lane, near the coast, a strip of alluvium runs in from quarter-sheet 90 N.E. It consists of a grey clay, apparently formed rather under lacustrine than fluviatile conditions. It rests in part upon the peat and in part on the Boulder clay, the former having been denuded away. At Back Lane, N.E. of Guinea Hall, it is five feet in thickness, resting on the Till. Farther east, in the same lane, it has thinned to two feet, resting on one foot of peat, which lies on the Boulder clay. A little beyond this point the grey clay thins out and seven feet of peat is seen resting on Lower Cyclas Clay, which deposit the present grey clay much resembles, having apparently been formed under lacustrine conditions. Still further west, beyond the margin of the sheet, it graduates into a marine clay, also resting on the peat, with *Tellina Balthica* and other shells. The sea is constantly making inroads over the whole of this lowlying tract, covering the peat with thick beds of marine silt, charged with *Scrobicularia piperata*, which cover a considerable area, and are therefore described separately as "Tidal Alluvium."

Tidal and Estuarine.—The estuary of the river Ribble occupies nearly ten square miles in this sheet, the whole of which, with the exception of the channels, being dry at low tide. From a little below Preston to Freckleton Point the river is confined between two high walls, about   feet apart. From their termination the river channel runs nearly straight to Lytham, and is joined by that of the Douglas (also artificially confined, for the purposes of navigation) a little to the west of Freckleton Point. Sand is brought down by the river, and also from the south, by the tidal current at flow tide, and is accumulated everywhere except along the lines of the channels, which scour away that already precipitated, and prevent by their motion the deposition of that held in mechanical suspension. The channels being thus confined to the north side of the estuary, immense banks of sand occur on the southern side, that above stretching in one continuous undulating plain from the coast at Hesketh Bank and Hutton Marsh to the Ribble channel at Hesketh Bank. Further to the west, in the direction of the open sea, this continuous bank is split up into detached portions, divided by channel kept open by tidal currents, the bottoms of which are never laid bare by the lowest spring tides. These sand banks together form a fan-shaped bar at the mouth of the Ribble, between Lytham and Southport.

Immense quantities of sand and mud are brought down by the Ribble from the 380 square miles of Glacial Drift country drained by it and its tributaries, a part of which is carried out seawards. Returning with the tide the sand is deposited on the banks, the mud from its lightness falling last, forming a thin coat on the sand near the shore as the tide recedes, while in the sand banks surrounded by water no mud is observable. The mud being at the surface of the water when the ebb takes place, a wedge of water is removed bodily from the bottom, running out seawards, carrying with it the principal part of the sand, while the more gradual retreat of the surface allows time for the precipitation. This takes place every spring tide, and to a less extent with each ordinary high tide, an extremely thin leaf-like layer being thrown, which dips towards the sea, the whole forming a laminated tidal clay reaching a thickness of 12 and even 20 feet. It is found all along the coast, on banks of the estuary of the Ribble from the western edge of the map, north of Brow Side House, to Penwortham, and from Freckleton Point to Old Chain House a large tract of it occurs, called Freckleton Marsh, bounded on the north by bluffs formed of the Upper Boulder Clay running by Newton, Clifton, Old and New Lea Hall. The whole of this marsh has been recently reclaimed up to the wall, on the north wall of the Ribble Navigation Company. In this tract the following section was obtained, by a boring of the Preston Farming Company, for the journal of which and the analysis of the water obtained, I have to thank Mr. W. Ascroft, of Preston :—

Sand, with salt water	24 feet.
Clay	12 ,,
Sand, salt water	4½ ,,
Clay	18 ,,
Sand, salt water	24 ,,
Boulder clay	57 ,,
Sand, fresh water	1 ,,
Gravel, conglomerate	4 ,,
Sand	4 ,,
Gravel	1 ,,
Clay (not penetrated, probably Keuper marls).	

The water of this well, by the analysis of Mr. J. G. Cope, of Preston, contained at first 40 grains of salts per gallon, but gradually became fresher, the residue consisting of

Bicarbonate of lime	40 per cent.
Sulphate of magnesia	10 ,, ,,
Chloride of magnesia	15 ,, ,,
Sulphate of lime	3 ,, ,,
,, of potassium	9 ,, ,,
Carbonate of potash and sodium	4 ,, ,,
Chloride of potassium	4 ,, ,,
	100

Tracts have also been reclaimed of this laminated clay on the opposite bank of the river, north of Dunkirk and Pipe Houses.

At Bonney Barn and Rydings a patch of marsh clay runs a little inland, resting on two feet of peat, lying on Boulder clay, which crops out on the shore for a little distance on either side of the Barn, dipping seawards under the laminated marsh clay. A little westward it is capped by pebbly sand, and soon after thins out along the line of a channel which flows to the N.W. With the exception of a thin band of it opposite Brow Side House, resting on peat and covered with sea sand, which also conceals the seaward extension of the peat, the latter also occurs on the shore to the eastward, opposite the lane by Cottams House, and there is no doubt that it is continuously connected with that occurring on the shore of the opposite coast, west of Freckleton Point, and which runs up the three small valleys under the tidal alluvium found in them, viz., west of the word "Bush," west of the word "Brow," and around the word "Brook Bridge" on the 1-inch map. In the peat, near low-water mark, opposite Bush, some very large horns of red deer were found in 1868. At Bonney Barn and Hesketh Marshes the clay is often overgrown to a considerable extent with tussock grass, thrift, Danish scurvy grass, and chrysanthemum, growing in little knolls, separated from each other by gutters which meander and ramify in different directions without having any direct outfall.

The plant surfaces are occasionally swamped during high tides by marine silt, and sometimes by freshwater alluvium brought down by the floods of the Douglas during the low tides of the sea; and when examining an artificial section of these beds I found it to consist of a succession of marine deposits (with *Scrobicularia piperata*, evidently occupying the position in which they lived and died,) of freshwater seams (with *Limnea*), and plant-growths, sometimes in one order of sequence, sometimes in another. Occasionally the mud seams (with *Scrobicularia*) contain more sand; in that case the shell *Tellina Balthica* is invariably very common, and the valves of the *Scrobicularia* are detached and appear to have been transported. The clay is found to run up the Douglas as far as Rufford, the ordinary high tide still flowing nearly a mile beyond that place; but in sections of the alluvium of this part of the river and in those of the Ribble above Preston the marine seams are the exception, and the freshwater deposits form seven eighths of the mass. At Rufford I found in one seam several shells of *Tellina Balthica* and *Cardium edule*. Both of them occur in the Ribble alluvium opposite Old Lea Hall, and the latter shell in that of Fishwick, above Preston, rather beyond the point to which tides now flow. The river Douglas receives the river Yarrow at a point opposite Sollom, and the latter river the Lostock north of the word "Fisherty Bridge" of the 1-inch map. The tide at springs flows up both these tributaries, the former to Little Wood and the latter to Croston Mill. This tract, especially west of Croston, is much subjected to floods, and the top of alluvium is essentially fluviatile, but tidal beds occur near the base. North of the junction the laminated clay is very thin and rests on peat.

All the alluvium between the western end of Longton is of a tidal character, running inland along the lines of the small brooks of Walmer Bridge, Tarra Carr Gutter, and that running between Longton and Hutton, the seaward alluvium of which is tidal, the landward fluviatile. Boulder clay occurs in two places at the bottom of the last-mentioned brook, and also under the Marsh Clay, near the end of the embankment north of Old Grange, where three feet of it is visible, and a little further is a low boulder cliff called Red Bank (Fig. 42), running as far as Farrer's Farm, where the alluvium again encroaches upon the land, in consequence of the Boulder Clay bluff trending inwards, running by Dungeon Farm to Hangman's Rack Wood, where a bed of rolled river gravel is seen capping the Pebble-beds at a level eight feet above that of the surface of the fluvio-marine alluvium, shown in Fig. 24.

Section South of Red Bank.

a. Aluvial sand, 2 ft.   *b.* Pebbles, 4 ins.; sand, 3 ins., with *Tellina Balthica.*
                         c. Boulder Clay, 4 ft.

In the centre of this tract the marsh clay is well seen in the sluice which runs from Mill Brook, dividing the parish of Howick from that of Hutton; it is there about 12 feet in thickness, its surface being much intersected by narrow channels and gutters, some of which have cut down nearly to the river level at low tide, forming steep banks eight or nine feet deep, and often a yard to four feet across. A space has been left in the wall where the sluice enters the river; the water enters here and connects itself with a channel, supplied by a spring, near the river-wall, about half a mile to the east; on the strip of alluvium between the wall and the channel (about 120 feet wide) sand and river shingle is piled up during floods of the river to a height of eight feet above the wall, or 16 feet above the ordinary river level, which deposit has entirely taken place during the last few years, since the building of the wall.

This patch of shingle, with the exception of a few loose stones, off the Bonney Barn (derived from the Boulder Clay there), is the only example of stones or shingle being found on the coast between the Ribble and Mersey, and there can be no doubt that these pebbles have been brought down by the Ribble, and have not been carried up by the tide. On the opposite side of the river the case is precisely opposite, and from Freckleton Point westwards a zone of pebble-beach invariably occurs, spring and ordinary high-tide marks, below which sand slopes to the water; here and there the peat has been denuded, and the underlying Boulder Clay crops to the surface, itself having been denuded, and its pebbles and boulders scattered over the shore, many of them still connected at the bottom with the clay below; this is well seen on the beach below the Boulder Clay cliff running eastwards from Bank House, in Warton, the clay forming a sort of bench or terrace at the foot of the cliff. A little west of this house the beach runs round a small inlet, formed by the mouth of a brook, which has cut a channel through it down to the *Scrobicularia* mud below, lying on the peat, which further west appears in force on the beach towards Lytham, and will be hereafter described in the "Explanation of Geological Survey Map, sheet 90, N.E."* And in Freckleton, Lea, and Ashton Marsh the laminated clay rests on a peaty bed, which is the undoubted representative

---

* Written 1870; Exp. 90 N.E. has been since published.

of the great sheet of peat of the Southport plains, so much of which overlies the lacustro-estuarine clay, which would appear to be the representative of the black silty sand found beneath the seam of peat underlying the laminated clay of Lea Marsh. At a point opposite the boundary brook (from Mill Brook) the following section is observable, the sand under the peat being nine feet in thickness :—

		feet.
a.	Laminated marsh clay	3
b.	Peaty clay, with trunks of oak trees	2
c.	Black loamy laminated sand dipping towards the river, apparently	9

Further east the sand under the peat has become clay, with very little sand, the following being the section :—

		feet.
a.	Laminated clay	3
b.	Peat	2
c.	Clay	9

Trunks of trees, chiefly of oaks, both at the top of the peat and at the bottom; the latter appear to have grown on the spot, while the former have evidently been brought down by floods.

In Preston Marsh the following artificial section was seen :—

		feet.
a.	Red sandy clay	7
a².	Grey ,, ,,	8
b.	Peaty clay	1
f.	River gravel	2
g.	Lower Boulder Clay	1 (+)

A little further west the Lower Till is known to rest on the Bunter Pebble Beds.

Upon the opposite bank of the river in Penwortham Marsh, east of the belt of shingle, already described, the following sections occur :—

a.	Laminated sand	4 feet.
g.	Lower Boulder Clay	12 ,,
h.	Pebble beds	(+) 2 ,, (+)*

Further to the east :—

a.	Laminated sand	1 foot.
g.	Lower Boulder Clay	10 feet.

Further east, opposite Marsh End :—

Alluvial clay	2 feet.
Lower Boulder Clay	8 ,,
Pebble beds	(+) 2 ,, (+)

Westward of this the alluvial beds are described under the head of ordinary alluvium; the succession is the same, but the tidal conditions cease by degrees, until the whole series becomes fluviatile.

Associated with pebbles, on the beach opposite Bonney Barn, before referred to, I observed in 1869, a considerable quantity of iron pyrites, precisely resembling that found in the Blue Lias of Dorsetshire. It is certain that the Keuper marls underlying the estuary of the Ribble must be extremely high in the series, and it is within the limits of possibility that the base of the Lias occurs there from which the pyrites may have

---

* In all the above sections the same lettering is used, to facilitate comparison ; (+) before a number signifies that it is below the usual water level ; (+) after a number that the base is not seen.

found its way into the Boulder Clay, from which it has been washed but on to the present beach.

Sequence of Post-Glacial Deposits.—The Shirdley Hill Sand has been shown to be a marine deposit, covering the plains up to a level of 30 feet, above which it occurs as "blown sand" in a line of old sand-dunes, masking an earlier cliff, the sand resting on a thin deposit of peat, both in the plain and under the dunes. It is therefore clear that the plain was formed by the sea, as far as the old cliff, before the deposition of the sand, an interval long enough for a thin growth of peat to have ensued having elapsed between the retreat and re-submergence. Two deposits are commonly found to rest upon the sand in the plain, viz., the Lower Cyclas and Scrobicularia Clays and the main peat.

Following the plain to the north, the peat was seen to be covered by tidal alluvium, which, in the entrance to the valley of the Ribble, overlies a thin deposit of peat, with fallen trees lying on silty clay, while further up the valley the tidal is replaced by fluviatile alluvium, peaty silt with trees invariably occurring at the base, the silt lying on a thin bed of gravel.

Still higher up the valley the low-level alluvium was seen to rest against an ancient cliff of old high-level alluvium, consisting of sand, gravel, peat, and clay, and occasionally gravel. It is clear that as the peat is of the same age as the base of the low-level alluvium, that the high-level alluvium must, like the Shirdley Hill Sand, be of older age than the main peat, and that it was formed by a river, after a period of obstruction of drainage, during which peat was formed in the bottom of the valley. It is therefore probable that both the thin peat deposit at the base of the Shirdley Hill Sand, and the high-level alluvium, were formed at the same time, and that the river that brought down the sand and gravel found above that of the latter flowed into the sea, on the margin of which the Shirdley Hill Sand was thrown up, and the shingle banks of Presall (near Morecambe Bay) were deposited. The sea during this period would appear to have stood about 15 feet higher than at present, which would prevent the Ribble cutting down its bed by that amount; and we find the top of this old alluvium in the valley of that river, about a mile beyond the point to which the highest spring tides now flow, to be about 36 feet above high-water level, which is nearly in proportion with the level of the recent alluvium, in regard to the existing sea-level. The land appears then to have risen, giving scope to fresh fluviatile denudation, represented by a bed of gravel found at the base of the low-level alluviums of rivers and brooks, after which an obstruction of drainage took place, trees were killed, and thick beds of peat overspread both the plains and the bottoms of all the great valleys.—C. E. R.

Agricultural.—In the extensive tract of peat-land, around Rufford, Croston, and Hesketh-Bank, and in the peat-covered slopes of Hutton, Horwick, and Lington, between the Douglas and Ribble, very large quantities of potatoes are raised, as well as some grain.

In the slightly undulating plains of Boulder Clay, forming an intermediate belt of country between the East Lancashire Fells and the peat mosses of the sea coast, the greater number of fields are laid down as permanent pasture, but a very large area produces grain crops, especially between Garstang and Preston, and around Kirkham, the area being a portion of the district known as the "Fylde of Lancashire."—C. E. R.

CHAPTER X.

RELATIVE AGES OF THE PENDLE AND PENINE CHAINS OF HILLS.

The Pendle Range of hills, understanding by that term the long line of parallel ridges and valleys which stretches from Parbold Hill to Colne, and which culminates in Pendle Hill, has hitherto been considered as nothing more than an offshoot of the Penine Chain, the name by which Conybeare and Phillips designated the mountainous tract extending from Derbyshire to the borders of Scotland.* In one sense indeed it is an offshoot, that is to say, as it appears upon a map, unto a casual observer; but when we come to examine the physical history of the two ranges, to compare the directions of the elevatory forces which gave them birth, and their relations to the Permian and Triassic formations where they come in contact with them, we find that such a view is a most superficial one, and will not stand the test of examination.

The Pendle Range, in fact, proves on examination to be a series of elevations, a miniature mountain chain, distinct both in age and direction from the Penine. It is really older, and had been upheaved in a series of parallel foldings when as yet the Penine chain lay in the lap of futurity; for it is easy to show that the age of the upheaval of the Pendle Range is post-Carboniferous, and that of the Penine Chain, by a less direct but nearly conclusive line of reasoning, post-Permian. In other words, the former range represents that period of disturbance and denudation which ensued upon the close of the Carboniferous and before the Permian period, and the latter the period of disturbance and denudation which marks the termination of the Permian age and Palæozoic epoch. As regards the relative directions of the lines of elevation, it will be apparent, even from the study of the district described in this memoir, that they are altogether different; that of the Pendle Range lying in an E.N.E. direction, that of the Penine in a direction nearly due north and south along the line of "the great anticlinal fault," already described (page 88). As the question is a theoretical one, and such as I am precluded from entering upon here, I must content myself with referring the reader to my paper "On the relative "ages of the principal Physical Features of Lancashire and the adjoining "parts of Yorkshire," in which he will find the subject fully discussed.†
—E. H.

Faults, and Lie of the Rocks, in the Triassic Plain.

The area within the Preston quarter-sheet geologically divides itself into three distinct tracts, divided from each other by two great faults, a central portion consisting of a trough of Bunter Sandstone, (2) an eastern, composed of Carboniferous and Permian rocks, and (3) a western, occupied by Keuper marl. Both these faults are downthrows to the west, and both appear to be of later date than the deposition of the Keuper marls; that to the west is a prolongation in a northerly direction of the Scarisbrick Fault, which I described in a previous memoir. It runs by Tarleton, Bretherton, Much Hool, and Longton, to near the "Saddle Inn," north of Lea Road Station, where it is believed to leave the margin of the map, and to run in a north-north-

---

* From the Roman name which appears to have been applied to it, "Alpes Penini," "Outlines of the Geology of England and Wales," p. 365. (1822.)
† Quart. Journ. Geol. Soc., vol. 24, p. 323.

westerly direction towards a point near Garstang Church in 91 S.E. From the Saddle Inn, southwards, to Longton, it throws the Keuper marls against the Pebble Beds, while further south, as in the Scarisbrick district (90 S.E. and 89 S.W.), it throws them against the Upper Mottled Sandstone. In the latter district a well was made some years ago, at Scarisbrick Hall, which failed in obtaining water, no beds corresponding to the Waterstones having apparently been reached. In endeavouring to estimate the throw of the Scarisbrick Fault, it becomes an interesting question whether the beds met with in this well, of which Mr. Binney gives the following section,* were Triassic beds at all, the more so, that in the deep well at Fleetwood, which was also unsuccessful, the grit met with under the Keuper marls was without the usual water-bearing marl:—

	ft.	in.
Soil	3	0
Brown sand	4	6
Brown clay	1	7
Variegated marl	230	8½
White grit pebble	0	7½
Variegated marl	18	0
Blue loam	4	2
Blue grit list (flagstone?)	19	3½
Brown strong rock list	1	1
Limestone list and limestone shale	7	4½
Brown flint kernel (chert?)	1	7
Blue grit list	2	0
Brown open grit	1	10
Blue shale	6	8½
Brown flint (chert?) pebble	0	4
Blue shale	0	5
	306	2½

But though the top bed of the Lower Keuper Sandstone, in the adjoining district, is a soft thin-bedded fine-grained sandstone alternating with bands of purple marls, yet further to the south it consists of an extremely hard yellowish compact sandstone with many rounded quartz pebbles, at Waterloo and Orrell almost forming a conglomerate similar to that invariably found at the base of the Keuper Sandstone, and it is not impossible that this hard grit bed may be that occurring at the Scarisbrick and Fleetwood well-borings.

Between the former place and the Carboniferous rocks to the east at Bispham and Newborough occur two faults (the Scarisbrick and the Euxton or boundary fault), both of which are downthrows to the west, the first being between the Red Marls and the Upper Mottled Sandstone, and the second between the latter and the Lower Coal-measures and Millstone Grit, forming the westerly margin of the Lancashire coalfield as far as the river Ribble, to Alston Hall, Skillaw Clough, Euxton, and at Roach Bridge cuts off the westerly extension of the Permian, which lies indiscriminately upon the denuded edges of various Carboniferous beds, one small outlier having been observed as far east as Hoghton, by Mr. Eccles, F.G.S., during the construction of the railway. It would appear that the Carboniferous deposits as a whole had received their south-easterly dip and suffered extensive denudation from the west and south-west before the deposition of the Permian strata, and that the faults with westerly downthrows running in a N.N.E. direction across this district not only came into existence after the Permian era, but long after the deposition of the Keuper Red Marls.

---

* Mem. Lit. Phil. Soc. Man., 2nd series, vol. xii., p. 250.

In regard to the Carboniferous strata which may underlie the Triassic country lying between Fleetwood and Ormskirk, it is probable that the beds which were there denuded before the Permian era must have belonged to the Millstone Grit, if not to the Yoredale series, for at several points patches of the former still form the surface of the country between the Lower Coal-measures and the coal-field boundary fault, which fault would have the effect of throwing the adjoining and slightly older grits to a lower level, though not of course to a depth corresponding to the thickness of all the formations between the Upper Mottled Sandstone and the Rough Rock, but only between the former and the Permian, which is probably thin, having been itself denuded before the deposit of the Lower Bunter Sandstones. Similarly the amount of the downthrow of the Scarisbrick fault, as affecting the hidden Lower Carboniferous rocks beneath the Trias, is probably at the most only the thickness of the beds between the top of the Red Marls and the bottom of the Upper Mottled Sandstone, supposing the latter to be present, and the beds between it and the bottom of the Red Marls.

Estimating the thickness of the Trias and Permian beds as follows :—

Keuper marls	1,500 feet.
,, sandstone	400
Upper mottled sandstone	600
Pebble beds	800
Lower mottled sandstone	100
Permian	100

the downthrow to the west of the coal-field boundary fault will be (estimating the Upper Mottled Sandstone at 50 feet), about 350 yards, that of the Scarisbrick fault about 716 yards. Supposing the Lower Bunter and Permian beds to be present, the Carboniferous beds would be 350 yards further from the surface than if they were absent. In this case the Carboniferous would be higher in the series than in the former ; but even if this tract stood higher than the adjoining country to the east, the extent of old denudation would strongly militate against the beds being sufficiently high to contain Coal-measures, nor does the described character of the Scarisbrick beds beneath the Keuper Marls strengthen such a view, as they appear to have a Lower Carboniferous facies.

The amount of westerly downthrow of the Euxton boundary fault appears to diminish northwards, as it is probably less at Roach Bridge, where the Permian has the same general dip as the Pebble Beds, than at Euxton and the country to the south towards Skillaw Clough. In the district north of Preston, between Garstang Junction and Garstang town, a new section has been recently exposed in the construction of the new railway to Knot End, Fleetwood, in which a considerable thickness of soft red current-bedded sandstone of Permian is seen, dipping to the west and north-west; and several other sections occur to the north, towards Morecambe Bay, and I also observed the red sandstone Permian rock cropping beneath the sands of that bay some distance from the coast.—C. E. R.

CHAPTER XI.

IGNEOUS ROCKS.

Trap Dykes.

All the rocks about which we have hitherto treated are sedimentary. The igneous rocks in this district are but poorly represented, indeed they were not known to exist before the survey commenced its investigations here.

In the summer of 1864 I found a dyke of dolerite, amygdaloidal at the edges, crossing two brooks above Caton near Lancaster. The black Carboniferous shales, a fissure in which it filled up, seemed to be hardened to some extent, but not otherwise much altered, for fossils in good preservation were found within two inches of the greenstone. I must defer a full description of this for a future memoir on that district. Meanwhile I may refer anyone seeking further information to a paper read by my friend, Mr. James Eccles, F.G.S., to the Manchester Geological Society in 1869, and published in their proceedings.*

I found another in the district now being described, three-quarters of a mile N. by W. from Grindleton, just where Grindleton Brook is crossed by the lane to Newton. Below the bridge is a small waterfall called Calf Lumb. This is formed by the dyke, which consists of dolerite. It is 10 feet in width, and might perhaps be profitably worked for road metal, or for stone "sets." It crosses the stream in a direction W. 25 N.; the rocks are much disturbed and contorted all about there, but there did not appear, so far as I could see, any shift of the beds along the fissure. It is only visible in the stream and on its banks, and cannot be traced in either direction.

Mr. Eccles and I found on a rubbish heap at the Whitewell Lead Mine† a fragment of basalt. The miners do not seem to be aware of having come across any such rock in their excavations. It is noteworthy that this locality is nearly in a straight line between the two dykes already mentioned, and not far out of the direction in which they appear to strike.

I found a fragment of dolerite on the surface near White Hill Ho, 2 miles N.W. of Tosside Chapel. It was too angular and unworn to have been an erratic, and it is very probable that there may be a dyke in that neighbourhood.

In a little brook near the village of Inglewhite, north of Preston, I saw what appeared to be a dyke, but there was too much water at the time of my visit for me to examine it carefully and come to a definite conclusion on the subject.

These are all the localities in Lancashire, east of the Sands, where I have as yet met with any traces of igneous action.

These W.N.Wly. dykes are probably of the same age as those of Scotland, considered by Prof. Archibald Geikie to be of the Miocene period, being similar both in direction and composition. If so, they are newer than any other formations in the North of England, except the superficial drifts.—R. H. T.

CHAPTER XII.

MINERALS, METAL-MINES, &c.

Minerals.—Although traces of several metallic ores occur in the district of Blackburn and Burnley, they are not found in sufficient quantity for profitable working. Barytes is not unfrequently to be found in veins and fissures of the Millstone Grits, and amongst other localities the following may be mentioned:—Lead Mine Clough, on Blackstone Edge Moor; amongst rocks at Little Hoar Edge; quarry on the eastern side of Counting Hill; quarry on Billington Moor near

---

* On Two Dykes recently found in North Lancashire. Trans. Man. Geol. Soc., vol. ix. (No. 1), pp. 26–8.
† or Brennand Mine.

Whalley; quarry at Stopping Hey near Ribchester Station; old Lead Mines on Anglezark Moor near Chorley.

Lead was formerly worked at several spots on Anglezark Moor. None of the pits appear to have been deep or the workings extensive. The principal mine occurs near the banks of the stream above Alance Bridge, near the large north and south fault. Mr. Standish of Duxbury Hall showed me some good specimens of galena from this mine. Another spot on which lead was worked occurs on the moors above High Bulloch. In the stuff thrown out from the old workings I found traces of galena, barytes, copper pyrites, and probably carbonate of lead. Several old mining heaps and traces of adits may be found along the dell of Grange Water, and one of the lodes which is in the line of a fault may be observed below Coppice Stile House.*

The most productive lode in the district seems to have been that which corresponds with the Thieveley fault already described (p. 85.)

The clay-ironstones of the Coal-measures are generally poor both in thickness and quality. The best beds of this mineral I have seen are unquestionably those in the Millstone Grit shales below Salmesbury in the river Darwen.—E. H.

Lead Mines of the Ribble Valley.

A memoir on the Geology of this district would hardly be complete without some mention of the lead mines of Skeleron, or Skelhorn, near Rimington. They were much worked in the reign of Queen Elizabeth by William Pudsay, Lord of Bolton-in-Bowland, who is said to have extracted silver from the lead which he obtained here at the rate of 26 lbs. per ton.† There are at least six distinct veins shown by the old workings. One runs N. 10 E., the remainder range between N. 10 W. and W. 35 N., and to speak in general terms are parallel to a great fault with a north-westerly direction, a little west of Stopper Lane, towards and beyond Rimington. Of this I shall speak again. Another vein containing lead and baryta lies on the S.W. side of the brook, and has a S.S.E. trend. These veins contain argentiferous galena, calamine, sulphate of baryta, carbonate of lime, and other minerals. Some of the limestones have that crystalline condition and brown colour which constitute what is called Dun Limestone. The oldest workings were not conducted by level, but entirely by shafts sunk from the surface.

An attempt to re-open these mines was made about 40 years ago. A level was carried up from the beck by the side of the old smelting house, about 380 yards to the N.E. towards Stopper Lane, but I believe without very great success. Still later, about 1867-8, a shaft was sunk in a field east of Stopper Lane, with the expectation of reaching the lode, but it was begun too far to the east, and ran entirely through micaceous clayey shale on the other side of the fault.

Mention has already been made of the fault running to the N.W., which would in mining phraseology be the lode in connection with these veins. It is interesting to see along nearly its whole length how it is associated with indications of mineral deposits. In the fault itself, where

---

* The mines on Anglezark Moor are mentioned in Conybeare and Phillips "Outlines of Geology of England and Wales," as copper mines producing carbonate of barytes; but though copper does occur in homœopathic quantities, lead and barytes were the chief minerals extracted.

† In "Ballads founded on Craven Legends" is the following note: "Webster (Johannes Hyphantes), in his Metallographia, says, that the Pudsay shillings were marked with an escallop, which was the Mint-mark of 1584-6, and adds that the lead ore in Skelhorn contains silver after the rate of 26 lbs. per ton."

it crosses the little brook coming down from Lower Gills, I found some little specks of galena and iron pyrites. Then we come further N.W. to the rich deposits at Skeleron on both sides of Ings Beck. Further down that beck, half a mile S.W. of the fault, is a baryta mine, the vein of which runs N. 43 W. On the left of the high road from Sawley to Gisburn, near Sawley Grange, is a limestone quarry, and in it a lead vein may be seen. It contains regular octohedral crystals (the less common form) of galena, several of which I obtained very perfect. Its direction is N. 10 W., and it appears to be running towards the fault which lies, as well as one may judge, about 200 yards N.E. of the quarry.

We next see the fault crossing the Ribble below the archery ground, at Bolton Hall, by Briery Bank Wood; it there contains a deposit of iron pyrites.

The fault is not seen again for two miles or more, the valley and its sides being covered up by superficial deposits, but several attempts at mining have been made along the west side of the Bolton Valley by Fat Hill, below Higher Heights, and in other places. We find a fault which is probably the same, but with a reversed throw (the downthrow being to the N.E. instead of to the S.W.) in Holden Clough at the Waterfall. Hard by are the Victoria Lead Mines. Three drifts or levels have been carried from the brook along veins in a S.Ely. direction. The nearest to the Waterfall was driven for 90 yards before coming to the fault, and then 20 yards along it, and I believe no lead worth speaking of was found in it. The next two were carried along through a bed of encrinital limestone 1 ft. 6 in. to 2 ft. thick, and lead was found along the cheeks of this limestone, but not in great abundance. The lead died out immediately on coming to the shale above and below, as is generally the case. Had this limestone been thicker, no doubt a better return would have accrued to the labour expended. But the surfaces for the deposit of lead being already very narrow, a maximum of about two feet, they were still further curtailed by there being a small fault along the middle of the vein, which shifted the beds a few inches, and threw limestone against shale above and below. On examining the brook section I found that this same encrinital limestone occurred higher up Fell Brook, but on the other side of the fault, and that the thick limestone which forms the Waterfall lies below it by about 60 or 70 feet. This is about the depth that the mine would have to be sunk on the vein to come to a thick limestone which *might*, but would not necessarily, contain a larger store of the metal. The encrinital limestone may be seen, as mentioned, higher up, at a southern curve of Fell Brook; a small lead vein crosses it running S.S.E. A miner extracted 2 or 3 cwt. of lead from it here.

St. Hubert Vein.—Four or five years ago a lead vein, was found crossing Far Field Brook, east of Harrop Fold. It was found on the west side, and a drift was carried along it on the opposite bank. It ran for about 50 fathoms from the brook in a direction E. 15 S., then turned at an angle to E. 40 S., a course which, with minor undulations, it held for 68 fathoms further, which was the extent of the workings at the time of my visit. I believe they have since been stopped. This vein was high up in the Bowland shales, and not very productive. The next hard rocks beneath it are the sandstones of the Lower Yoredale Grit. If we take 7° as the lowest average of dip from the top of those beds, where the brook crosses Smalden Lane to the mine, we find that there is probably a thickness of 250 feet of these shales below the mine to these sandstones, and then it is by no means certain that the vein would bear ore.

Two veins were worked many years ago below the Waterfall in a brook running down from Harris's Lathe, near Roddel Chapel. I was unable to procure much information about them. They were in rocks close under the Lower Yoredale Grit, and therefore the bearing beds were probably the same as those at the Victoria Mine, Holden Clough.

At Lane-side, a farm N.N.W. of Grindleton, some large nuggets of lead ore were found in draining, on the left bank of the brook, a few years ago. A trench was cut with a view to finding the vein from which they had come, but to my mind not in the direction most likely to give that result. Considering the disturbed condition of the rocks here, and the dyke hard by, in connection with the probable existence of a trap dyke at the Brennand Mine along the same line of country (*see* p. 172), I think it very likely that a search properly conducted might bring to light a good vein.

Fifty yards west of the fault, which brings down the Permian Sandstones against the Limestone below Waddow Hall, in the bed of the stream, may be seen a small vein containing galena, and "heavy spar." It lies in the Permian Sandstone, and is of interest as occurring in rock of Post-Carboniferous age. A man, by name Furness, many years ago drove a level along it from the north bank of the river in sandstone all the way.

It is a matter for consideration whether it might not be worth while to sink a shaft on the vein through the sandstone to the underlying limestones and shales, which are probably not far below; but no doubt the difficulty of keeping out the water is an important element in the problem.

South of Nick of Pendle a fault crosses the Pendle Range with a W.N.Wly. direction from Sabden to Audley Reservoir. At the top of the ridge above Parsley Barn, and about 500 yards from the Nick, are some rubbish-heaps, the old workings of a lead vein. Apparently the vein is parallel to the fault, about W. 20 N. From all accounts the vein was not a rich one.

A mile and a quarter south of Gisburn, east of the road to Colne, a vein of lead-ore has been worked at some time below Todber Nursery, From the old workings it seems that the vein ran N. 40 W. for 50 yards, and then W.N.W. for 130 more, down to the stream. I was unable to obtain any information about it. It probably has some connection with the Great Barnoldswick Fault.—R. H. T.

For other notices of minerals, see pp. 29, 32, 47-51.—W. G.

Fire-clays, &c.

Clay for fire-bricks and pottery is worked at Darwen in the Cliviger Valley, in Dulesgate, at Littleborough, and at Cherry Tree and Great Harwood near Blackburn.—E. H.

Fire-clays are also largely worked at the Townley Collieries, Burnley, where they are manufactured into drain, water, and sewer-pipes, as well as into fire-bricks for lining furnace floors. The clay, which is the under-clay of the mountain mines, is full of ferns, and other vegetable matter; is brought out to the surface and left to temper in the air and rain some time before it is used; that which is required for specially resisting great heat is burnt before being placed in the pug-mill.

Clays associated with coals on a similar horizon, were until lately worked at Charnock Richard, near Coppul, west of Chorley. Fire-clay is raised to the surface by an incline shaft at Eller Beck, near Adlington; and at Brinscall, on the Chorley and Blackburn Railway, a bed five feet thick is worked.—C. E. R.

APPENDIX I.

The accompanying list of Fossils has been prepared in the Palæontological Department from collections made in the field by Mr. Tiddeman whilst surveying portions of the area embraced in this Memoir. The Geological Survey is much indebted to Mr. James Eccles for the use of his fine collection of Carboniferous Fossils, from which to catalogue the names of species collected in the same area, without which the list would have been very incomplete.

I have also made extensive use of those species named by Prof. J. Phillips, in his "Geology of Yorkshire," where the localities coincided.

These three sources combined have enabled me to construct the following Zoological Table of the Fauna occurring in the district.

R. ETHERIDGE.

SPECIES.		Carboniferous Limestone.														
		Clitheroe.	Bollard.	Waddow Hall.	Chatburn.	Coplow, Clitheroe.	Raw-Cause, Clitheroe.	Withcll.	Salt Hill, Clitheroe.	Worston, Clitheroe.	Downham.	Twiston.	Ravensholme.	Wybrowy, Bolton in Bowland.	Blue Butts, Slaidburn.	Great Dunnow, Slaidburn.
1. Calamites	"			+	+	+	+	+				+				
2. Amplexus coralloides	Sby.			+								+				
3. Cyathophyllum regium	Phil.	+														
4. " crenulare	Phil.			+							+					
5. " sp.																
6. Cyathopsis fungites	Fleme.		+													
7. Favosites dentifera	Phil.		+													
8. " incrustans	Phil.		+													
9. " parasitica	Phil.		+													
10. Michelinia tenuisepta-	Phil.		+													
11. Lonsdaleia floriforme	Mart.		+													
12. Lithostration ensifer	M'Edw.															
13. Michelinia megastoma	Phil.		+													
14. Syringopora geniculata	Phil.	+														
15. " ramulosa	Goldf.		+													
16. " reticulata	Goldf.					+										
17. Zaphrentis sp.																
18. Actinocrinus atlas	M'Coy.		+	+			+		+	+						
19. " bursa	Phil.	+	+													
20. " Gilbertsoni	Mil.	+	+													
21. " globosus	Phil.		+													
22. " mammillaris	Phil.		+													
23. " polydactylus	Mill.	+	+				+			+						
24. " tessellatus	Phil.	+														
25. " triacontadactylus	Mill.	+	+	+			+		+			+	+		+	
26. " sp.								+		+					+	
27. Archæocidaris triserialis	M'Coy.								+							
28. " urei	Flem.															
29. " sp. spines and plates				+		+										
30. Cladocrinus sp.						+										
31. Codonaster acutus	M'Coy.															
32. Cyathocrinus calcaratus	Phil.		+													
33. " conicus	Phil.	+	+													
34. " distortus	Phil.		+													
35. " mammillaris	Phil.		+	+												
36. " ornatus	Phil.	+	+													
37. Dichocrinus elongatus	Phil.		+													
38. Eurycrinus concavus	Phil.		+													
39. Forbesiocrinus nobilis	De Kon.		+													
40. Pentremites acutus	Sby.		+													
41. " angulatus	Sby.		+									+				
42. " Derbiensis	Sby.											+				
43. " ellipticus	Sby.		+													
44. " inflatus	Sby.		+													
45. " oblongus	Sby.		+													
46. " pentagonalis	Sby.															
47. " orbicularis	Sby.		+									+				
48. " sp.					+											
49. Palechinus gigas	M'Coy.	+														
50. " sp. plate						+		+			+					
51. Platycrinus contractus	Phil.		+													
52. " ellipticus	Phil.		+													
53. " gigas	Phil.		+													
54. " granulatus	Mull.		+													

APPENDIX 1.—FOSSILS.

	SHALES-WITH-LIMESTONES.							BOWLAND SHALES.				
	Ashnot, Knowlmere.	Brockhall, Ribble.	Primrose Print Works.	Worsaw End House.	Hodder Bridge, Ribble.	Bashall Brook, Clitheroe.	Gisburn.	Forest Becks.	Fat Hill, Holden.	Harrowfield, Knowlmere.	Dinckley Hall, Ribble.	Wiswell, Clitheroe.	Near Stanch n Hall.	Beck above Little Mearley Hall.
1.		·	·	·	·	·	·		·		·	·	·	·
2.	+													
3.														
4.														
5.	·	·	+											
6.														
7.														
8.														
9.														
10.														
11.														
12.														
13.														
14.														
15.														
16.														
17.	·	·	·	+										
18.														
19.														
20.														
21.														
22.														
23.														
24.														
25.	·	·	·	·	·	·	·		;	+				
26.														
27.														
28.														
29.														
30.														
31.														
32.														
33.														
34.														
35.														
36.														
37.														
38.														
39.														
40.														
41.														
42.														
43.														
44.														
45.														
46.														
47.														
48.														
49.														
50.														
51.														
52.														
53.														
54.														

Species.		Clitheroe.	Bellman.	Waddow Hall.	Chatburn.	Coplow, Clitheroe.	Race Course, Clitheroe.	Withgill.	Salt Hill, Clitheroe.	Worston, Clitheroe.	Downham.	Twiston.	Ravenholme.	Wybersey, Bolton in Bowland.	Blue Butts, Slaidburn.	Great Dunnow, Slaidburn.
1. Platycrinus laciniatus	Phil.		+													
2. " laevis	Mill.		+													
3. " (stem)				+												
4. " megastylus	Phil.	+	+													
5. " microstylus	Phil.		+													
6. " plicatus	Goldf.	+														
7. " ? tuberculatus	Mill.		+													
8. " rugosus	Mill.	+	+								+					
9. Poteriocrinus quinquangularis	Mill.															
10. " conicus	Phil.		+													
11. " crassus	Mill.								+		+					
12. " granulosus	Phil.		+													
13. " sp.																
14. Rhodocrinus calcaratus	Phil.		+													
15. " stem									+		+					
16. Synbathocrinus conicus	Phil.		+													
17. Serpula hexicarinata	M'Coy.	+														
18. Brachymetopus ouralicus ?	Vern.	+														
19. " sp.					+	+										
20. Endomychus scouleri	M'Coy.	+			+		+			+						
21. Griffithides globiceps	Port.															
22. Phillipsia Derbiensis	Mart.	+	+													
23. " Bronnartii	Fisch.		+													
24. " gemmulifera	Phil.		+													
25. " seminifera	Phil.															
26. " truncata	Phil.					+										
27. " sp.																
28. Ceriopora rhombifera	Phil.		+													
29. Fenestella crassa	M'Coy.													+		
30. " cuneida	M'Coy.				+											
31. " flabellata	Phil.		+													
32. " membranacea	Phil.															
33. " Morrisii	M'Coy.	+														
34. " plebeia ?	M'Coy.							+						+	+	
35. " multiporata	M'Coy.													+		
36. " quadridecimalis	M'Coy.	+									+		+	+	+	
37. " sp.																
38. Glauconome sp.																
39. " pluma	Phil.	+	+													
40. Gorgonia ?																
41. Polypora verrucosa	M'Coy.	+														
42. Ptilopora flustriformis	Phil.		+													
43. " nodulosa	Phil.		+													
44. " undulata	Phil.															
45. Retepora undata	M'Coy.		+											+		
46. Athyris ambigua	Sby.		+													
47. " expansa	Phil.		+													
48. " lamellosa	Lev.					+										
49. " planosulcata	Phil.		+													
50. " Royssi	Lev.	+	+			+		+					+	+	
51. " sp.						+			+		+			+	+	
52. " globularis	Phil.															
53. Camarophoria crumena	Mart.	+														
54. " globulina	Phil.		+													
55. Choneies Dalmanniana	D'Kon.					+										
56. " Hardrensis	Phil.					+							+			
57. " papilionacea	Phil.		+													
58. " sp.								+								
59. Crania sp.																
60. Lingula squamiformis	Phil.		+													
61. Meristella tumida	Dalm.		+													
62. Orthis antiquata	Sby.		+													
63. " michelini	Mart.							+		+		+				
64. " resupinata	Lev.	+	+	+	+	+		+			+		+			+
65. Productus aculeatus	Mart.	+									+					
66. " arenarius ? ?		+														
67. " Cora	D'Orb.	+	+					+			+					
68. " costatus	Sby.		+													
69. " fimbriatus	Sby.	+	+													
70. " giganteus	Mart.				+											
71. " humerosus	Sby.															
72. " longispinus	Sby.	+														

APPENDIX I.—FOSSILS.

	SHALES-WITH-LIMESTONES.						BOWLAND SHALES.								MILLSTONE GRIT SHALES.		
	Ashnot, Knowlmere.	Brockhall, Ribble.	Primrose Print Works.	Worsaw End House.	Hodder Bridge, Ribble.	Bashall Brook, Clitheroe.	Gisburn.	Forest Becks.	Pat Hill, Holden.	Harrowfield, Knowlmere.	Dinckley Hall, Ribble.	Wiswell, Clitheroe.	Near Standen Hall.	Beck above Little Mearley Hall.	Fellside, Slaidburn.	Champion.	Sales Wheel.	Root.	Alston Hall.
1.																			
2.																			
3.																			
4.																			
5.																			
6.																			
7.																			
8.																			
9.																			
10.																			
11.																			
12.																			
13.	.	.	+																
14.																			
15.																			
16.																			
17.																			
18.																			
19.					+														
20.	+	.	.																
21.																			
22.																			
23.																			
24.	.	.	.		.			+											
25.																			
26.	+																		
27.	.	.			+														
28.																			
29.																			
30.																			
31.																			
32.																			
33.																			
34.																			
35.																			
36.																			
37.	+																		
38.	.		+																
39.																			
40.																			
41.																			
42.																			
43.																			
44.																			
45.																			
46.																			
47.																			
48.																			
49.																			
50.	+																		
51.																			
52.																			
53.																			
54.																			
55.																			
56.	.	.	.	.			+												
57.																			
58.	.	.	.	.	+														
59.																			
60.																			
61.																			
62.																			
63.	+	.	.	.	+														
64.																			
65.																			
66.																			
67.																			
68.																			
69.																			
70.																			
71.																			
72.																			

Species.		Clitheroe.	Bolland.	Waddow Hall.	Chatburn.	Coplow, Clitheroe.	Race Course, Clitheroe.	Withgill.	Salt Hill, Clitheroe.	Worston, Clitheroe.	Dornham.	Twiston.	Ravensholme.	Wyberset, Bolton in Bowland.	Blue Butts, Slaidburn.	Great Dunnow, Slaidburn.
						Carboniferous Limestone.										
1. Productus mesolobus	Phil.	+	+			+										+
2. ,, margaritaceus?	Phil.		+													
3. ,, muricatus	Phil.								+			+				
4. ,, punctatus	Mart.	+	+					+				+				
5. ,, pustulosus	Phil.	+	+		+		+	+		+		+				
6. ,, semireticulatus	Mart.	+	+					+		+	+	+			+	+
7. ,, scabriculus	Mart.	+	+				+	+				+			+	
8. ,, spinulosus	Sby.	+	+		+											
9. ,, striatus	Fisch.		+									+				
10. ,, sublævis	Kon.		+													+
11. ,, undatus	Def.	+										+				
12. Retzia radialis	Phil.															
13. Rhynchonella acuminata	Mart.	+	+													
14. ,, acuminata var mesogonia					+			+								
15. ,, cordiformis	Sby.							+		+						
16. ,, flexistria	Phil.	+	+									+				
17. ,, pleurodon	Phil.	+	+	+				+				+				+
18. ,, pugnus	Mart.	+	+					+				+				
19. ,, proava	Phil.		+													
20. ,, reniformis	Sby.		+				+									
21. Spirifera bisulcata	Sby.														+	
22. ,, convoluta	Phil.	+	+					+				+			+	
23. ,, cuspidata	Mart.	+	+		+										+	
24. ,, crassa	D. Kon.															
25. ,, distans	Sby.	+	+													
26. ,, duplicicosta	Phil.	+	+													
27. ,, fusiformis	Phil.		+													
28. ,, glabra	Mart.	+	+		+	+		+				+				
29. ,, globularis	Phil.															
30. ,, integricostata	Phil.		+													
31. ,, lineata	Mart.		+		+			+		+		+				
32. ,, ovalis	Phil.		+					+								
33. ,, planata	Phil.		+													
34. ,, pinguis	Sby.	+	+									+				
35. ,, rhomboidea	Phil.	+	+									+				
36. ,, striata	Mart.	+	+	+	+					+		+		+	+	+
37. ,, triangularis	Mart.		+													
38. ,, trirndialis	Phil.															
39. Spiriferina cristata		+														
40. ,, insculpta	Phil.		+			+										
41. Streptorhynchus crenistria	Phil.		+					+	+						+	+
42. Strophomena analoga	Phil.		+	+				+							+	
43. Terebratula hastata	Sby.	+	+				+	+							+	
44. ,, sacculus	Mart.							+								
45. ,, sp.																
46. Avicula cyeloptera	Phil.															
47. ,, lunulata	Phil.		+													
48. ,, squamosa	Phil.		+													
49. ,, laminosa	Phil.		+													
50. ,, lævigata?	M'Coy															
51. Aviculopecten circularis	Kon.					+										
52. ,, concentricus	M'Coy															
53. ,, margaritoides	M'Coy									+		+				
54. ,, hemisphericus	Phil.		+													
55. ,, tessellatus	Phil.	+	+													
56. ,, ellipticus	Phil.															
57. ,, dissimilis	Flemg.															
58. ,, arenosus	Phil.		+													
59. ,, interstitialis	Phil.		+													
60. ,, papyraceus	Sby.		+													
61. ,, granosus	Sby.		+													
62. ,, radiatus	Phil.		+													
63. ,, sp.															+	
64. Posidonomya Becheri	Gold.	+														
65. ,, Gibbsoni	Brown															
66. ,, vetusta	Sby.		+													
67. ,, sp.																
68. Pterinea sp.																
69. Pinna costata	Phil.														+	
70. ,, flabelliformis	Mart.		+													
71. ,, flexicostata	M'Coy	+														
72. ,, sp.					+										+	

APPENDIX I.—FOSSILS.

	Shales-with-Limestones								Bowland Shales								Millstone Grit Shales			
	Ashnot, Knowlmere.	Brook Hall, Ribble.	Primrose Print Works.	Worsaw End House.	Hodder Bridge, Ribble.	Bashall Brook, Clitheroe.	Gisburn.	Forest Becks.	Pat Hill, Holden.	Harrowfield, Knowlmere.	Dinckley Hall, Ribble.	Wiswell, Clitheroe.	Near Standen Hall.	Beck above Little Mearley Hall.	Fellside, Slaidburn.	Champion.	Sales Wheel.	Root.	Alston Hall.
1.	+																		
2.	++																		
3.																			
4.																			
5.	+	+																	
6.			+																
7.																			
8.																			
9.																			
10.																			
11.																			
12.																			
13.																			
14.	+																		
15.																			
16.																			
17.	+																		
18.	++																		
19.																			
20.																			
21.	.	+																	
22.																			
23.																			
24.																			
25.																			
26.																			
27.																			
28.	+	+																	
29.																			
30.																			
31.	+																		
32.																			
33.																			
34.																			
35.																			
36.	+																		
37.																			
38.																			
39.																			
40.																			
41.	.	+	+																
42.																			
43.																			
44.																			
45.																			
46.	+																		
47.																			
48.																			
49.																			
50.																			
51.																			
52.																			
53.																			
54.	+																		
55.																			
56.																			
57.																			
58.																			
59.																			
60.																			
61.																			
62.																			
63.																			
64.																+	+		
65.													.	+	+	.	+	.	
66.		.	.	.	.	.	.	.				.	.	+	+	.	+	.	
67.		.	.	.	.	.	.	.			++	.	.	+	.	.	.	.	
68.		.	.	.	.	.	.	.						.	.		.	.	
69.																			+
70.																			
71.																			
72.																			

GEOLOGY OF THE BURNLEY COALFIELD, ETC.

SPECIES.		Clitheroe.	Bolland.	Wadlow Hall.	Chatburn.	Coplow, Clitheroe.	Raws Course,Clitheroe.	Withcill.	Salt Hill, Clitheroe.	Winston, Clitheroe.	Downham.	Twiston.	Ravensholme.	W'rhersey; Bolton in Bowland.	Blue Butts, Shaldburn.	Great Dunnow, Shaldburn.
				Carboniferous Limestone.												
1. Modiola clongata	Phil.		+													
2. „ squamifera	Phil.		+													
3. „ granulosa	Phil.		+													
4. Myalina sp.																
5. Myalina lamellosa	Kon.							+								
6. Gervillia lunulata ?	Phil.															
7. Cucullæa obtusa	Phil.		+					+								
8. „ arguta	Phil.		+													
9. „ allied to M'Coyanum								+								+
10. Pullastra elliptica	Phil.										+					
11. Concardium minax	Phil.		+								+					
12. „ rostratum	Mart.		+													
13. „ trigonalis	Phil.		+													
14. „ koninckii	Baily.	+						+								
15. „ sp.			+													
16. Ctenodonta cuneata	Phil.		+													
17. „ attenuata			+													
18. „ gibbosa	Flem.		+													
19. „ undulata	Phil.		+													
20. „ claviformis	Phil.		+													
21. „ luciniformis	Phil.		+													
22. „ tumida	Phil.															
23. Sanguinolites angustatus	Phil.		+													
24. Cardiomorpha oblonga	Sby.		+													
25. „ laminata	Phil.		+													
26. „ ornata			+								+					
27. Cypricardia globrata	Phil.															
28. „ parallela	Phil.															
29. „ rhombea	Phil.		+													
30. Edmondia sulcata	Phil.							+				+				
31. „ oblonga								+								
32. „ orbicularis	M'Coy		+								+					
33. „ unioniformis	Phil.		+					+								
34. Arca sp.																
35. Myacites tumidus	Phil.		+													
36. Leptodomus senilis	Phil.		+													
37. Lutraria prisca	M'Coy							+								
38. Acroculia trilobus	Phil.		+									+				
39. „ tubifer	Sby.		+													
40. „ striatus	Phil.		+													
41. „ neritoides	Phil.		+													
42. „ vetustus ?	Sby.		+	+		+					+	+				
43. „ angustus	Phil.		+													
44. Metoptoma elliptica	Phil.		+									+				
45. „ pileus	Phil.		+													
46. „ oblonga	Phil.		+													
47. „ imbricata	Phil.		+													
48. „ sulcata	Phil.		+													
49. „ trilobata	Phil.		+													
50. „ sp.							+		+		+					
51. Cirrus acutus	Sby.		+													
52. Murchisonia angulata	Phil.															
53. „ fusiformis	Phil.		+													
54. „ vittata	Phil.		+													
55. Loxonema sinuosa	Phil.															
56. „ constricta	Sby.		+													
57. „ sulculosa	Phil.		+													
58. „ scaluroidea	Phil.		+													
59. „ tumida	Phil.		+													
60. Macrocheilus parallelus	Phil.		+													
61. „ acutus	Phil.		+													
62. „ curvilineus	Phil.		+													
63. „ rectilineus	Phil.	+	+													
64. „ imbricatus	Sby.		+										+			
65. „ sigmalineus	Phil.		+													
66. „ globularis	Phil.		+													
67. Natica ampliata	Phil.		+													
68. „ lirata	Phil.		+													
69. „ elliptica	Phil.		+													
70. „ planispira	Phil.		+													
71. „ variata	Phil.		+													
72. „ plicistria	Phil.		+					+	+							

APPENDIX I.—FOSSILS.

	SHALES-WITH-LIMESTONES.						BOWLAND SHALES.							MILLSTONE GRIT SHALES.		
	Ashnot Knowlmere.	Brock Hall, Ribble.	Primrose Print Works.	Worsaw End House.	Hodder Bridge, Ribble.	Bashall Brook, Clitheroe.	Gisburn.	Forest Beds.	Fat Hull, Holden.	Harrowfield, Knowlmere.	Dinckley Hall.	Wiswell, Clitheroe.	Near Stansfield Hall.	Beck above Little Mearley Hall.	Fellside, Slaidburn.	Champion.	Sales Wheel.	Root.	Alston Hall.
1.																			
2.																			
3.																			
4.																			
5.	+	.				.			.					.	.		+	+	
6.																			
7.																			
8.																			
9.																			
10.																			
11.																			
12.																			
13.																			
14.																			
15.																			
16.	+	.	.	.	.	.		+											
17.																			
18.																			
19.																			
20.																			
21.																			
22.																			
23.																			
24.	.	+																	
25.																			
26.																			
27.																			
28.																			
29.																			
30.																			
31.																			
32.																			
33.																			
34.																			
35.																			
36.																			
37.																			
38.																			
39.																			
40.																			
41.																			
42.																			
43.																			
44.																			
45.																			
46.																			
47.																			
48.																			
49.	+																		
50.																			
51.																			
52.																			
53.																			
54.																			
55.																			
56.	+																		
57.																			
58.																			
59.																			
60.																			
61.																			
62.																			
63.																			
64.																			
65.																			
66.																			
67.																			
68.																			
69.																			
70.																			
71.																			
72.																			

GEOLOGY OF THE BURNLEY COALFIELD, ETC.

SPECIES.		Clitheroe.	Bolland.	Waddow Hall.	Clauburn.	Coplow, Clitheroe.	Race Course, Clitheroe.	Withgill.	Salt Hill, Clitheroe.	Worston, Clitheroe.	Downham.	Twiston.	Ravensholme.	Wyloscey, Bolton in Bowland.	Blue Butts, Slaidburn.	Great Dunnow, Slaidburn.
						Carboniferous Limestone.										
1. Natica elongata	Phil.	.	+													
2. ,, tabulata	Phil.	+	+													
3. ,, sp.		.	.													
4. ,, elliptica	Phil.	.	.													
5. Patella scutiformis	Phil.	.	+													
6. ,, sinuosa	Phil.	.	+													
7. ,, mucronata	Phil.	.	+													
8. ,, curvata	Phil.	.	+													
9. ,, retrorsa	Phil.	.	+													
10. ,, lateralis	Phil.	.	+													
11. Turbo tiara	Phil.	.	+													
12. ,, semisulcatus	Phil.	.	+													
13. ,, biserialis	Phil.	.	+													
14. Turritella suturalis	Phil.	.	+													
15. Pleurotomaria atomaria	Phil.	.	+													
16. ,, abdita	Phil.	.	+													
17. ,, acuta	Phil.	.	+													
18. ,, carinata	Sby.	.	+													
19. ,, crenato-striata	Sandb.	.	.													
20. ,, conica	Phil.	.	+													
21. ,, depressa	Phil.	.	+													
22. ,, gemmulifera	Phil.	.	+													
23. ,, inconspicua	Phil.	.	+													
24. ,, interstrialis	Phil.	.	+													
25. ,, expansa	Phil.	.	+													
26. ,, lirata	Phil.	.	+													
27. ,, limbata	Phil.	.	+													
28. ,, excavata	Phil.	.	+													
29. ,, monilifera	Phil.	.	+													
30. ,, strialis	Phil.	.	+													
31. ,, sculpta	Phil.	.	+													
32. ,, squamula	Phil.	.	+													
33. ,, sulcata	Phil.	.	+													
34. ,, sulcatula	Phil.	.	+													
35. ,, spiralis	Phil.	.	+										+			
36. ,, suturalis (Eulima)	Phil.	.	+													
37. ,, undulata	Phil.	.	+													
38. ,, tornatilis	Phil.	.	+													
39. ,, vittata	Phil.	.	+													
40. ,, rotundata	Sby.	.	+													
41. ,, tumida	Phil.	.	+													
42. ,, sp.		.	.													
43. Platyschisma glabrata	Phil.	.	+													
44. ,, helicoides	Phil.	.	+													
45. ,, ovoidea	Phil.	.	+		+											
46. Pleurothus cristatus	Phil.	.	+													
47. Euomphalus catillus	Sby.	.	.					+				+				+
48. ,, calyx	Phil.	.	+	+												
49. ,, pugilis	Phil.	.	+													
50. ,, pentangulatus	Sby.	.	+													
51. ,, pileopsideus	Phil.	.	+													
52. ,, pentagonalis	Phil.	.	+													
53. ,, tabulatus	Phil.	+	+													
54. ,, sp.		.	.													
55. Trochus Yvanii	Lév.	.	+													
56. Bellerophon costatus	Sby.	.	+							+						
57. ,, tangentialis	Phil.	.	+		.	+		.	.	+					
58. ,, hiulcus	Sby.	.	+		+	+		+								
59. ,, tenuifascia	Sby.	.	+					.								
60. ,, cornu-arietis	Sby.	.	.					•								
61. ,, sp.		.	.													
62. Porcellia Woodwardii	Sby.	.	+	+												
63. Dentalium ingens	D. Kon.	.	+													
64. Actinoceras giganteum	Sby.	.	.						.	+		+				
65. Discites carineferus	Sby.	.	.							+						
66. ,, tetragonus	Phil.	.	+							+		+				
67. ,, sp.		.	.		+											
68. Goniatites bilinguis	Salt.	+	+		+									+		
69. ,, Henslowii	Sby.	+														
70. ,, sphaericus	Mart.	+												+		
71. ,, truncatus	Phil.	+														
72. ,, striatus	Sby.	.	+													

APPENDIX I.—FOSSILS.

	SHALES-WITH-LIMESTONES.						BOWLAND SHALES.								MILLSTONE GRIT SHALES.		
	Ashnod, Knowlmere.	Brock Hall, Ribble.	Primrose Print Works.	Worsaw End House.	Hodder Bridge, Ribble.	Bashall Brook, Clitheroe.	Gisburn.	Forest Becks.	Fat Hill, Holden.	Harrowfield, Knowlmere.	Duckley Hall, Ribble.	Wiswell, Clitheroe.	Near Standen Hall.	Beck above Little Mearley Hall.	Fellside, Sladburn.	Champion.	Sales Wheel.	Root.	Alston Hall.

No.																			
1.																			
2.																			
3.	+																		
4.																			
5.																			
6.																			
7.																			
8.																			
9.																			
10.																			
11.																			
12.																			
13.																			
14.																			
15.																			
16.																			
17.																			
18.																			
19.			+																
20.																			
21.																			
22.																			
23.																			
24.																			
25.																			
26.																			
27.																			
28.																			
29.																			
30.																			
31.																			
32.																			
33.																			
34.																			
35.																			
36.																			
37.																			
38.																			
39.																			
40.																			
41.									+			+							
42.																			
43.																			
44.																			
45.																			
46.																			
47.	+																		
48.																			
49.																			
50.																			
51.																			
52.																			
53.																+				
54.					.		.	.	.	.	.	.	.	.	+				
55.																			
56.																			
57.																			
58.																			
59.																			
60.																			
61.																			
62.																			
63.																			
64.														.	+	.	+	+	+
65.	.	.	.	.	.	.		.	.	.	.	.	.	+	.	+	+	
66.	.	.	+	+	+	.	.	.	.	.	.	.	.	.	.	.	.	.
67.																			
68.													+	.	.	.	++		
69.	.	*	.	.	.	.	.												
70.																			
71.																			
72.																			

Species		Clitheroe	Bolland	Waddow Hall	Chatburn	Coplow, Clitheroe	Race Course, Clitheroe	Wribeill	Salt Hill, Clitheroe	Worston, Clitheroe	Downham	Twiston	Ravenshohne	Wytterser, Bolton in Bowland	Blue Butts, Slaidburn	Great Dunnow, Slaidburn
1. Goniatites cyclolobus	Phil.		+													
2. „ carina	Phil.		+													
3. „ intercostalis	Phil.		+													
4. „ mixolobus	Phil.		+													
5. „ rotiformis	Phil.		+													
6. „ vesica	Phil.		+													
7. „ vittager	Phil.		+													
8. „ sp.																
9. Nautilus discus			+													
10. „ dorsalis	Phil.		+													
11. „ bistrialis	Phil.		+													
12. „ biangulatus	Sby.		+													
13. „ globatus	Sby.		+													
14. „ ? goniolobus	Phil.		+													
15. „ subsulcatus	Phil.		+													
16. „ sp.																
17. Orthoceras cinctum	Sby.		+			+										
18. „ planoseptatum																
19. „ Steinhauerii	Sby.		+													+
20. „ sulcatulum	M'Coy.															
21. „ unguis	Phil.		+													
22. „ undulatum	Sby.		+													
23. „ inæquiseptum	Phil.		+													
24. „ anguläre	Fleng.	+	+													
25. „ Gesneri	Mart.		+													
26. „ reticulatum	Phil.		+													
27. „ sp.						+					+		+			
28. Phragmoceras sp.																
29. Poterioceras fusiforme	Sby.		+													
30. Holoptychius (scale)																
31. ? (scale)																

APPENDIX I.—FOSSILS.

	Shales-with-Limestones.							Bowland Shales.							Millstone Grit Shales.		
	Ashnot, Knowmere.	Brock Hall, Ribble.	Primrose Print Works.	Worsaw End House.	Hodder Bridge, Ribble.	Eshhall Brook, Clitheroe.	Gisburn.	Forest Becks.	Fat Hill, Holden.	Harrowfield, Knowlmere.	Dinckley Hall, Ribble.	Wiswell, Clitheroe.	Near Standen Hall.	Beck above Little Mearley Hall.	Fellside, Standburn.	Champion.	Sales Wheel.	Root.	Alston Hall.
1.																			
2.																			
3.																			
4.																			
5.																			
6.																			
7.			+		+	.	.		.		+				+		+		
8.			+																
9.		.		+															
10.																			
11.																			
12.																			
13.																			
14.																			
15.					.														
16.				.		.		.			+	.		+					
17.											+								
18.		+																	
19.			+				.		.		+								
20.																			
21.																			
22.																			
23.																			
24.																			
25.																			
26.							.		.		+	.	.		+				
27.													+						
28.																			
29.											.	.	.		+				
30.		.				+													
31.																			

APPENDIX II.

A List of Works and Papers relating to the Geology of Lancashire and some of the adjacent Country.

By W. Whitaker and R. H. Tiddeman.

NOTICE.

The accompanying list of papers is meant to be useful for reference to any one who is interested in the geological structure and bearings of the district of which the memoir treats. It is not confined to Lancashire, but includes many important papers in adjacent counties which may be brought to bear upon the district. The list owes its existence in the first instance to the untiring and disinterested labours of my friend and colleague Mr. Whitaker, to whom I am indebted for the chief part of the titles, and I am glad to have this opportunity of expressing my gratitude. I am also glad to acknowledge the kind assistance of Mr. Edward Best, Mr. Louis, C. Miall, and other friends. I do not profess to have read all the papers and books mentioned, and the list as a first edition will necessarily be somewhat imperfect, but I shall be very grateful for corrections or additions with a view to a future issue.

The arrangement is as follows: First comes a list of authors in alphabetical order, with the numbers attached to their papers in the list of works appended. Then follows the list arranged in the first place by the years of publication. Each year is arranged in the alphabetical sequence of the authors' names. In the case of a joint work it is inserted under the name of the author whose name appears first in the title although in the index there is a reference to each author. In the case of a number of works by an officer of the Geological Survey in the same year I have arranged them under his name in the following order: Official works—1-inch maps, 6-inch maps, sections, memoirs—Private works.

R. H. Tiddeman.

ALPHABETICAL LIST OF AUTHORS.

A.

Aikin, J., 17.
Aitken, John, 269, 291, 315-6, 367-373, 422-3, 443-4.
Anon, 15, 31, 36, 47, 49, 50, 51, 53, 56, 78, 165, 292, 487-8.
Aveline, W. T., 374, 424, 489, 518.

B.

Baines, S., 166, 181.
Bakewell, R., 28.
Barker, R., 117.
Beddoe, Dr., 546a.
Bewick, J., 394.
Binney, E. W., 65, 93-6, 105-6, 109-111, 114, 117-8, 122, 124-6, 129, 134, 137, 143-4, 153-4, 157, 163b, 182-4, 217-225, 248-50, 270-2, 293-6, 317-322, 344-8, 375-9, 395-6, 445-6, 490-1.
Black, Dr., 79.
Bostock, Dr., 42.
Bostock, R., 397, 425.
Boult, J., 164, 323, 519.
Boulton, J., 398, 447.
Bowman, J. E., 88, 97-8.
Bramall, A., 273.
Brigg, John, 547.
Brockbank, W., 324, 520.
Brown, Capt. T., 99.
Bunbury, Sir C. J. F., 168.
Burn, F., 80.
Busk, Prof. G., 521, 548-9.
Butterworth, J., 325, 399.

C.

Cameron, A. G., 448-9.
Campbell, Dr., 22, 24.
Carruthers, W., 449-50, 492.
Chambers, R., 127.
Charnock, J. C., 128.
Coal Commission, 451.
Creyke, R., 115.
Cumberland, G., 34.
Cunningham, J., 119, 123.
Curry, J., 349.

D.

Dakyns, J. R., 383, 402, 424, 432, 452-3, 461-3, 482, 523, 550.
Dalton, James, 55.
Dalton, W. H., 471.
Darbyshire, R. D., 297, 551.
Davidson, T., 169, 170, 172-3, 197, 251.
Dawkins, W. B., 298, 426, 454, 522, 552-3.
De Koninck, Prof. L., 161, 171, 427.
Denham, Capt., 81.
Denny, H., 152b, 184a.
De Rance, C. E., 400-1, 428, 434, 455-9, 468, 495, 524-6, 554.

Dickinson, J., 138, 185.
Dickinson, Jos., 252, 274, 326.
Duncan, T., 147.
Dunn, Matth., 112.

E.

Earwaker, J. P., 429.
Eddy, S., 116, 174.
Eccles, James, 380-1, 430.
Everett, Prof., 460.

F.

Fairbairn, Sir W., 226.
Farey, 25-7, 35.
Farrer, J. W., 130, 327.

G.

Gages, A., 186.
Garnett, W. J., 131.
Geinitz, Dr., 275.
George, E. J., 41.
George, E. S., 72.
Gilbertson, W., 53, 61.
Gray, Thomas, 13.
Gomersall, W., 554a.
Green, A. H., 198, 256, 278, 282, 328, 332, 382-3, 402, 424, 461-3.
Greenwell, G. C., 329.
Gunn, W., 405, 464, 466, 496.

H.

Hall, Elias, 57, 64.
Hardwick, C., 330.
Harkness, Prof. R., 114, 156.
Harmer, F. W., 486.
Haughton, Revd. Dr. S., 403.
Hawkshaw, John, 100-1, 120.
Henderson, A., 164.
Henry, Dr. W., 32.
Hewlett, A., 330a.
Heywood, Jas., 82, 104.
Hibbert, Dr., 39.
Higgins, Rev. H. H., 465, 497.
Hodgkinson, E., 89, 102.
Hodgson, Miss E., 253, 276, 350, 404, 431.
Hooker, Dr. J. D., 157.
Hopkins, J., 9.
Housman, J., 20, 21.
Howson, W., 145.
Hughes, Prof. T. McK., 331, 351, 374, 384, 527, 555.
Hull, E., 136, 158, 175, 187-9, 190-2, 199-208, 227-239, 254-260, 277-82, 299-302, 332-4, 352-361, 385-6, 405, 405a, 406-12, 432-6, 466-9, 498-9, 528-30, 556.
Hume, Rev. Dr., 335.
Hurtley, F., 16.

I.

Inglis, G., 30.

J.

Jackson, Joseph, 113a.
Jackson, S. B., 261.
Jars, M., 14.
Jones, Prof. T. R., 240.
Jopling, C. P., 107.

K.

Kerr, —, 470.
King, Prof. W., 135.
Kirkby, J. W., 241.
Kirwan, R., 18.
Knowles, A., 242.
Knowles, J., 437.

L.

Leigh, C., 8.
Leigh, J., 65.
L. E. O., 47.
Lister, M., 2-5, 7.
Looney, F., 66.
Lucas, J., 471.

M.

Mackintosh, D., 303, 413-5, 438, 472, 531.
Maleverer, Dr., 7.
Mallet, Dr. J. W., 146, 148.
Marrat, F. P., 500.
Maw, G., 416.
Miall, L. C., 501-2, 532.
Moon, M. A., 387.
Mordacque, Rev. S. H., 193.
Morris, J. P., 388.
Morrison, Walter, 473.
Morton, G. H., 162, 194, 209-212, 243-5, 262-3, 304, 336-7, 362, 474-5, 503-5, 533.
Murchison, Sir R. I., 58, 62.
Mushet, D., 19.

N.

Neild, J., 338.
Nicholson, Dr. H. A., 417.

O.

Ormerod, G. W., 132, 139, 140.
Owen, Professor R., 363.

P.

Paley, F. A., 364.
Paterson, 305.
Peace, Wm., 83.
Peacock, R.A., 365.
Percy, Dr. J., 283.

Phillips, Prof. J., 52, 59, 63, 67, 73-4, 86, 113, 149, 150, 152c, 159, 176-7, 306-8, 339.
Pigott, Dr., 163.
Plant, J., 340, 389-391, 476, 557.
Platt, J., 12.

R.

Ramsay, Prof. A. C., 309, 506.
Reade, T. M., 507-9, 534-7, 558.
Rennie, G., 164.
Richardson, T., 84.
Richmond, L., 11.
Ricketts, Dr. C., 418, 439, 477, 559.
Roberts, G. E., 298, 310.
Roberts, Isaac, 538.
Roberts, J., 419, 478-9.
Rofe, J., 284, 311-2, 341, 420, 440, 480, 539.
Russell, R., 424, 463, 482, 530, 560.
Ruthven, J., 160.

S.

Sainter, J. D., 213, 264, 313.
Salmon, H. C., 214.
Salter, J. W., 265.
Sedgwick, Prof. A., 68-70, 121, 146a.
Sharpe, D., 108.
Sharpe, Rev. W., 102a.
Short, Dr. T., 10.
Shirley, T., 1.
Smith, Ecroyd, 314.
Smith, J. T., 481.
Smith, Wm., 29, 33, 46.
Smith, Roach, 113a.
Smyth, W. W., 136.
Sorby, H. C., 151, 178-9.
Sowerby, G. B., 40, 48, 60.
Sowerby, J., 38, 41, 43.
Stark, J., 44.
Strangways, C. F., 432, 471.
Sturdie, J., 6.

T.

Talbot, J. H., 396.
Tate, A. N., 266.
Taylor, J., 195, 246-7, 285.
Taylor, J. E., 366.
Teale, T. P., 163a, 179a.
Tiddeman, R. H., 374, 393, 405, 405a, 406, 433, 467, 469, 510-11, 541-4, 556, 561.
Thornber, Rev. W., 75.
Thorp, Rev. W., 90, 91, 102a.
Thorpe, T. E., 392.
Tomlinson, C., 196, 215.
Tooke, A. W., 76.
Trimmer, Jos., 141, 152.
Turner, W., 180.

V.

Vaux, F., 133.

ALPHABETICAL LIST OF AUTHORS.

A.

Aikin, J., 17.
Aitken, John, 269, 291, 315-6, 367-373, 422-3, 443-4.
Anon, 15, 31, 36, 47, 49, 50, 51, 53, 56, 78, 165, 292, 487-8.
Aveline, W. T., 374, 424, 489, 518.

B.

Baines, S., 166, 181.
Bakewell, R., 28.
Barker, R., 117.
Beddoe, Dr., 546a.
Bewick, J., 394.
Binney, E. W., 65, 93-6, 105-6, 109-111, 114, 117-8, 122, 124-6, 129, 134, 137, 143-4, 153-4, 157, 163b, 182-4, 217-225, 248-50, 270-2, 293-6, 317-322, 344-8, 375-9, 395-6, 445-6, 490-1.
Black, Dr., 79.
Bostock, Dr., 42.
Bostock, R., 397, 425.
Boult, J., 164, 323, 519.
Boulton, J., 398, 447.
Bowman, J. E., 88, 97-8.
Bramall, A., 273.
Brigg, John, 547.
Brockbank, W., 324, 520.
Brown, Capt. T., 99.
Bunbury, Sir C. J. F., 168.
Burn, F., 80.
Busk, Prof. G., 521, 548-9.
Butterworth, J., 325, 399.

C.

Cameron, A. G., 448-9.
Campbell, Dr., 22, 24.
Carruthers, W., 449-50, 492.
Chambers, R., 127.
Charnock, J. C., 128.
Coal Commission, 451.
Creyke, R., 115.
Cumberland, G., 34.
Cunningham, J., 119, 123.
Curry, J., 349.

D.

Dakyns, J. R., 383, 402, 424, 432, 452-3, 461-3, 482, 523, 550.
Dalton, James, 55.
Dalton, W. H., 471.
Darbyshire, R. D., 297, 551.
Davidson, T., 169, 170, 172-3, 197, 251.
Dawkins, W. B., 298, 426, 454, 522, 552-3.
De Koninck, Prof. L., 161, 171, 427.
Denham, Capt., 81.
Denny, H., 152b, 184a.
De Rance, C. E., 400-1, 428, 434, 455-9, 468, 495, 524-6, 554.

Dickinson, J., 138, 185.
Dickinson, Jos., 252, 274, 326.
Duncan, T., 147.
Dunn, Matth., 112.

E.

Earwaker, J. P., 429.
Eddy, S., 116, 174.
Eccles, James, 380-1, 430.
Everett, Prof., 460.

F.

Fairbairn, Sir W., 226.
Farey, 25-7, 35.
Farrer, J. W., 130, 327.

G.

Gages, A., 186.
Garnett, W. J., 131.
Geinitz, Dr., 275.
George, E. J., 41.
George, E. S., 72.
Gilbertson, W., 53, 61.
Gray, Thomas, 13.
Gomersall, W., 554a.
Green, A. H., 198, 256, 278, 282, 328, 332, 382-3, 402, 424, 461-3.
Greenwell, G. C., 329.
Gunn, W., 405, 464, 466, 496.

H.

Hall, Elias, 57, 64.
Hardwick, C., 330.
Harkness, Prof. R., 114, 156.
Harmer, F. W., 486.
Haughton, Revd. Dr. S., 403.
Hawkshaw, John, 100-1, 120.
Henderson, A., 164.
Henry, Dr. W., 32.
Hewlett, A., 330a.
Heywood, Jas., 82, 104.
Hibbert, Dr., 39.
Higgins, Rev. H. H., 465, 497.
Hodgkinson, E., 89, 102.
Hodgson, Miss E., 253, 276, 350, 404, 431.
Hooker, Dr. J. D., 157.
Hopkins, J., 9.
Housman, J., 20, 21.
Howson, W., 145.
Hughes, Prof. T. McK., 331, 351, 374, 384, 527, 555.
Hull, E., 136, 158, 175, 187-9, 190-2, 199-200, 227-239, 254-260, 277-82, 299-302, 332-4, 352-361, 385-6, 405, 405a, 406-12, 432-6, 466-9, 498-9, 528-30, 556.
Hume, Rev. Dr., 335
Hurtley, F., 16.

I.

Inglis, G., 30.

J.

Jackson, Joseph, 113a.
Jackson, S. B., 261.
Jars, M., 14.
Jones, Prof. T. R., 240.
Jopling, C. P., 107.

K.

Kerr, —, 470.
King, Prof. W., 135.
Kirkby, J. W., 241.
Kirwan, R., 18.
Knowles, A., 242.
Knowles, J., 437.

L.

Leigh, C., 8.
Leigh, J., 65.
L. E. O., 47.
Lister, M., 2-5, 7.
Looney, F., 66.
Lucas, J., 471.

M.

Mackintosh, D., 303, 413-5, 438, 472, 531.
Maleverer, Dr., 7.
Mallet, Dr. J. W., 146, 148.
Marrat, F. P., 500.
Maw, G., 416.
Miall, L. C., 501-2, 532.
Moon, M. A., 387.
Mordacque, Rev. S. H., 193.
Morris, J. P., 388.
Morrison, Walter, 473.
Morton, G. H., 162, 194, 209-212, 243-5, 262-3, 304, 336-7, 362, 474-5, 503-5, 533.
Murchison, Sir R. I., 58, 62.
Mushet, D., 19.

N.

Neild, J., 338.
Nicholson, Dr. H. A., 417.

O.

Ormerod, G. W., 132, 139, 140.
Owen, Professor R., 363.

P.

Paley, F. A., 364.
Paterson, 305.
Peace, Wm., 83.
Peacock, R.A., 365.
Percy, Dr. J., 283.

Phillips, Prof. J., 52, 59, 63, 67, 73-4, 86, 113, 149, 150, 152c, 159, 176-7, 306-8, 339.
Pigott, Dr., 163.
Plant, J., 340, 389-391, 476, 557.
Platt, J., 12.

R.

Ramsay, Prof. A. C., 309, 506.
Reade, T. M., 507-9, 534-7, 558.
Rennie, G., 164.
Richardson, T., 84.
Richmond, L., 11.
Ricketts, Dr. C., 418, 439, 477, 559.
Roberts, G. E., 298, 310.
Roberts, Isaac, 538.
Roberts, J., 419, 478-9.
Rofe, J., 284, 311-2, 341, 420, 440, 480, 539.
Russell, R., 424, 463, 482, 530, 560.
Ruthven, J., 160.

S.

Sainter, J. D., 213, 264, 313.
Salmon, H. C., 214.
Salter, J. W., 265.
Sedgwick, Prof. A., 68-70, 121, 146a.
Sharpe, D., 108.
Sharpe, Rev. W., 102a.
Short, Dr. T., 10.
Shirley, T., 1.
Smith, Ecroyd, 314.
Smith, J. T., 481.
Smith, Wm., 29, 33, 46.
Smith, Roach, 113a.
Smyth, W. W., 136.
Sorby, H. C., 151, 178-9.
Sowerby, G. B., 40, 48, 60.
Sowerby, J., 38, 41, 43.
Stark, J., 44.
Strangways, C. F., 432, 471.
Sturdie, J., 6.

T.

Talbot, J. H., 396.
Tate, A. N., 266.
Taylor, J., 195, 246-7, 285.
Taylor, J. E., 366.
Teale, T. P., 163a, 179a.
Tiddeman, R. H., 374, 393, 405, 405a, 406, 433, 467, 469, 510-11, 541-4, 556, 561.
Thornber, Rev. W., 75.
Thorp, Rev. W., 90, 91, 102a.
Thorpe, T. E., 392.
Tomlinson, C., 196, 215.
Tooke, A. W., 76.
Trimmer, Jos., 141, 152.
Turner, W., 180.

V.

Vaux, F., 133.

W.

Ward, J. C., 383, 402, 421, 424, 434, 452-3, 461-3, 482.
Watson, H. H., 51.
West, T., 152a.
Whitaker, J., 216, 286, 343.
Würzburger, P., 512.
Wild, G., 287.
Wilkinson, T. T., 216, 267-8, 288, 342.
Williamson, W. C., 71, 77, 85, 87, 142, 483, 513-14, 545.

Wollaston, G. H., 441, 484.
Wood, E., 298.
Wood, S. V., jun., 442, 485-6.
Woodward, H., 289, 515-17, 546.
Woolnoth, J. C., 37.
Wynne, R. H., 290.

Y.

Yates, J. B., 92, 103.

A LIST OF WORKS AND PAPERS RELATING TO THE GEOLOGY OF LANCASHIRE AND SOME OF THE ADJACENT COUNTRY.

1667.

1. SHIRLEY, T.   The Description of a Well and Earth in Lancashire taking fire by a Candle approached to it. *Phil. Trans.*, *vol.* ii. (*No.* 26), *p.* 482.

1674.

2. LISTER, M.   A description of certain stones figured like plants, and by some observing men esteemed to be plants petrified. (Craven.) *Phil. Trans.*, *vol.* viii. (*No.* 100), *p.* 6181.
3. ———   Of Trochitæ and Entrochi. *Ibid.*
4. ———   On a Subterranean Fungus and a Mineral Juice. *Ibid. p.* 6179.

1675.

5. ———   Of an odd figured Iris. ("Rainsborough Scar Rible" and "Eshton Tarn.") *Phil. Trans.*, *vol.* ix. (*No.* 110), *p.* 222.

1693.

6. STURDIE, J.   Extracts of some Letters concerning Iron Ore and Iron, particularly of the Hæmatites, wrought into iron at Milnthorp Forge. (Lancashire.) *Phil. Trans.*, *vol.* xvii. (*No.* 199), *p.* 695.

1699.

7. LISTER, DR. M., and MALEVERER.   Of Coal Borings. *Phil. Trans.*, *vol.* xxi. (*No.* 250), *p.* 73.

1700.

8. LEIGH, C.   Natural History of Lancashire, Cheshire, and the Peak in Derbyshire, &c. *Fol. Oxford.*

1732.

9. HOPKINS, J.   An Extract of a Letter concerning an extraordinary large horn of the Stag kind taken out of the sea on the coast of Lancashire. (Raven's Barrow Hole, Holker Old Park.) *Phil. Trans.*, *vol.* xxxvii., *p.* 257.

1740.

10. SHORT, DR. THOMAS - An Essay towards a Natural, Experimental, and Medicinal History of the Principal Mineral Waters of Cumberland, Northumberland, Westmoreland, Bishop-prick of Durham, Lancashire, &c., particularly those of —— Cartmall, &c. 4to. Sheffield.

1745.

11. RICHMOND, L. A Letter concerning a moving Moss in the neighbourhood of Church Town in Lancashire. Phil. Trans., vol. xliii. (No. 475), p. 282.

1751.

12. PLATT, J. A Letter concerning a flat spheroidal Stone, having Lines regularly crossing it (Ardwick.) Phil. Trans., vol. xlvi. (No. 496), p. 534.

1769.

13. GRAY, THOMAS Journal of a Tour in the North of England (Letters).

1774.

14. JARS, M. Voyages Métallurgiques. Tome 1. Quarto. Lyons. (Lancashire, p. 251. Yorkshire, p. 255.)

1781.

15. ANON. Tour to the Caves in the Environs of Ingleborough and Settle in the West Riding of Yorkshire, their Geology, &c. 8vo.

1786.

16. HURTLEY, F. Account of Natural Curiosities at Malham in Craven, Yorkshire. 8vo.

1795.

17. AIKIN, J. Description of the Country from 30 to 40 miles round Manchester. 4to.

1798.

18. KIRWAN, R. Experiments on the Composition and Proportion of Carbon in Bitumen and Mineral Coal. Journ. of Nat. Phil. Chem. and Arts. 4to. vol. i., p. 487.

1799.

19. MUSHET, D. On Primary Ores of Iron. Phil. Mag., (Ser. i.), vol. iii., p. 350.

1800.

20. HOUSMAN, J. - Descriptive Tour and Guide to the Lakes, Caves, Mountains, &c., in Cumberland, Westmoreland, and Lancashire. 8vo. Carlisle.

21. ——— A Topographical Description of Cumberland, Westmoreland, Lancashire, and a part of the West Riding of Yorkshire. (Map of the Soils. Accounts of Caves, Soils, Minerals, Stone.) 8vo. Carlisle.

1810.

22. CAMPBELL, DR. — Remarks upon the Inferior Strata of the Earth occurring in Lancashire, with some miscellaneous observations arising from the subject. With a Geol. Map of Lancashire, by DR. WILKINSON. *Correspondence of the Bath and W. of England Soc.*, vol. xii., p. 85.

23. FAREY, J. A List of about 500 Collieries in and near to Derbyshire. *Phil. Mag.*, vol. xxxv., p. 431.

1811.

24. CAMPBELL, DR. — Remarks upon the Inferior Strata of the Earth occurring in Lancashire, with some Observations arising from the subject. *Phil. Mag.*, vol. xxxviii., p. 268.

25. FAREY, J. — Geological Remarks and Queries on Dr. Campbell's Map, and account of the Stratification of Lancashire. *Ibid.*, p. 336.

1812.

26. FAREY, J. Geological Observations in correction and addition to the Paper on the Great Derbyshire Denudation, and the Report on Derbyshire, &c. *Phil. Mag.*, Ser. i., vol. xxxix., p. 93.

1813.

27. FAREY, J. — Cursory Geological Observations lately made in Shropshire, Wales, Lancashire, Scotland, Durham, Yorkshire N. R., and Derbyshire. (Some Observations on Mr. Bakewell's Geological Map, and on the supposed identity of the Derbyshire Peak and the Craven Limestone Rocks.) *Phil. Mag.*, vol. xlii., p. 53.

1814.

28. BAKEWELL, R. An Account of the Coalfield at Bradford, near Manchester. *Trans. Geol. Soc.*, Ser. i., vol. ii., p. 282.

1815.

29. SMITH, WILLIAM A Delineation of the Strata of England and Wales with part of Scotland (16 sheets in case).

1817.

30. INGLIS, G. — On the cause of Ebbing and Flowing Springs. *Phil. Mag.*, vol. ii., p. 81.

1818.

31. ANON. Blue Iron Earth (Knots-hole, near Liverpool). *Ann. of Phil.* vol. xi., p. 391.

32. HENRY, DR. W. — Analysis of a Siliciferous Subsulphate of Alumine found in considerable quantity in a Coal Mine near Oldham. *Ibid.*, p. 432.

1821.

33. SMITH, WILLIAM — A Geological Map of Yorkshire (4 sheets).

1823.

34. CUMBERLAND, G. On the Origin of the Accumulations of Bones in the Caves of the Vale of Pickering, in Yorkshire, and in other places. *Thomsons' Ann. Phil.*, Ser. ii., vol. v., p. 127-9.

35. FAREY, J. On the unconformable position of the Pontefract Rock of Sandstone with respect to the subjacent Coal Measures, &c. *Ibid. p.* 270.

1824.

36. ANON. Fossils (Stag's Horns, Wallasey Pool). *Phil. Mag., vol.* lxiv., *p.* 315.
37. WOOLNOTH, Lieut. J. C. Analysis of the Holy Well Water, near Cartmell, Lancashire. *Ibid. p.* 394.

1824 ?-5.

38. SOWERBY, J. The Mineral Conchology of Great Britain, *vol.* v. 8vo., London. [*Lancashire, pp.* 34–5, 83. *Yorkshire, pp.* 23, 36, 109, 119, 160.]

1825.

39. HIBBERT, DR. On the Dispersion of Stony Fragments remote from their native beds, as displayed in a Stratum of Loam near Manchester. (Roy. Soc. Edin.) *Edin. Journ. Sci., vol.* ii., *p.* 208.
40. SOWERBY, G. B. Description of a new Species of *Pentremites*. *Zool. Journ., vol.* ii., *p.* 316.
41. SOWERBY, J. D. C., and GEORGE, E. J. Additional Observations upon a Fossil found in Coal Shale, and the description of a Palate found in Coal near Leeds. *Ibid., p.* 22.

1826.

42. BOSTOCK, DR. Notice respecting the Pebbles in the Bed of Clay which covers the New Red Sandstone in the S.W. of Lancashire. *Trans. Geol. Soc., Ser.* 2, *vol.* ii., *p.* 138.

1826 ?-9.

43. SOWERBY, J. The Mineral Conchology of Great Britain, *vol.* vi. 8vo. London.

1826.

44. STARK, J. Notice regarding the discovery of Live Cockles in a Peat Moss at a great distance from the Sea. *Edin. Journ., Sci., vol.* iv. *p.* 142.
45. WITHAM, A. On the discovery of Live Cockles in Peat Moss at a great distance from the Sea, and much above its present level. (Wernerian Soc.) *Annals of Phil., Ser.* 2, *vol.* xi. *p.* 464.

1827.

46. SMITH, W. On retaining Water in Rocks for summer use. *Phil. Mag., Ser.* 2, *vol.* i. *p.* 415.

1828.

47. ANON; L. E. O. Granite found North of the Humber. *Mag. Nat. Hist., vol.* i. *p.* 596.
48. SOWERBY, G. B. On some new species of *Pentatrematites*. *Zool. Jour., vol.* iv. *p.* 89.

1829.

49. ANON; R. C. T[AYLOR?] Granite found North of the Humber. *Mag. Nat. Hist., vol.* ii. *p.* 120.
50. —————————— Coal Fossils from Clifton, near Manchester. *Mag. Nat. Hist., vol.* ii. *pp.* 106, 299.

51. ANON. B. — Drawings of others. *Ibid.* p. 486–7.
52. PHILLIPS, PROF. J. — On a group of Slate Rocks, ranging E.S.E. between the Rivers Lune and Wharfe, from near Kirkby Lonsdale to near Malham, and on the attendant Phenomena. *Trans. Geol. Soc. Ser.* 2, *vol.* iii. *p.* 1.

1830.

53. GILBERTSON, W., and R. C. T[AYLOR] — A Collection of Shells from Preston. *Mag. Nat. Hist.,* vol. iii. p. 170.
54. WATSON, H. H. — Sulphate of Barytes, forming a vein in Cannel Coal. *Phil. Mag., Ser.* 2, *vol.* viii. p. 304.

1832.

55. DALTON, JAMES — Observations, chiefly Chemical, on the Nature of the Rock Strata in Manchester and its Vicinity. *Mem. Lit. and Phil. Soc., Manchester, Ser.* 2, *vol.* v. *pp.* 148–153.

1831.

56. ANON.: (A. R. B.) — Observations on the Limestone District of Yealand, near Lancaster, and on the Vegetable Phenomena displayed by the growth of Plants on one of its rocky hills. *Mag. Nat. Hist., vol.* v. *p.* 550.
57. HALL, E. — Mineralogical and Geological Map of the Coalfields of Lancashire, with parts of Yorkshire, Cheshire, and Derbyshire. *Manchester.*

1833.

58. MURCHISON, SIR R. I. — Observations on certain accumulations of Clay, Gravel, Marl, and Sand around Preston, in Lancashire, which contain Marine Shells of existing species. *Brit. Assoc., Rep. for* 1831, p. 82.
59. PHILLIPS, PROF. J. — Account of the Geology of Yorkshire. *Brit. Assoc. Rep. for* 1831, *p.* 56.

1834?

60. SOWERBY, G. B. — On *Pentatrematites Orbicularis, acuta* and *pentangularis. Zool. Journ., vol.* v., *p.* 456.

1835.

61. GILBERTSON, W. — On Marine Shells of recent Species, at considerable Elevations near Preston. *Rep. Brit. Assoc. for* 1834, p. 654.
62. MURCHISON, SIR R. — The Gravel and Alluvia of S. Wales, and Siluria as distinguished from a northern drift covering Lancashire, Cheshire, N. Salop, and parts of Worcestershire and Gloucestershire. *Proc. Geol. Soc., vol.* ii. p. 230.
63. PHILLIPS, JOHN — Illustrations of the Geology of Yorkshire, &c. *London.* 2 vols. 4to.

1836.

64. HALL, ELIAS — Introduction to the Mineral and Geological Map of the Coalfield of Lancashire, with a part of Yorkshire, Derbyshire, and Cheshire (pp. iv., 28). 8vo. *Manchester.*

65. LEIGH, J., and BINNEY, E. W. — Observations on a patch of Red and Variegated Marls containing Fossil Shells at Collyhurst, near Manchester. *Proc. Geol. Soc., vol.* ii. p. 391.
66. LOONEY, FRANCIS — List of Organic Remains, &c., and where found; to accompany Mr. Elias Hall's Introduction and Map. *Manchester?*.
67. PHILLIPS, JOHN — Guide to Geology. London. 12mo.
68. SEDGWICK, REV. PROF. A. — Introduction to the General Structure of the Cambrian Mountains, with a description of the great Dislocations by which they have been separated from the neighbouring Carboniferous Chain. *Trans. Geol. Soc., Ser.* 2, *vol.* iv., *p.* 47.
69. ——— — Description of a series of Longitudinal and Transverse Sections through a portion of the Carboniferous Chain between Penigent and Kirkby Stephen. *Ibid.*, *p.* 69.
70. ——— — On the New Red Sandstone Series in the Basin of the Eden and North-western Coasts of Cumberland and Lancashire. *Ibid.*, *p.* 383.
71. WILLIAMSON, W. C. — On the Limestones found in the Neighbourhood of Manchester. *Phil. Mag., Ser.* 3, *vol.* ix., pp. 241, 348.

1837.

72. GEORGE, E. S. - On the Yorkshire Coalfield. *Trans. Leeds Phil. and Lit. Soc., vol.* i.
73. PHILLIPS, PROF. J. - A Report on the Probability of the Occurrence of Coal and other Minerals in the vicinity of Lancaster. 4to. *Lancaster.*
74. ——— - On certain Limestones and associated Strata in the Neighbourhood of Manchester. *Rep. Brit. Assoc. for* 1836, *Trans. of Sections*, *p.* 86.
75. THORNBER, REV. W. - History of Blackpool (with Geological References).
76. TOOKE, A. W. - The Mineral Topography of Great Britain. (Lancashire, p. 46, Yorkshire, p. 53.) *Mining Review, No.* 9, *p.* 39.
77. WILLIAMSON, W. C. - On Fossil Fishes in the Lancashire Coalfield. *Proc. Geol. Soc., vol.* ii., *p.* 571.

1838.

78. ANON. Geological Curiosities [Trees at Rochdale Gasworks], from the *Birmingham Daily Post.* *Mining Review, No.* 6, *vol.* iv., *p.* 80.
79. BLACK, DR. - On a Fossil Stem of a Tree recently discovered near Bolton-le-Moors. *Proc. Geol. Soc., vol.* ii., *p.* 670.
80. BURN, FREDERICK Notice of the Localities and General Features of the Slate Quarries of Great Britain. *Mining Review, No.* xii., *vol.* iv., *pp.* 177, 178 (4to.).
81. DENHAM, CAPTAIN On the Tidal Capacity of the Mersey Estuary, the proportion of Salt held in Solution . the excess of Deposit upon each Influx, and the consequent effect . . . . . *Rep. Brit. Assoc. for* 1837. *Trans. of Sec.*, *p.* 85.
82. HEYWOOD, JAMES On the Geology of the District of South Lancashire. *Ibid.*, *p.* 77.
83. PEACE, WM. On the Dislocations of the Coal Strata in Wigan and the Vicinity. *Ibid. p.* 82.
84. RICHARDSON, THOMAS Researches upon the Composition of Coal. *Trans. Nat. Hist. Soc., Northumberland, Durham, and Newcastle-upon-Tyne, vol.* ii., *part* 2, *pp.* 401–411; *Phil. Mag., vol.* xiii., *p.* 121; *and Mining Review, vol.* iv., *No.* 8, *pp.* 124–6.

85. WILLIAMSON, W. C.   On the Coal Measures of West Lancashire. *Rep. Brit. Assoc. for 1837, Trans. of Sections,* p. 82.

1839.

86. PHILLIPS, JOHN   A Treatise on Geology. London, 8vo.
87. WILLIAMSON, W. C.   A Notice of the Fossil Fishes of the Yorkshire and Lancashire Coalfields. *Proc. Geol. Soc.,* vol. iii., p. 153.

1840.

88. BOWMAN, J. E. -   On the Character of the Fossil Trees lately discovered near Manchester on the line of the Manchester and Bolton Railway, and on the Formation of Coal by gradual subsidence. *Proc. Geol. Soc.,* vol. iii., p. 270.
89. HODGKINSON, EATON   On the Temperature of the Earth in the Deep Mines of Lancashire and Cheshire. *Rep. Brit. Assoc. for 1839, Trans. of Sections,* p. 19.
90. THORP, REV. W. -   On the proposed Line of Section between the Coalfield of Yorkshire and that of Lancashire. *Proc. Geol. and Polyt. Soc., W. R. Yorks.,* vol. i. (No. 2), p. 7.
91. ————— -   On the Disturbances in the District of the Valley of the Don. *Ibid.* (No. 4), p. 4.
92. YATES, JOS. B. -   - On the Rapid Changes which take place at the Entrance of the River Mersey, and the Means adopted for establishing an Easy Access to Vessels resorting thereto. *Rep. Brit. Assoc. for 1839, Trans. of Sections,* pp. 77-78.

1841.

93. BINNEY, E. W.   Sketch of the Geology of Manchester and its Vicinity. *Trans. Man. Geol. Soc.,* vol. i., p. 35.
94. ————— -   Observations on the Lancashire and Cheshire Coalfield. *Ibid.,* p. 67.
95. —————   - Remarks on the Marine Shells found in the Lancashire Coalfield. *Ibid.,* p. 80.
96. ————— -   On the Fossil Fishes of the Pendleton Coalfield. *Ibid.,* p. 153.
97. BOWMAN, J. E. -   - Observations on the Characters of the Fossil Trees lately discovered in the Line of the Bolton Railway, near Manchester. *Ibid.,* p. 112.
98. ————— -   - On the Fossil Trees found on the Line of the Bolton Railway at Dixon Fold, near Manchester, and the light they throw on several points still undecided among geologists. *Edinburgh New Phil. Journ.,* vol. xxxi., p. 150.
99. BROWN, CAPTAIN T. -   Description of some new species of Fossil Shells, found chiefly in the Vale of Todmorden, Yorkshire. *Trans. Man. Geol. Soc.,* vol. i., p. 212.
100. HAWKSHAW, JOHN -   Description of the Fossil Trees found in the Excavations for the Manchester and Bolton Railway. *Trans. Geol. Soc.,* Ser. 2, vol. vi., p. 173.
101. —————   Further Observations on ditto. *Ibid.,* p. 177.
102. HODGKINSON, EATON -   On the Temperature of the Earth in the Deep Mines in the neighbourhood of Manchester. *Rep. Brit. Assoc. for 1840, Trans. of Sections,* p. 17.
102a. THORP, REV. W.   On the Agriculture of the West Riding of Yorkshire considered Geologically. *Proc. West Riding Geol. Soc.,* vol. i.

103. YATES, JAMES - Account of the Footsteps of Extinct Animals observed in a Quarry in Rathbone Street, Liverpool. *Rep. Brit. Assoc. for* 1840, *Trans. of Sections, pp.* 99–100.

1842.

104. HEYWOOD, JAMES - Remarks on the Coal District of South Lancashire. *Mem. Lit. and Phil. Soc., Manchester,* Ser. 2, *vol.* vi., *pp.* 426–463.

1843.

105. BINNEY, E. W. On the Fossil Fishes of the Manchester Coalfield. *Geologist, p.* 15.
106. ——————— Notes on the Lancashire and Cheshire Drift. *Geologist, p.* 112. (*Man. Geol. Soc.*)
107. JOPLING, CHARLES P. - Sketch of Furness and Cartmel.
108. SHARPE, DAN. On the Silurian Rocks of the South of Westmoreland and North of Lancashire. *Proc. Geol. Soc., vol.* iv., *p.* 23.

1844.

109. BINNEY, E. W. - Report on the Excavation made at the junction of the Lower New Red Sandstone with the Coal Measures at Collyhurst, near Manchester. *Rep. Brit. Assoc. for* 1843, *p.* 241.
110. ——————— On the remarkable Fossil Trees lately discovered near St. Helen's. *Phil. Mag., Series* 3, *vol.* xxiv., *p.* 165.
111. ——————— On the Fossil Trees found standing upright in the Coal Measures. *Proc. Geol. and Polyt. Soc., W. R. Yorks., vol.* ii., *p.* 186.
112. DUNN, MATTH. An Historical, Geological, and Descriptive View of the Coal Trade of the North of England. 8vo.
113. PHILLIPS, PROF. J. - Memoirs of William Smith. 8vo. *London.*
113a. SMITH, ROACH, and JACKSON, JOSEPH. On Victoria Cave. *Collectanea Antiqua, vol.* i., No. 5.

1845.

114. BINNEY, E. W., and HARKNESS, PROF. R. An Account of the Fossil Trees found at St. Helen's. *Phil. Mag., Ser.* 3, *vol.* xxvii., *p.* 241.
115. CREYKE, R. Some Account of the Process of Warping. *Journ. Roy. Agri. Soc., vol.* v., *p.* 398.
116. EDDY, S. - Account of the Grassington Lead Mine. *Rep. Brit. Assoc. for* 1844, *Trans. of Sect., p.* 52.

1846.

117. BINNEY, E. W. - The relation of the New Red Sandstone to the Carboniferous Strata in Lancashire and Cheshire. *Quart. Journ. Geol. Soc., vol.* ii., *p.* 12.
118. ——————— Description of the Dukinfield Sigillaria. *Ibid., p.* 390.
119. CUNNINGHAM, JOHN - On some Footmarks and other impressions observed in the New Red Sandstone quarries of Storton, near Liverpool. *Ibid., p.* 410.
120. HAWKSHAW, JOHN Plan of part of the Yorkshire Coalfield.
121. SEDGWICK, REV. PROF. A. On the Classification of the Fossiliferous Slates of Cumberland, Westmoreland, and Lancashire. *Quart. Journ. Geol. Soc., vol.* ii., *p.* 106.

1847.

122. BINNEY, E. W. On Fossil Calamites standing in an erect position in the Carboniferous Strata near Wigan. *Phil. Mag., Ser.* 3, *vol.* xxxi., *p.* 259.

123. CUNNINGHAM, JOHN — On the Geological Conformation of the Neighbourhood of Liverpool, as respects the Supply of Water. *Proc. Lit. and Phil. Soc., Liverpool,* No. iii., pp. 58–74.

1848.

124. BINNEY, E. W. — On the Origin of Coal. *Mem. Lit. and Phil. Soc., Manchester,* Ser. 2, vol. ix., pp. 148–194.
125. ——— Sketch of the Drift Deposits of Manchester and its Neighbourhood. *Ibid.,* pp. 195–234.
126. ——— A Glance at the Geology of Low Furness, Lancashire. *Ibid.,* pp. 423–445.
127. CHAMBERS, ROBT. — Ancient Sea Margins as Memorials of Changes in the relative level of Sea and Land. 8vo. *Edin. and Lond.*
128. CHARNOCK, J. C. — The Farming of the West Riding of Yorkshire. *Journ. Roy. Agric. Soc., vol.* ix., p. 284.

1849.

129. BINNEY, E. W. — Description of a Mineral Vein in the Lancashire Coalfield, near Skelmersdale. *Mem. Lit. and Phil. Soc., Manchester,* Ser. 2, vol. ix., pp. 115–124.
130. FARRER, J. W. — On Ingleborough Cave, *Quart. Journ. Geol. Soc., vol.* v., p. 49.
131. GARNETT, W. J. — The Farming of Lancashire (with Notes on the Geology). *Journ. Roy. Agric. Soc., vol.* x., p. 1.
132. ORMEROD, G. W. — On the Drainage of a Portion of Chat-moss. *Rep. Brit. Assoc. for* 1848, *Trans. of Sections,* p. 72.
133. VAUX, FREDERICK — Ultimate Analysis of some Varieties of Coal. *Quart. Journ. Chem. Soc., vol.* i., p. 318, (Lancashire), p. 320.

1850.

134. BINNEY, E. W. — Remarks on Sigillaria, and some Spores found imbedded in the inside of its roots. [Wigan, &c.] *Quart. Journ. Geol. Soc., vol.* vi., p. 17.
135. KING, PROF. WILLIAM — A Monograph of the Permian Fossils of England. *Palæontograph. Soc.* 4to. *Lond.*
136. SMYTH, W. W., HULL, E. — Quarter Sheet 79 N.E. 1-inch map. *Geol. Survey.*

1851.

137. BINNEY, E. W. — A Description of some supposed Meteorites found in Seams of Coal. *Mem. Lit. and Phil. Soc., Manchester,* Ser. 2, vol. ix., pp. 306–320.
138. DICKINSON, JOSEPH — The Flora of Liverpool (with an Introductory Notice "On the Physical Geography of Liverpool and Wirral," with Remarks on the Geology). 8vo. *Lond.;* and *Proc. Lit. and Phil. Soc., Liverpool* (No. 6).
139. ORMEROD, G. W. — On the gradual Subsidence of a portion of the Surface of Chat-moss, in Lancashire, by Drainage. *Rep. Brit. Assoc. for* 1850, *Trans. of Sections,* p. 101.
140. ——— Section and Analysis of Permian Beds at Astley, Lancashire. *Quart. Journ. Geol. Soc.,* vol. vii., p. 268.

141. TRIMMER, JOSHUA	-	On the Erratic Tertiaries bordering the Penine Chain between Congleton and Macclesfield, and on the scratched detritus of the Till. Part I. *Quart. Journ. Geol. Soc., vol.* vii., *p.* 201.
142. WILLIAMSON, PROF. W. C.		On the Structure and Affinities of the Plants hitherto known as Sternbergiæ. *Mem. Lit. and Phil. Soc., Manchester,* Ser. 2, *vol.* ix., *pp.* 340–358.

1852.

143. BINNEY. E. W.	-	Notes on the Drift Deposits found near Blackpool. *Mem. Lit. and Phil. Soc., Manchester, Ser.* 2, *vol.* x., *pp.* 121–135.
144. ————	-	On some Trails and Holes found in Rocks of the Carboniferous Strata, with Remarks on the Microconchus Carbonarius. *Ibid., pp.* 181–201.
145. HOWSON, W.	-	Illustrated Guide to the Curiosities of Craven.
146. MALLET, DR. J. W.		On a new Fossil Resin. *Phil. Mag.*, Ser. 4, Vol. iv., *p.* 261.
146a. SEDGWICK, PROF. A.		On the Lower Palæozoic Rocks at the base of the Carboniferous Chain between Ravenstonedale and Ribblesdale. *Quart. Journ. Geol. Soc., vol.* viii.

1853.

147. DUNCAN, THOMAS	-	Description of the Liverpool Corporation Waterworks (with Geological Notices. Analyses of Water, p. 502). *Proc. Inst. Civ. Eng., vol.* xii., *p.* 460.
148. MALLET, J. W.	-	Schlerotinite, a new Fossil Resin from the Coal Measures of Wigan, England. (*From American Journ. Sci. and Art, vol.* xv., *p.* 433.) *Edin. New Phil. Journ., vol.* lv., *p.* 359.
149. PHILLIPS, JOHN	-	The Rivers, Mountains, and Sea Coast of Yorkshire, with Essays on the Climate, Scenery, and Ancient Inhabitants of the Country. *Lond.* 8vo.
150. ————	-	A Map of the Principal Features of the Geology of Yorkshire. *York*.
151. SORBY, H. C.		On the Microscopical Structure of some British Tertiary and Post Tertiary Fresh-water Marls and Limestones. *Quart. Journ. Geol. Soc., vol.* ix., *p.* 344.
152. TRIMMER, JOSHUA		On the Erratic Tertiaries bordering the Penine Chain, &c. Part II. *Ibid. p.* 352.
152a. WEST, T.	-	On certain Cavities in the Limestone District of Craven. *Proc. West Riding Geol. Soc., vol.* iii., 1854.
152b. DENNY, H.		On the Discovery of Hippopotamic and other remains in the neighbourhood of Leeds. *Proc. West Riding Geol. Soc., vol.* iii.
152c. PHILLIPS, PROF. J.		On the Dispersion of Erratic Rocks at higher levels than their parent rock in Yorkshire. *Brit. Assoc. Reports for* 1853.

1855.

153. BINNEY, E. W.	-	On the Origin of Ironstones, and more particularly the newly-discovered Red Stone at Ipstones, near Cheadle, Staffordshire; with some account of the Ironstones of South Lancashire. *Mem. Lit. and Phil. Soc., Manchester, Ser.* 2, *vol.* xii., *pp.* 31–45.
154. ————	-	On the Permian Beds of the North-West of England. *Ibid., pp.* 209–269.

APPENDIX II.—LIST OF WORKS AND PAPERS. 201

155. BRAITHWAITE, FREDERICK. On the Infiltration of Salt Water into the Springs of Wells under London and Liverpool. *Proc. Inst. Civ. Eng., vol.* xiv., *pp.* 507–523.
156. HARKNESS, PROF. R. On Mineral Charcoal. *Edin. New Phil. Journ.*, Ser. 2, *vol.* i., *p.* 73.
157. HOOKER, DR. J. D., and BINNEY, E. W. On the Structure of certain Limestone-nodules enclosed in Seams of Bituminous Coal, with a description of some Trigonocarpons contained in them. *Phil. Trans., vol.* cxlv., *p.* 149.
158. HULL, E. - The Geology of the Country around Prescot, Lancashire. 2nd ed, *Geol. Survey Memoirs* (on Sheet 80, N.W.)
159. PHILLIPS, JOHN - Manual of Geology, practical and theoretical: illustrated. *Lond.* 8vo.
160. RUTHVEN, J. A Geological Map of the Lake District. *Kendal* 1.

1856.

161. DE KONINCK On the Mountain Limestone and Old Red of Britain. *Quart. Journ. Geol. Soc., vol.* xiii., *pt.* 2, *p.* 12.
162. MORTON, G. H. On the Sub-divisions of the New Red Sand Stone between the River Dee and the Up-throw of the Coal Measures east of Liverpool. *Proc. Lit. and Phil. Soc., Manchester,* No. 10, *pp.* 68–76.
163. PIGOTT, DR. On the Harrogate Spas. *8vo.*
163a. TEALE, T. P. On some Works of Man associated with the remains of Extinct Mammals in the Aire Valley Deposit, *Proc. West Riding Geol. Soc.*
163b. BINNEY, E. W. - Additional Observations on the Permian Beds of the North West of England. *Mem. Lit. and Phil. Soc., Manchester, Ser.* 2, *vol.* xiv., *pp.* 101–120.

1857.

164. RENNIE, G., BOULT, J., and HENDERSON, A. Report from the Committee appointed to investigate and report upon the effects produced upon the Channels of the Mersey by the alterations which within the last 50 years have been made on its banks. *Rep. Brit. Assoc. for* 1856, *p.* 1.

1858.

165. ANON.: (G. W.) - On Fish Remains in the Yorkshire Coalfield. *Geologist, vol.* i. *p.* 51.
166. BAINES, S. - On the Yorkshire Flagstone with Fossils. (*Brit. Assoc.*) *Geologist, vol.* i. *p.* 537.
167. BRUNLEES, JAMES Description of the Iron Viaducts erected across the Tidal Estuaries of the Rivers Leven and Kent, in Morecambe Bay. (Note of the Estuary Sands.) *Proc. Inst. Civ. Eng., vol.* xvii. *p.* 442.
168. BUNBURY, Sir CHAS. - On a remarkable specimen of Neuropteris, with Remarks on the Genus. *Quart. Journ. Geol. Soc., vol.* xiv. *p.* 243.
169. DAVIDSON, THOMAS - A Monograph of the British Fossil Brachiopoda, Part iv. The Permian Brachiopoda. *Palæontograph. Soc.* 4to. *Lond.*
170. ——————— Do. Do. The Carboniferous Brachiopoda. No. 1. *Ibid.*
171. DE KONINCK, PROF. L. On some new Palæozoic Crinoids from England and Scotland. *Geologist, vol.* i. *pp.* 146–178.

1859.

172. DAVIDSON, THOMAS — A Monograph of the British Fossil Brachiopoda. Part v. The Carboniferous Brachiopoda. No. 2. (Lancashire, Pl. 16.) *Palæontograph. Soc. 4to. London.*

173. ———————— Gems from private collections. Spirifera Convoluta, (from the Carb. Lime. of Thorneley) N. E. of Preston. *Geologist, vol.* ii., *p.* 313.

174. EDDY, STEPHEN — On the Lead Mining Districts of Yorkshire. *Brit. Assoc. Reports for* 1858, *p.* 167.

175. HULL, E. — Quarter Sheet. 80 N.W. 1-inch Map. *Geol. Survey.*

176. PHILLIPS, PROF. J. — Anniversary Address to the Geological Society of London. (With note on Iron Ore of West Cumberland and North Lancashire, p. xliii.) *Quart. Jour. Geol. Soc. vol.* xv.

177. ———————— and R. BARKER — On the Hæmatite Iron Ores of North Lancashire and West Cumberland. *Rep. Brit. Assoc. for* 1858, *Trans. of Sections, p.* 106.

178. SORBY, H. C. - - The Structure and Origin of the Millstone Grit of South Yorkshire. *E. Baines and Sons, Leeds; and Geol. and Polyt. Soc. West Riding, Yorkshire, vol.* iii., *pp.* 669–675.

179. ———————— On the currents present during the deposition of the Carboniferous and Permian Strata in South Yorkshire and North Derbyshire. *Rep. Brit. Assoc. for* 1858, *Trans. of Sections, p.* 108.

179a. TEALE, T. P. - On the superficial Deposits of the Aire at Leeds. *Brit. Assoc. Reports for* 1858.

180. TURNER, W. On some Fossil Bovine Remains found in Britain. *Edin. New Phil. Journ., Ser.* 2, *vol.* x. *p.* 31.

1860.

181. BAINES, S. - - On some of the Differences of the Deposition of Coal. *Proc. Geol. and Polyt. Soc., W. Riding, Yorks., vol.* iv. *p.* 77.

182. BINNEY, E. W. - Excursion to Burnley. *Trans. Manchester, Geol. Soc., vol.* ii. (*Part* 5), *pp.* 49–52.

183. ———— Observations on the Fossil Shells of the Lower Coal Measures. *Ibid.* (*Part* 7), *pp.* 72–84.

184. ————————— Excursion to Brooks-bottom and the Neighbourhood. *Ibid.* (*Part* 8), *pp.* 85–89.

184a. DENNY, H. - - On the Geological and Archæological contents of the Victoria and Dowkerbottom Caves in Craven. *Proc. West Riding Geol. Soc., vol.* iv.

185. DICKINSON, JOSEPH - Sulphate of Lime Crystals from Haughton, near Denton. *Trans. Man. Geol. Soc., vol.* ii., *part* 4, *pp.* 47–8.

186. GAGES, A. Report on the result obtained by the Mechanico-Chemical Examination of Rocks and Minerals. *Rep. Brit. Assoc. for* 1859, *p.* 65.

187. HULL, E. Quarter Sheet 89 S.W. 1-inch Map. *Geol. Survey.*

188. ———————— Sheet 92. Lancashire 6-inch Map. Neighbourhood of Bickerstaffe. *Ibid.*

189. ———————— Sheet 100. Lancashire 6-inch Map. Neighbourhood of Knowsley, Rainford. *Ibid.*

190. ———————— Sheet 107. Lancashire 6-inch Map. Neighbourhood of Prescott. *Ibid.*

191. ———————— Sheet 108. Lancashire 6-inch Map. Neighbourhood of St. Helens, Burton Wood. *Ibid.*

192. HULL, E.		On the New Subdivision of the Triassic Rocks. *Trans. Man. Geol. Soc.*, vol. ii. (Part 3), pp. 22-34.
193. MORDACQUE, S. H.	REV.	On a Stalactite found in Flagstone Rock near Haslingden. *Proc. Geol. Assoc.*, vol. i., p. 46.
194. MORTON, G. H.		Traces of Icebergs near Liverpool. *Proc. Lit. and Phil. Soc., Liverpool.*
195. TAYLOR, JOHN		Letter on Fossil Fruit from Upper Coal Measures near Bolton. *Geologist*, vol. iii., p. 198.
196. TOMLINSON, C		On the Action of Heat on certain Sandstones of Yorkshire (near Huddersfield). *Proc. Geol. Assoc.*, vol. i., p. 50.

1861.

197. DAVIDSON, THOMAS	A Monograph of the British Fossil Brachiopoda. Part V. The Carboniferous Brachiopoda (No. 3, pp. 81-116). *Palæontograph. Soc.* 4to. London.
198. GREEN, A. H. -	On the Faults of the Lancashire Coalfield. *Geologist*, vol. iv., p. 538 (*Brit. Assoc.*).
199. HULL, E.	Quarter-sheet 80, N.E. 1-inch Map. *Geol. Survey.*
200. —————	Sheet 81. Lancashire 6-inch Map. Neighbourhood of Wardle. *Ibid.*
201. —————	Sheet 84. Lancashire 6-inch Map. Neighbourhood of Ormskirk, St. John's. *Ibid.*
202. —————	Sheet 85. Lancashire 6-inch Map. Neighbourhood of Standish. *Ibid.*
203. —————	Sheet 93. Lancashire 6-inch Map. Neighbourhood of Wigan, Up Holland. *Ibid.*
204. —————	Sheet 101. Lancashire 6-inch Map. Neighbourhood of Billinge, Ashton. *Ibid.*
205. —————	Sheet 27. Vertical Sections. *Ibid.*
206. —————	The Geology of the Country around Altrincham, Cheshire (80 N.E.). *Geol. Survey Memoirs.*
207. —————	The Coalfields of Great Britain. *London.*
208. —————	Notes on the Glacial Phenomena of Wastdale, Cumberland. *Geologist*, vol. iv., p. 478.
209. MORTON, G. H.	On the Pleistocene Formation of the District around Liverpool. *Ibid.*, p. 211.
210. —————	On the Coal Measures in the Neighbourhood of Liverpool, and the probability of their extension beneath the town. *Proc. Lit. and Phil. Soc., Liverpool.*
211. —————	On the Basement Bed of the Keuper Formation in Wirral and the South-west of Lancashire. *Proc. Liverpool Geol. Soc., Sessions* 1 *and* 2, pp. 4, 5.
212. —————	On the Pleistocene Deposits of the District around Liverpool. *Ibid.*, pp. 12-14.
213. SAINTER, J. D.	A Salt Spring in a Coal Mine (at Dukinfield). *Geologist*, vol. iv., p, 398.
214. SALMON, H. C.	New Caves in Yorkshire (Letter on). *Ibid.*, p. 312.
215. TOMLINSON, C.	On the Efflorescence which succeeds the Action of Heat on certain Sandstones of Yorkshire. *Proc. Geol. Assoc.*, vol. i., p. 158.
216. WHITAKER, J., and WILKINSON, T.T.	The Burnley Coalfield. *Geologist*, vol. iv., p. 508. (*Brit. Assoc.*)

1862.

217. BINNEY, E. W. - -	A Succinct Account of the Geological Features of the Neighbourhood of Manchester. *Rep. Brit. Assoc.* 1861, *Trans. of Sections*, p. 109.

218. BINNEY, E. W. - On the Geology of Manchester. *Geologist, vol. v., p. 463. (Man. Geol. Soc.)*
219. ——— - Additional Observations on the Permian Beds of South Lancashire. *Mem. Lit. and Phil. Soc. Manchester, Ser. 3, vol. ii., p. 29.*
220. ——— On some Fossil Plants, showing Structure, from the Lower Coal Measures of Lancashire. *Quart. Journ. Geol. Soc., vol. xviii., p. 106.*
221. ——— Observations on Down Holland Moss. *Trans. Man. Geol. Soc., vol. iii., p. 9.*
222. ——— - On Jelly Peat, found at Church Town, near Southport. *Ibid., p. 19.*
223. ——— An Account of the Excursion to Todmorden. *Ibid., p. 325.*
224. ——— Geology of Manchester and its Neighbourhood. *Ibid., p. 350.*
225. ——— On Drift near Manchester. *Proc. Lit. and Phil. Soc. Manchester, vol. iii., No. 3, pp. 15, 16.*
226. FAIRBAIRN, SIR WM. - Remarks on the Temperature of the Earth's Crust as exhibited by Thermometrical Returns obtained during the sinking of the Deep Mine at Dukinfield. *Rep. Brit. Assoc. for 1861, Trans. of Sections, p. 53.*
227. HULL, E. - Quarter-sheet 89 S.E., 1-inch Map. *Geol. Survey.*
228. ——— - Sheet 86. Lancashire, 6-inch Map. Neighbourhood of Haddington, Horwick. *Ibid.*
229. ——— - Sheet 94. Lancashire, 6-inch Map. Neighbourhood of West Houghton, Hindley, Atherton. *Ibid.*
230. ——— - Sheet 95. Lancashire, 6-inch Map. Neighbourhood of Radcliffe, Peel, Swinton. *Ibid.*
231. ——— - Sheet 102. Lancashire, 6-inch Map. Neighbourhood of Leigh Lowton. *Ibid.*
232. ——— - Sheet 103. Lancashire, 6-inch Map. Neighbourhood of Astley, Eccles. *Ibid.*
233. ——— - Sheet 104. Lancashire, 6-inch Map. Neighbourhood of Manchester and Salford. *Ibid.*
234. ——— - The Geology of the Country around Bolton Lancashire, (89 S.E.). *Geol. Survey Memoirs.*
235. ——— The Geology of the Country around Wigan (89 S.W.). *Ibid.*
236. ——— The Burnley Coalfield, Lancashire. *Mining and Smelting Mag., vol. i., p. 163.*
237. ——— - The Lancashire and Cheshire Coalfields. *Ibid., p. 85.*
238. ——— - Explanation of the Lancashire Maps of the Geological Survey (with Sections). *Trans. Manchester Geol. Soc., vol. iii., p. 23.*
239. ——— - Marine Shells at Dukinfield (in Coal). *Ibid., p. 348.*
240. JONES, PROF. T. R. - A Monograph of the Fossil Estheriæ. *Palæontograph. Soc.* 4to, Lond. (*Lancashire, pp. 24, 28–30, 32, 115, &c.; Yorkshire, p. 101.*)
241. KIRKBY, J. W. - On some Remains of Chitons from the Mountain Limestone of Yorkshire. *Quart. Journ. Geol. Soc., vol. xviii., p. 233.*
242. KNOWLES, A. - On the Bank Top and Hagside Pits, and the proving of Faults. *Trans. Manchester Geol. Soc., vol. iii., p. 190.*
243. MORTON, G. H. On Glacial Surface Markings on the Sandstone near Liverpool. *Quart. Journ. Geol. Soc., vol. xviii., p. 377.*

244. Morton, G. H. - The Geology of the Country around Liverpool.
245. ———— - On the Pleistocene Deposits of the District around Liverpool. *Rep. Brit. Assoc. for* 1861, *Trans. of Sections,* p. 120.
246. Taylor, John Pleistocene Deposits on the Stockport and Woodley Railway. *Trans. Manchester Geol. Soc., vol.* iii., p. 147.
247. ———— - On the Geology of the Railway between Hyde and Marple. *Ibid.,* p. 296.

1863.

248. Binney, E. W. - [Note on the Permian Beds.] *Proc. Lit. and Phil. Soc., Manchester, vol.* iii., No. 6, pp. 35–37.
249. ———— - [Note on Drift with remains of human work.] *Ibid.,* No. 10, pp. 74–77.
250. ———— - Section of the Drift near Rainford. *Geologist, vol.* vi., p. 307 (*Man. Geol. Soc.*).
251. Davidson, Thos. - A Monograph of the British Fossil Brachiopoda. Part V. The Carboniferous Brachiopoda, No. 5. *Palæontograph. Soc.* 4to. London.
252. Dickinson, Joseph - On the Coal Strata of Lancashire. *Geologist, vol.* vi., p. 261.
253. Hodgson, Miss E. - On a Deposit containing Diatomaceæ, Leaves, &c., in the Iron Ore Mines near Ulverston. *Quart. Journ. Geol. Soc., vol.* xix., p. 19.
254. Hull, E. - Quarter-sheet, 88 S.W. 1-inch Map. *Geol. Survey.*
255. ———— - Sheet 87. Lancashire, 6-inch Map. Neighbourhood of Bolton-le-Moors. *Ibid.*
256. Hull, E., and Green, A. H. Sheet 88. Lancashire 6-inch Map. Neighbourhood of Bury, Heywood. *Ibid.*
257. Hull, E. - Sheet 96. Lancashire, 6-inch Map. Neighbourhood of Middleton, Prestwich. *Ibid.*
258. ———— - Sheet 97. Lancashire, 6-inch Map. Neighbourhood of Oldham. (Revised 1866.) *Ibid.*
259. ———— - Sheet 113. Lancashire, 6-inch Map. Part of Liverpool, &c. *Ibid.*
260. ———— - The Geology of the Country around Oldham. (88 S.W.) *Geol. Survey Memoirs.*
261. Jackson, S. B. - On the Wigan Coalfield. *Proc. Liverpool Geol. Soc., Session* iv., pp. 9–13. *Geologist, vol.* vi., p. 463.
262. Morton, G. H. On the Surface Markings near Liverpool, supposed to have been caused by Ice. *Proc. Liverpool Geol. Soc., Session* 3, pp. 9–11.
263. ———— On the Thickness of the Bunter and Keuper Formations around Liverpool. *Ibid.,* p. 15.
264. Sainter, J. D. - Bones at Macclesfield. *Geologist, vol.* vi., p. 185.
265. Salter, J. W. - On some Species of Eurypterus and allied forms. *Quart. Journ. Geol. Soc., vol.* xix., p. 81.
266. Tate, A. N. On the Composition of Black Sandstone occurring in the Trias round Liverpool. *Proc. Liverpool Geol. Soc., Session* 4, p. 16.
267. Wilkinson, T. T. On the Drift Deposits near Burnley. *Geologist, vol.* vi., p. 192. (*Man. Geol. Soc.*)
268. ———— Section of Strata near Worsthorne, Burnley. *Proc. Lit. and Phil. Soc. Manchester, vol.* iii. (No. 5), pp. 31, 32.

1864.

269. AITKEN, JOHN — Notice of Bones from the Valley of the Irwell, near Rawtenstall. *Trans. Man. Geol. Soc., vol.* iv. (No. 15), *pp.* 333–5.

270. BINNEY, E. W. - The Geology of Manchester. *Ibid., vol.* iv., *p.* 217.

271. ————— Notice of Fossils from the Lancashire Coalfield. *Ibid. vol.* iv. (No. 14), *p.* 317.

272. ————— A few Remarks on the Lancashire and Cheshire Drift. *Proc. Lit. and Phil. Soc. Manchester, vol.* iii. (No. 10), *pp.* 214–216, and *Geologist, vol.* vii., *p.* 140.

273. BRAMALL, A. On Sinking through Drift Deposits (Section at Rainford, p. 202 ; also p. 212). *Trans. Manchester Geol. Soc., vol.* iv., *p.* 194.

274. DICKINSON, J. On the Coal Strata of Lancashire. *Ibid., vol.* iv., *p.* 155.

275. GEINITZ, DR. The Dyas or Permian Formation in England. Translated from "Dyas, &c.," by E. W. BINNEY and J. W. KIRKBY. *Ibid., vol.* iv., *p.* 120.

276. HODGSON, MISS E. On Helix and perforated Limestone (near Ulverston). *Geologist* 1864, *p.* 42.

277. HULL, E. Quarter-sheet 81, N.W. 1-inch Map. *Geol. Survey.*

278. HULL, E., and GREEN, A. H. Sheet 89. Lancashire, 6-inch Map. Neighbourhood of Rochdale. *Ibid.*

279. HULL, E. Sheet 112. Lancashire, 6-inch Map. Neighbourhood of Stockport, &c. *Ibid.*

280. ————— Sheet 105. Lancashire, 6-inch Map. Neighbourhood of Ashton-under-Lyne. Revised 1866. *Ibid.*

281. ————— Sheet 64. Horizontal Sections. *Ibid.*

282. HULL, E., and GREEN, A. H. On the Millstone Grits of North Staffordshire, and the adjoining parts of Derbyshire, Cheshire, and Lancashire. *Quart. Journ. Geol. Soc., vol.* xx., *p.* 242.

283. PERCY, DR. J. Metallurgy (vol. 2). Iron and Steel. 8vo. *Lond.*

284. ROFE, J. - On some recent Marine Shells found in the excavations for railway works at Preston. *Proc. Geol. Assoc., vol.* i., *p.* 321.

285. TAYLOR, J. Geological Essays and Sketch of the Geology of Manchester and the Neighbourhood. 8vo.

286. WHITAKER, J. On some Teeth and a Flint found in the Drift at Barrowfield near Burnley. *Trans. Manchester Geol. Soc., vol.* iv., *p.* 176.

287. WILD, G. - On the Fulledge Section of the Burnley Coalfield. *Ibid., vol.* iv., *p.* 179.

288. WILKINSON, T. T. On the Drift Deposits near Burnley. *Trans. Man. Geol. Soc., vol.* iv., *p.* 108.

289. WOODWARD, HENRY On the Eurypteridæ (Manchester). *Geol. Mag., vol.* i., *p.* 239.

290. WYNNE, R. H. - On Coal Mining in Lancashire. *Trans. S. Wales Inst. of Engineers, vol.* iv., *p.* 20.

1865.

291. AITKEN, JOHN On certain appearances of Glacial Action on Rock Surfaces near Clitheroe. (Man. Geol. Soc.) *Geol. Mag., vol.* ii., *p.* 179.

292. ANON. - - On a bed of Lower Boulder Clay at Heaton Mersey near Manchester. *Ibid., vol.* ii., *p.* 236.

293. BINNEY, E. W. Additional Observations on the Permian Beds of South Lancashire. *Mem. Lit. and Phil. Soc., Manchester, Ser.* 3, *vol.* ii., *pp.* 29–47.

294. BINNEY, E. W. A few Remarks on Mr. Hull's Additional Observations on the Drift Deposits in the neighbood of Manchester. *Ibid., pp.* 462–4.
295. ——— Further Observations on the Permian and Triassic Strata of Lancashire. *Mem. Lit. and Phil. Soc., Manchester*, vol. iv., No. 13, *pp.* 134–139.
296. ——— - A description of some Fossil Plants showing Structure found in the Lower Coal Measures of Lancashire and Yorkshire. *Phil. Trans., vol.* clv., *p.* 579.
297. DARBISHIRE, R. D. - On the Genuineness of certain Fossils from the Drift Bed at Macclesfield. *Geol. Mag., vol.* ii,, *p.* 293.
298. DAWKINS, W. B., WOOD, E., and ROBERTS, G. E. On the Mammalian Remains found near Richmond, Yorkshire. With an introductory note on the deposit in which they were found. *Quart. Journ. Geol. Soc., vol.* xxi., *p.* 493.
299. HULL, E. Sheets 65, 66, 67, 68. Horizontal Sections. *Geol. Survey.*
300. ——— New Red Conglomerate near Manchester (Letter on). *Geol. Mag., vol.* ii., *p.* 429.
301. ——— - On the New Red Sandstone and Permian Formations as Sources of Water Supply for Towns. *Mem. Lit. and Phil. Soc., Manchester, Ser.* 3, *vol.* ii., *pp.* 256–276.
302. ——— Additional Observations on the Drift Deposits and more recent Gravels in the neighbourhood of Manchester. *Ibid., pp.* 449–461.
303. MACKINTOSH, D. Marine Denudation illustrated by the Brimham Rocks. *Geol. Mag., vol.* ii., *p.* 154.
304. MORTON, G. H. - The Geology of the Country around Liverpool. *8vo. Liverpool.*
305. PATERSON, — On the Geology of Warrington. (Warrington Field Nat. Soc.) *Geol. Mag., vol.* ii., *p.* 376.
306. PHILLIPS, PROF. JOHN Note on the Geology of Harrogate. *Quart. Journ. Geol. Soc., vol.* xxi., *p.* 232.
307. ——— On the Formation of Valleys near Kirkby Lonsdale. *Brit. Assoc. Rep. for* 1864 *Trans. of Sections, p.* 63.
308. ——— On the Distribution of Granite Blocks from Wastdale Craig. *Ibid., p.* 65.
309. RAMSAY, PROF. A. C. - The Ice Drifted Conglomerates of the Old Red Sandstone. *Reader, vol.* vi., *p.* 186.
310. ROBERTS, G. E. Geological Notes on the Mountain Limestone of Yorkshire. *Geol. Mag., vol.* vi., *p.* 163.
311. ROFE, JOHN Description of a new species of *Actinocrinus* from the Mountain Limestone of Lancashire (near Clitheroe). *Ibid., p.* 12.
312. ——— Notes on some Echinodermata from the Mountain Limestone, &c. *Ibid., p.* 245.
313. SAINTER, JOHN On the Macclesfield Drift Beds. *Ibid., p.* 365.
314. SMITH, ECROYD (On the Victoria Cave.) *Trans. Historic Soc., Cheshire.*

1866.

315. AITKEN, J. - - On certain Appearances of Glacial Action on Rock Surfaces near Clitheroe, and incidentally a brief description of the chief Geological Features of that locality. *Trans. Man. Geol. Soc., vol.* v., *p.* 84.

316. AITKEN, J. - On the Union of the Gannister and Higher-Foot Coal Mines at Bacup, together with some remarks on the circumstances under which it occurs. *Ibid., p.* 185.
317. BINNEY, E. W. - A few Remarks on the so-called Lower New Red Sandstone of Central Yorkshire. *Geol. Mag., vol.* iii., *p.* 49.
318. ——— - The so-called Lower New Red Sandstone of Plumpton, Yorkshire. *Ibid., p.* 473.
319. ——— - On Calcareous Nodules, with wood, from Coal near Oldham. *Proc. Lit. and Phil. Soc., Manchester, vol.* v., *No.* 13, *pp.* 113–115.
320. ——— - [On Specimens from Birkdale Park, Southport.] *Ibid., vol.* vi., *No.* 3, *pp.* 17, 18.
321. ——— - Account of Calcareous Nodules from the Lower Coal-seams of Lancashire and Yorkshire, full of Fossil Wood. *Proc. Lit. and Phil. Soc., Manchester, vol.* v., *No.* 7, *pp.* 61–64.
322. ——— - Remarks on Specimen from Wigan. (Manchester Geol. Soc.) *Geol. Mag. vol.* iii., *p.* 271.
323. BOULT, JOS. Further Observations on the alleged Submarine Forests on the Shores of Liverpool Bay and the River Mersey. *Trans. Hist. Soc., Lancashire and Cheshire, vol.* vi.
324. BROCKBANK, W. - Notes on a Section of Chat Moss near Astley Station. *Proc. Lit. and Phil. Soc., Manchester, vol.* v., *No.* 9, *pp.* 91–95.
325. BUTTERWORTH, J. Fossil Wood (Lancashire Coal). *Science Gossip*, No. 23, *pp.* 250–1.
326. DICKINSON, JOSEPH - On some of the Leading Features of the Lancashire Coalfield. *Trans. N. Inst. Mining Eng., vol.* xv., *pp.* 13–17.
327. FARRER, J. W. - Further Explorations in the Dowker-bottom Caves in Craven. *Proc. W. Riding Geol. Soc., vol.* iv.
328. GREEN, A. H. On the River Denudation of Valleys. *Geol. Mag., vol.* iii., *p.* 523.
329. GREENWELL, G. C. On an Impression of Lepidodendron Sternbergii found at Poynton. *Trans. Man. Geol. Soc., vol.* v., *p.* 194.
330. HARDWICK, C. - - A few Thoughts on Geology in its relation to Archæology. *Ibid., p.* 201.
330a. HEWLETT, A. - On the Wigan Coalfield. *Journ. Liverpool Polytechnic Soc.*
331. HUGHES, T. McK. Note on the Silurian Rocks of Casterton Low Fell, Kirkby Lonsdale, Westmoreland. *Geol. Mag. vol.* iii., *p.* 206.
332. HULL, E., and GREEN, A. H. The Geology of the Country around Stockport, Macclesfield, Congleton, and Leek (81 N.W. and S.W.). *Geol. Survey Memoirs.*
333. HULL, E. River Denudation of Valleys. *Geol. Mag., vol.* iii., *p.* 474 and 570.
334. ——— - - Modern Views of Denudation. *Pop. Science Rev., vol.* v., *p.* 453.
335. HUME, REV. DR. On the Changes in the Sea Coast of Lancashire and Cheshire. *Trans. Hist. Soc. Lanc. and Chesh., vol.* vi.
336. MORTON, G. H. On the Geology of the Country bordering the Mersey and Dec. *Liverpool Naturalist's Journal*, No. 1, *p.* 15; *Proc. Liverpool Geol. Soc., Session* 7, *pp.* 37–42.
337. ——— On the Position of the Public Wells for the Supply of Water in the Neighbourhood of Liverpool. *Proc. Liverpool Geol. Soc., Session* 7, *pp.* 27–30; and *Geol. Mag., vol.* iii., *p.* 81.

338. NEILD, J. — — Sandstone Markings. *Science Gossip*, No. 23, pp. 258, 259.
339. PHILLIPS, PROF. J. On Glacial Striation [Wastdale]. *Rep. Brit. Assoc. for* 1865, *Trans. of Sections*, p. 71.
340. PLANT, J. Alluvial Deposits on Travis Isle, Collyhurst. *Trans. Man. Geol. Soc.*, vol. v., p. 56.
341. ROFE, J. — — Notes on the Starr Hills of the Lancashire Coast. *Proc. Geol. Assoc.*, vol. iii., p. 13.
342. WILKINSON, T. T. Additional Notes on the Drift Deposits in Burnley and the Neighbourhood. *Trans. Man. Geol. Soc.*, vol. v. p. 45.
343. WHITAKER, J. — On the Outcrop of the Lower Coal-measure Rocks on Boulsworth and Gorple, together with Observations on the Origin of some "Rock-Basins" therein. *Ibid.* p. 94.

1867.

344. BINNEY, E. W. — [On Two Fossils from the Coal Measures, near Huddersfield.] *Proc. Lit. and Phil. Soc. Man.*, vol. vi., No. 8, p. 59.
345. ——————— — On a Section at Ardwick. *Ibid.*, No. 12, pp. 119–121.
346. ——————— Note on the Age of the Hæmatite Iron Deposits of Furness. *Ibid.* vol. vii., No.5, pp. 55–61. Discussion by BROCKBANK, W., pp. 59–61.
347. ——————— — On the Upper Coal Measures of England and Wales. *Trans. Man. Geol. Soc.*, vol. vi., p. 38.
348. ——————— On the Drift of the Western and Eastern Counties. *Geol. Mag.*, vol. iv., p. 231.
349. CURRY, J. — On the Drift of the North of England. *Quart. Journ. Geol. Soc.* vol. xxiii., p. 40.
350. HODGSON, MISS E. The Moulded Limestones of Furness. *Geol. Mag.*, vol. iv., p. 401.
351. HUGHES, T. McK. — On the Break between the Upper and Lower Silurian Rocks of the Lake District, as seen between Kirkby Lonsdale and Malham, near Settle. *Ibid.* p. 346.
352. HULL, E. — — Sheet 71. Lancashire, 6-inch Map. Neighbourhood of Haslingden. *Geol. Survey*.
353. ——————— — Sheet 72. Lancashire, 6-inch Map. Neighbourhood of Cliviger, Bacup. *Ibid.*
354. ——————— · Sheet 73. Lancashire, 6-inch Map. Neighbourhood of Todmorden. *Ibid.*
355. ——————— — Sheet 78. Lancashire, 6-inch Map. Neighbourhood of Belmont. *Ibid.*
356. ——————— · Sheet 79. Lancashire, 6-inch Map. Neighbourhood of Entwistle. *Ibid.*
357. ——————— Sheet 80. Lancashire, 6-inch Map. Neighbourhood of Tottington. *Ibid.*
358. ——————— · Sheet 109. Lancashire, 6-inch Map. Neighbourhood of Winwick. *Ibid.*
359. ——————— · — Faults in Drift (Rochdale). *Geol. Mag.*, vol. iv., p. 182.
360 ——————— On the Parallelism of the Drift Deposits in Lancashire and the Eastern Counties. *Ibid.* p. 183.
361. ——————— — — On a Section of the Drift Deposit in the Banks of the Ribble, near Balderston Hall. *Proc. Lit. and Phil. Soc., Manchester*, vol. vi., No. 13, pp. 136–140.
362. MORTON, G. H. — On the presence of Glacial Ice in the Valley of the Mersey during the Post-pliocene Period. *Proc. Liv. Geol. Soc.*, Session 8, p. 4.

363. OWEN, PROF. R.   On the Mandible and Teeth of Cochliodonts. *Geol. Mag.*, vol. iv., p. 59.
364. PALEY, F. A.   On a Series of Elevated Sea Terraces on Hampsfell, near Cartmell, Lancashire. *Proc. Camb. Phil. Soc., Parts* v., vi. *pp.* 107, 108.
365. PEACOCK, R. A.   Gradual change of Form and Position of Land in the South End of the Isle of Walney. *Rep. Brit. Assoc. for* 1866. *Trans. of Sections*, p. 66.
366. TAYLOR, J. E.   On the Parallellism of the Drift Deposits in Lancashire and Norfolk. *Geol. Mag., vol.* iv., p. 281.

1868.

367. AITKEN, JOHN   Notes on the Origin and Structure of a Flint Pebble found in the Drift on Holcombe Hill. *Geol. and Nat. Hist. Rep., vol.* ii., p. 192. (*Trans. Man. Geol. Soc.* .)
368. ————   On Mr. Hull's Horizontal Section of Mid-Lancashire. *Ibid., vol.* ii., p. 192. (*Ibid.* .)
369. ————   The Geology of Rossendale. A Chapter in Thos. Newbigging's "History of the Forest of Rossendale." London.
370. ————   Excursion of the Manchester Geological Society to Bacup and Todmorden. *Trans. Man. Geol. Soc., vol.* vi. p. 22.
371. ————   ——— from Stubbins to Bacup. *Ibid.,* p. 67.
372. ————   ——— to Clitheroe and Pendle Hill. *Ibid., vol.* vii., p. 15.
373. ————   Remarks on an Outlier of Drift Gravel on Holcombe Hill. *Ibid., vol.* vii. p., 57, and p., 80.
374. AVELINE, W. T., HUGHES, T. McK., TIDDEMAN, R. H.   Quarter-sheet 98, S.E. 1-inch Map. *Geol. Survey.*
375. BINNEY, E. W.   Observations on the Structure of Fossil Plants found in the Carboniferous Strata (Part I. Calamites and Calamo-dendron). *Mem. Palæontog. Soc.* 4to. London, pp. 1–32.
376. ————   Notes on the Excursion of the Members of the Manchester Geological Society to Disley. *Trans. Man. Geol. Soc., vol.* vi. p. 103.
377. ————   On the Upper Coal Measures of England and Scotland. *Ibid.,* p. 38.
378. ————   [Note on Pholas borings (Furness). *Proc. Lit. and Phil. Soc. Manchester, vol.* vii. (No. 7), pp. 76, 77.
379. ————   Description of a Dolerite at Gleaston, in Low Furness. *Ibid.* (No. 12), pp. 148–154.
380. ECCLES, JAMES   On some Instances of the Superficial Curvature of Inclined Strata near Blackburn. *Trans. Man. Geol. Soc., vol.* vii., p. 20.
381. ————   On the Excursion to Holcombe, &c. *Ibid.,* p. 36.
382. GREEN, A. H.   Sea Cliffs and Escarpments (Letter on). *Geol. Mag., vol.* v., p. 40.
383. GREEN, A. H., DAKYNS, J. R., WARD, J. C.   Quarter-sheet 88 S.E. 1-inch Map. *Geol. Survey.*
384. HUGHES, T. McK.   Notes on the Geology of Parts of Yorkshire and Westmoreland. *Proc. West Riding Geol. Soc., vol.* iv.
385. HULL, E.   Quarter-sheet 89 N.E. 1-inch Map. *Geol. Survey.*

386. HULL, E. — Observations on the Relative Ages of the leading Physical Features and Lines of Elevation of the Carboniferous District of Lancashire and Yorkshire. *Quart. Journ. Geol. Soc.*, vol. xxiv., p. 323.
387. MOON, M. A. — Geological Notes on Iron Ores with special reference to West Cumberland and North Lancashire Hæmatite Deposits. *Whitehaven.*
388. MORRIS, J. P. — On the so-called Fossils from the Iron Ore of Furness and Millom. 8vo. *From the Furness, Cartmel, and Grange Visitor, July 11, 1868.*
389. PLANT, J. On the Glacial Groovings on the Bunter Sandstone at Orsdale Clough, Salford. *Trans. Man. Geol. Soc.*, vol. vii., p. 40 and vol. vi., p. 120.
390. ———— Remarks on a Stone Axe found in the Valley of the Mersey at Flexton, near Manchester, in 1846 (with Section). *Ibid.*, p. 46, and p. 65. (*Geol. and Nat. Hist. Rep.*, vol. ii., p. 187.)
391. ———— The Alluvial Gravels at Ordsall Lane, Salford. *Ibid.*, p. 95. (*Ibid.*, vol. ii., p. 277.)
392. THORPE, T. E. Analysis of the Water of the Holy Well, a Medicinal Spring at Humphrey Head, N. Lancashire. *Quart. Journ. Chem. Soc.*, Ser. 2, vol. vi., pp. 19–25.
393. TIDDEMAN, R. H. - The Valleys of Lancashire (Letter on). *Geol. Mag.*, vol. v., p. 39.

1869.

394. BEWICK, J. On Mining in the Mountain Limestone of the North of England. *Trans. N. Inst. Mining. Eng.*, vol. xviii., pp. 163–182. Discussion on ditto. Vol. xix., pp. 92–95, 102–111 (1870).
395. BINNEY, E. W. Notes on the Lancashire and Cheshire Drifts. (Read in 1842.) *Trans. Man. Geol. Soc.*, vol. viii., p. 30.
396. ———— and TALBOT, J. H. On the Petroleum found in the Down Holland Moss near Ormskirk. (Read in 1843.) *Ibid.*, p. 41.
397. BOSTOCK, R. - The New Red Sandstone as a Source of Water Supply. *Proc. Liv. Geol. Soc., Session 10*, p. 58.
398. BOULTON, JOHN Geological Fragments. *Ulverston.* 8vo.
399. BUTTERWORTH, J. Entomostraca in Shale (Bradford). *Science Gossip*, No. 53, pp. 111–112.
400. DE RANCE, C. E. Quarter-sheet, 90 S.E. 1-inch Map. *Geol. Survey.*
401. ———— - The Geology of the Country between Liverpool and Southport (90 S.E.) *Geol. Survey Memoirs.*
402. GREEN, A. H., DAKYNS, J. R., and WARD, J. C. Geology of part of the Yorkshire Coalfield, 88 S.E.
403. HAUGHTON, REV. DR. S. On the Theory of Secondary Joints, as illustrated by Mr. Hull's Paper on the Lines of Elevation of the Pendle Hills in Lancashire. *Journ. Roy. Geol. Soc., Ireland*, vol. ii. (Part 2), p. 163.
404. HODGSON, MISS E. - The Coast of Furness (Letter on). *Geol. Mag.*, vol. vi., p. 286.
405. HULL, E., TIDDEMAN, R. H., and GUNN, W. Sheet 48. Lancashire. 6-inch Map. Neighbourhood of Colne. *Ibid.*
405a. HULL, E., and TIDDEMAN, R. H. Sheet 56. Lancashire. 6-inch Map. Neighbourhood of Haggate. *Ibid.*
406. ———— Sheet 62. Lancashire. 6-inch Map. Neighbourhood of Balderstone. *Ibid.*

407. HULL, E. Sheet 63. Lancashire. 6-inch Map. Neighbourhood of Accrington. *Ibid.*
408. ———— Sheet 64. Lancashire. 6-inch Map. Neighbourhood of Burnley. *Ibid.*
409. ———— Sheet 65. Lancashire. 6-inch Map. Neighbourhood of Stiperden Moor. *Ibid.*
410. ———— Sheet 70. Lancashire 6-inch Map. Neighbourhood of Blackburn. *Ibid.*
411. ———— Sheet 33. Vertical Sections. *Ibid.*
412. ———— On the Evidence of a Ridge of Lower Carboniferous Rocks crossing the Plain of Cheshire beneath the Trias, &c. *Quart. Journ. Geol. Soc., vol. xxv., p.* 171.
413. MACKINTOSH, D. On the Correlation, Nature, and Origin of the Drifts of N.W. Lancashire and a part of Cumberland, with Remarks on Denudation. *Ibid., p.* 407.
414. ———— Apparent Lithodomous Perforations in Northwest Lancashire. *Ibid. p.* 280.
415. ———— Lithodomous Borings 667 feet above the Sea. *Geol. Mag., vol.* vi, *p.* 96.
416. MAW, G. On some raised Shell-beds on the Coast of Lancashire. *Ibid., p.* 72.
417. NICHOLSON, DR. H. A. Notes on the Green Slates and Porphyries of the Neighbourhood of Ingleton. *Ibid., p.* 213.
418. RICKETTS, DR. C. The Geology of the Neighbourhood of Ingleborough. *Proc. Liverpool Geol. Soc.*, Session 10, *p.* 34.
419. ROBERTS, J. On the Wells and Water of Liverpool. *Ibid., p.* 84.
420. ROFE. J. Note on the Enlargements on some Crinoidel Columns. *Geol. Mag., vol.* 6, *p.* 351.
421. WARD, J. C. On Beds of supposed Rothliegende Age near Knaresborough. *Quart. Journ. Geol. Soc., vol.* xxv., *p.* 291.

1870.

422. AITKEN, JOHN Notice of a Fossil (Bones) from the Lancashire Flag Rock. *Trans. Man. Geol. Soc., vol.* ix., No. 2, *pp.* 39, 40.
423. ———— On the Pholas-boring Controversy (Furness). *Ibid.,* No. 3, *pp.* 31-41.
424. AVELINE, W. T., GREEN, A. H., DAKYNS, J. R., WARD, J. C., RUSSELL, R. The Geology of the Carboniferous Rocks N. and E. of Leeds, and the Permian and Trias Rocks about Tadcaster. (93 S.W.) *Geol. Survey Memoirs.*
425. BOSTOCK, R. The Mersey and the Dee; their former Channels and Change of Level. *Proc. Liverpool Geol. Soc., &c.*, Session 11, *p.* 41.
426. DAWKINS, W. B. Excursion to Ardwick. *Trans. Man. Geol. Soc., vol.* ix. (No. 3), *pp.* 14-15.
427. DE KONINCK, PROF. On some new and remarkable Echinoderms from the British Palæozoic Rocks. *Geol. Mag., vol.* vii., *p.* 258 (translated from *Bull. Roy. Acad. Bruxelles*, 2me. Ser, *t.* xxviii., *p.* 57).
428. DE RANCE, C. E. Notes on the Geology of the Country around Liverpool. *Nature, vol.* ii., No. 46, *pp.* 391-394.
429. EARWAKER, J. P. Geological Discovery in Liverpool (Thatto Heath). *Ibid., p.* 397.
430. ECCLES, JAMES On Two Dykes recently found in North Lancashire. *Trans. Man. Geol. Soc., vol.* ix. (No. 1), *pp.* 26-8.

431. HODGSON, MISS E. The Granite Drift of Furness. *Geol. Mag.*, vol. vii., p. 328.
432. HULL, E., DAKYNS, J. R., WARD, J. C., STRANGWAYS, C. F. Quarter-sheet 88 N.W. 1-inch Map. *Geol. Survey*
433. HULL. E., TIDDEMAN, R. H. Quarter-sheet, 89 N.E. 1-inch Map. New Edition. *Geol. Survey.*
434. HULL, E., DE RANCE, C. E. Sheet 62. Horizontal Sections. *Ibid.*
435. HULL, E. - - - Sheet 34. Vertical Sections. *Ibid.*
436. ——— Recent Observations on Underground Temperature, or the causes of variation in different localities. *Quart. Journ. of Science*, vol. vii., p. 207.
437. KNOWLES, JOHN - Observations on the Temperature at the Pendleton Colliery. *Trans. Man. Geol. Soc.*, vol. ix., No. 2, pp. 72-83.
438. MACKINTOSH, D. - On the Nature, Correlation, and Mode of Accumulation of the Drift Deposits of the West Riding of Yorkshire. *Geol. and Polyt. Soc., W. Riding, Yorks.*
439. RICKETTS, DR. C. The Sections of Strata exposed on the St. Helens and Huyton Branch Railway. *Proc. Liverpool Geol. Soc.*, Session 11, p. 56.
440. ROFE, J. - - On some supposed Lithodomous Perforations in Limestone Rocks (near Ulverston). *Geol. Mag.*, vol. vii., p. 1
441. WOLLASTON, G. H. On the Erratic Blocks of the Skiddaw District. [Note.] *Ibid.*, p. 587.
442. WOOD, S. V., JUN. Observations on the Sequence of the Glacial Beds. *Ibid.*, pp. 17-61.

1871.

443. AITKEN, JOHN - The President's Address. (Notes on Sections at Stockport, Furness, and Settle, pp. 26-8.) *Trans. Man. Geol. Soc.*, vol. x., No. 1, p. 8.
444. ——— - Notice of Specimens from Drift, Holcombe Hill. *Ibid.*, No. 2, pp. 115-6.
445. BINNEY, E. W. - Lepidostrobus and some allied ones. *Palæontog. Soc.*, 4to. Lond., pp. 33-62.
446. ——— - Notes on some of the High Level Drifts in the Counties of Chester, Derby, and Lancaster. *Proc. Lit. and Phil. Soc. Man.*, vol. x., p. 66.
447. BOULTON, JOHN - Particulars of a First Exploration of the extensive and newly-discovered Cavern at Stainton, Low Furness. 12mo. *Ulverston.*
448. CAMERON, A. G. - Description of the recently discovered Caverns at Stainton (in Furness). *Geol. Mag.*, vol. viii., p. 312.
449. CARRUTHERS, W. Remarks on the Fossils from the Railway section at Huyton. *Rep. Brit. Assoc.*, 1870. *Trans. of Sections*, p. 71.
450. ——— - On the Sporangia of Ferns from the Coal Measures. *Rep. Brit. Assoc.*, 1870, *Trans. of Sections*, p. 71.
451. COAL COMMISSION Report of the Commissioners appointed to inquire into the several matters relating to Coal in the United Kingdom. Vol. i. General Report and 22 Sub-Reports. Vol. ii. General Minutes and Proceedings of Committees. *Lond.*
452. DAKYNS, J. R.; WARD, J. C. Sheet 260. Yorkshire, 6-inch Map. Neighbourhood of Honley. *Geol. Survey.*
453. ——— Sheet 272. Yorkshire, 6-inch Map. Neighbourhood of Holmfirth. *Ibid.*

454. DAWKINS, W. B. On the Formation of the Caves round Ingleborough. *Trans. Man. Geol. Soc., vol. x. (Part 2), pp.* 106-114.
455. DE RANCE, C. E. Quarter-sheet 90 N.E. 1-inch Map. *Geol. Survey.*
456. ——— Quarter-sheet 91 S.W. 1-inch Map. *Ibid.*
457. ——— On the Two Glaciations of the Lake District *Geol. Mag., vol. viii., p.* 107.
458. ——— On the Glacial Phenomena of Western Lancashire and Cheshire. *Quart. Journ. Geol. Soc., vol.* xxvi., *p.* 641.
459. ——— On the Post-Glacial Deposits of Western Lancashire and Cheshire. *Ibid., p.* 655, both published *in first No. of vol.* xxvii.
460. EVERETT, PROF. - Third Report of the Underground Temperature Committee (Rose Bridge Colliery, Ince, Wigan). *Rep. Brit. Assoc. for* 1870, *p.* 29.
461. GREEN, A. H., DAKYNS, J. R., WARD, J. C. Sheet 273. Yorkshire 6-inch Map. Neighbourhood of Penistone. *Geol. Survey.*
462. ——— Sheet 281. Yorkshire 6-inch Map. Neighbourhood of Langsett. *Ibid.*
463. GREEN, A. H., DAKYNS, J. R., WARD, J. C., and RUSSELL, R. The Geology of the Neighbourhood of Dewsbury, Huddersfield, and Halifax. 88 N.E. *Geol. Survey Memoirs.*
464. GUNN, W. - Sheet 49. Lancashire 6-inch Map. Neighbourhood of Laneshaw Bridge. *Geol. Survey.*
465. HIGGINS, REV. H. H. - On some Specimens supposed to be the Fossils of a Plant named Pycnophyllum, Flabellaria, Næggerathia, Cordaites, in the Ravenhead Collection of Fossils, Free Public Museum, Liverpool. *Proc. Liverpool Geol. Soc., Session* 12, *p.* 71.
466. HULL, E., and GUNN, W. Sheet 57. Lancashire 6-inch Map. Neighbourhood of Winewall. *Geol. Survey.*
467. HULL, E., and TIDDEMAN, R. H. Sheet 62. Lancashire 6-inch Map. (New Edition with alterations.) Neighbourhood of Balderstone. *Ibid.*
468. HULL, E., DE RANCE, C. E. Sheet 63. Horizontal Sections. *Ibid.*
469. HULL, E., and TIDDEMAN, R. H. Sheet 85. Horizontal Sections. *Ibid.*
470. KERR, — On Traces of Glacial Phenomena in the Valley of the River Irwell and its tributaries in Rossendale. *Trans. Man. Geol. Soc., vol. x. (No. 2), pp.* 116, 126.
471. LUCAS, J., STRANGWAYS, C. F., and DALTON, W. H. Sheet 201. Yorkshire 6-inch Map. Neighbourhood of Bingley. *Geol. Survey.*
472. MACKINTOSH, D. On the Drifts of the West and South Borders of the Lake District, and on the Three Great Granitic Dispersions. *Geol. Mag., vol.* viii., *pp.* 250, 303.
473. MORRISON, W. On the Exploration of the Settle Caves, Yorkshire (Abstract). *Transactions of the Plymouth Institution, vol. iv., Part* II.
474. MORTON, G. H. Anniversary Address. *Proc. Liverpool Geol. Soc., Session* 12, *p.* 3.
475. ——— On the Glaciated Condition of the Triassic Rocks around Liverpool. *Rep. Brit. Assoc. for* 1870, *Trans. of Sections, p.* 81.
476. PLANT, J. On a Flint-flake Core, found in the Upper Valley Gravel at Salford, Manchester. *Rep. Brit. Assoc. for* 1870, *Trans. of Sect. p.* 156.
477. RICKETTS, DR. C. On Sections of Strata between Huyton and St. Helen's. *Ibid., p.* 85.

478. ROBERTS, J. — Effect produced by Red Sandstone upon Salt Water. *Proc. Liverpool Geol. Soc., Session* 12. *p.* 66.
479. ———— — Section of the Boulder Clay at the Gasworks, Linacre, near Liverpool. *Ibid., p.* 68.
480. ROFE, J. - Notes on the Crinoidea (Clitheroe). *Geol. Mag., vol.* viii., *p.* 241.
481. SMITH, J. T. — Iron Ores Committee (Cumberland and Lancashire). *Journ. Iron. and Steel Institute, vol.* ii., *p.* 2-6.
482. WARD, J. C., RUSSELL, R., DAKYNS, J. R. — Sheet 246. Yorkshire 6-inch Map. Neighbourhood of Huddersfield. *Geol. Survey.*
483. WILLIAMSON, PROF. W. C. — On the Organization of Wolkmannia Dawsoni, an undescribed Verticillate Strobilus from the Lower Coal Measures of Lancashire. *Mem. Lit. and Phil. Soc. Man., Ser.* 3, *vol.* v., *p.* 28.
484. WOLLASTON, G. H. — Glaciation of the Lake District. *Geol. Mag., vol.* viii., *p.* 143.
485. WOOD, S. V., JUN. — Mr. Croll's Hypothesis of the Formation of the Yorkshire Boulder Clay. *Ibid., p.* 92.
486. WOOD, S. V., and HARMER, F. W. — On the Palæontological Aspects of the Middle Glacial Formations of the East of England, and their bearing upon the Age of the Middle Sands of Lancashire. *Rep. Brit. Assoc. for* 1870, *Trans. of Sect., p.* 90.

1872.

487. ANON. [? H. WOODWARD.] — Eurypterus (Arthropleura), Mammatus, Salter. *Geol. Mag., vol.* ix., *p.* 432.
488. ANON. — Discovery of Fossil Oysters [in Lancashire]. *The Earth, No.* 3, *p.* 71.
489. AVELINE, W. T., CAMERON, A. G. — Quarter-sheet 91 N.W. 1-inch Map. *Geol. Survey.*
490. BINNEY, E. W. — Observations on the Structure of Fossil Plants found in the Carboniferous Strata. Part iii. Lepidodendron, *pp.* 63-96. Plates XIII.-XVIII. *Palæontograph, Soc.* 4to. *London.*
491. ———— — On some Specimens of a Fossil Plant from the Upper-Foot Coal Seam near Oldham. *Proc. Lit. and Phil. Soc., Manchester, vol.* xi., *No.* 7, *p.* 69, and *No.* 10, *p.* 99.
492. CARRUTHERS, W. — Notes on some Fossil Plants. *Geol. Mag., vol.* ix., *p.* 49.
493. DAKYNS, J. R. — On the Glacial Phenomena of the Yorkshire Uplands. *Quart. Journ. Geol. Soc., vol.* xxviii., *p.* 382-8.
494. DALTON, W. H. — On the Geology of Craven. *Proc. West Riding Geol. Soc., New Series, Part* I.
495. DE RANCE, C. E. — The Geology of the Country around Southport, Lytham, and South Shore (90 N.E.). *Geol. Survey Memoirs.*
496. GUNN, W. — Sheet 184. Yorkshire 6-inch Map. Neighbourhood of Kelbrook. *Geol. Survey.*
497. HIGGINS, REV. H. H. — On some Fossil Ferns in the Ravenhead Collection, Free Public Museum, Liverpool. *Proc. Liverpool Geol. Soc., Session* 13, *p.* 94.
498. HULL, E. — Sheet 64. Lancashire 6-inch Map. Neighbourhood of Burnley. *Geol. Survey, new edition, with additions by* C. E. DE RANCE.
499. ———— - - On a remarkable Fault in the New Red Sandstone of Rainhill, Lancashire. *Journ. Roy. Geol. Soc., Ireland, vol.* iii., *Part* II., *pp.* 73-75.
(Further Observations on the Well at St Helen's.) *Ibid., p.* 86.

500. MARRAT, F. P. - On the Fossil Ferns in the Ravenhead Collection. *Proc. Liverpool Geol. Soc., Session* 13, pp. 97–134.
501. MIALL, L. C. Further Experiments and Remarks on Contortion of Rocks. *Rep. Brit. Assoc. for* 1871, *Trans. of Sections*, p. 106.
502. ——— - Descriptive Guide to the Mineral Collection in the Leeds Museum. *Leeds*.
503. MORTON, G. H. Minerals that occur in the Neighbourhood of Liverpool, with the Localities, &c. *Proc. Liv. Geol. Soc., Session* 13, p. 91.
504. ——— Shells found in the Glacial Deposits around Liverpool, with the Localities, &c. *Ibid.*, pp. 92–93.
505. ——— - Minerals that occur in the Neighbourhood of Liverpool, with the Localities, &c. *Ibid.*, pp. 91–93.
506. RAMSAY, PROF. A. C. The Physical Geology and Geography of Great Britain. 3rd edition. *Lond.* 8vo. (1st edition, 1863.
507. READE, T. M. - The Geology and Physics of the Post-Glacial Period, as shown in the Deposits and Organic Remains in Lancashire and Cheshire. *Proc. Liv. Geol. Soc., Session* 13, pp. 36–88.
508. ——— - The Post-Glacial Geology and Physiography of West Lancashire and the Mersey Estuary. *Geol. Mag., vol. ix.*, p. 111.
509. ——— - [Letter on boring for Coal.] *Liverpool Daily Post, Sept.* 16. Noticed in *Nature, vol.* vi., No. 151, p. 421.
510. TIDDEMAN, R. H., and DE RANCE, C. E. Sheet 86. Horizontal Sections. *Geol. Survey.* Sheet 87. Horizontal Sections. *Ibid.*
511. TIDDEMAN, R. H. - On the Evidence for the Ice-sheet in North Lancashire and adjacent parts of Yorkshire and Westmoreland (with Map). *Quart. Journ. Geol. Soc., vol. xxviii.*, pp. 471–491.
512. WÜRZBURGER, P. - The Hæmatite Iron Ore Deposits of Furness. *Journ. Iron and Steel Institute, vol.* i., pp. 135–142.
513. WILLIAMSON, PROF. W. C. On the Organisation of the Fossil Plants of the Coal Measures. Part III. Lycopodiaceæ *(continued)*. *Phil. Trans.*, pp. 283–318.
514. ——— On the Structure of the Dictyoxylons of the Coal Measures. *Rep. Brit. Assoc. for* 1871, *Trans. of Sections*, p. 111.
515. WOODWARD, HENRY - A Monograph of the British Fossil Crustaceans belonging to the Order *Merostomata*. Part IV. (*Lancashire*, p. 163–8.) *Palæont. Soc.* 4to. *Lond.*
516. ——— - On a new Arachnid from the Coal Measures of Lancashire. *Geol. Mag., vol. ix.*, p. 385.
517. ——— - Notes on some British Palæozoic Crustacea belonging to the Order *Merostomata*. *Ibid.*, pp. 433–441.

1873.

518. AVELINE, W. T. - The Geology of the Southern part of the Furness District in North Lancashire (91 N.W.). *Geol. Survey Memoirs.*
519. BOULT, JOSEPH - The Mersey as known to the Romans. *Proc. Lit. and Phil. Soc., Liverpool, No.* xxvii., p. 249.
520. BROCKBANK, W. (On the Victoria Cave). *Proc. Lit. and Phil. Soc., Manchester, March* 1873.

APPENDIX II.—LIST OF WORKS AND PAPERS.   217

521. Busk, Prof. George — Human Skull and Fragments of Bones of the Red Deer, &c., found at Birkdale, near Southport, Lancashire. *Journ. Anthrop. Inst., vol.* iii. (No. 1), *pp.* 104, 105.
522. Dawkins, W. B. — Report on the Victoria Cave. The Archæological and Zoological Results. *Rep. Brit. Assoc. for* 1872, *Trans. of Sections, p.* 178.
523. Dakyns, J. R. — On some points connected with the Drift of Derbyshire and Yorkshire. *Geol. Mag., vol.* x., *p.* 62.
524. De Rance, C. E. — On the occurrence of Lead, Zinc, and Iron Ores in some Rocks of Carboniferous Age in the North-west of England. *Geol. Mag., vol.* x., *pp.* 64–74.
525. ——————— - The Cyclas Clay of West Lancashire. *Ibid., pp.* 187–189.
526. ——————— - The Lower Scrobicularia and Lower Cyclas Clays of the Mersey and the Ribble. *Ibid., pp.* 287, 288.
527. Hughes, Prof. T. McK. — On a Series of Fragments of Chert collected below a chert-bearing Limestone in Yorkshire (Ingleborough). *Rep. Brit. Assoc. for* 1872, *Trans. of Sections, p.* 189.
528. Hull, E., Tiddeman, R. H., De Rance, C. E. — Quarter-sheet 89 N.W. 1-inch Map. *Geol. Survey.*
529. Hull, E. - - - Quarter-sheet 89, S.E. 1-inch Map. Superficial Deposits. *Ibid.*
530. ——————— - The Coalfields of Great Britain. Third edition, enlarged. *London.*
531. Mackintosh, D. - Observations on the more remarkable Boulders of the North-west of England and the Welsh Borders. *Quart. Journ. Geol. Soc., vol.* xxix., *pp.* 351–360.
532. Miall, L. C. - Descriptive Guide to the Fossil Collection in the Leeds Museum (with Bibliography). *Leeds.*
533. Morton, George H. - The Strata below the Trias in the Country around Liverpool, and the Probability of Coal occurring at a Moderate Depth. *Proc. Lit. and Phil. Soc., Liverpool, No.* xxvii., *pp.* 157–174.
534. Reade, T. M. - Glacial Striæ at Miller's Bridge, Bootle, near Liverpool. *Proc. Liverpool Geol. Soc., Session* 14, *pp.* 31–32.
535. ——————— - The Buried Valley of the Mersey. *Ibid., pp.* 42–65.
536. ——————— - Cyclas Clay. *Geol. Mag., vol.* x., *p.* 139.
537. ——————— - Formby and Leasowe Marine Beds, or the so-called "Cyclas Clay." *Ibid., pp.* 238, 239.
538. Roberts, Isaac - Section of Strata above the Boulder Clay at Whitechapel. *Proc. Liverpool Geol. Soc., Session* 14, *pp.* 32–34.
539. Rofe, John — Further Notes on Crinoidea (Lancashire, &c.). *Geol. Mag., vol.* x., *pp.* 262–267.
540. Russell, R. - Sheet 232. Yorkshire 6-inch Map. Neighbourhood of Birstal. *Geol. Survey.*
541. Tiddeman, R. H. - Report on the Victoria Cave. The Physical History of the Deposits. *Rep. Brit. Assoc. for* 1872, *Trans. of Sections, p.* 179.
542. ——————— - The Older Deposits in the Victoria Cave, Settle. *Geol. Mag., vol.* x., *p.* 11.
543. ——————— - The Age of the North of England Ice-sheet. *Ibid., p.* 140.
544. ——————— - The Relation of Man to the Ice-sheet in the North of England. *Nature, vol.* ix., No. 210, *p.* 14.

545. WILLIAMSON, PROF. W. C. — On the Organisation of the Fossil Plants of the Coal Measures. Part iv. Dictyoxylon, Lyginodendron, and Heterangium. *Phil. Trans.*, vol. 163, pp. 377–408.

546. WOODWARD, HENRY — On some supposed Fossil Remains of Arachnida (?) and Myriapoda from the English Coal Measures. *Geol. Mag.*, vol. x., pp. 104–112.

1874.

546a. BEDDOE, DR. — The Anthropology of Yorkshire. *Brit. Assoc. Report for* 1873, *Trans. of Section*, p. 134.

547. BRIGG, JOHN — The Industrial Geology of Bradford. Sco. Leeds, and *Ibid.*, p. 76.

548. BUSK, PROF. — On a Human Fibula of unusual form discovered in the Victoria Cave, Settle. *Journ. Anthrop. Inst.*, vol. iii., p. 392.

549. ——— — (Remarks on the same.) Presidential address. *Ibid.*, p. 516–7.

550. DAKYNS, J. R. — On the Geology of part of Craven. *Brit. Assoc. Reports for* 1873, *Trans. of Sections*, p. 78.

551. DARBISHIRE, R. D. — On a Deposit of Middle Pleistocene Gravel in the Worden Hall Pits, Leyland, Lancashire. *Quart. Journ. Geol. Soc.*, vol. xxx., p. 38.

552. DAWKINS, W. B. — Report of the Committee for assisting in the Exploration of the Settle Caves. *Brit. Assoc. Report for* 1873, p. 250.

553. ——— — Observations on the Rate at which Stalagmite is being accumulated in the Ingleborough Cave. *Ibid.*, *Trans. of Sections*, p. 80.

554. DE RANCE, C. E. — The Geology of the Country between Blackpool and Fleetwood (91 S.W.). *Geol. Survey Memoirs.*

554a. GOMERSALL, W. — On the round Boulder Hills of Craven. *Brit. Assoc. Report for* 1873, *Trans. of Sections*, p. 80.

555. HUGHES, PROF. T. Mc. K. — Exploration of Cave Ha, near Giggleswick. *Journ. Anthrop. Inst.*, vol. iii., No. 3, p. 383.

556. HULL, E., and TIDDEMAN, R. H. — Sheet 55. Lancashire 6-inch Map. Neighbourhood of Whalley. *Geol. Survey.*

557. PLANT, J. — Notice of Mammoth Bones discovered in a cave in Lothersdale near Skipton. *Trans. Man. Geol. Soc.*, vol. xiii., part 3, p. 54.

558. READE, T. MELLARD — The Drift Beds of the North-west of England. Part I. Shells of the Lancashire and Cheshire Low-level Boulder Clay and Sands. *Quart. Journ. Geol. Soc.*, vol. xxx., p. 27.

559. RICKETTS, CHAS. — Is the Mersey filling up? *Liverpool Daily Courier, May 15th,* 1874.

560. RUSSELL, R. — Geology of the Country round Bradford, Yorkshire. *Brit. Assoc. Reports for* 1873, *Trans. of Sections*, p. 88.

561. TIDDEMAN, R. H. — Sheet 47. Lancashire 6-inch Map. The Neighbourhood of Clitheroe. *Geol. Survey.*

INDEX.*

A.

Abbot House, 24.
Abbot Stone, 112.
Accrington, 63, 64.
Admarsh, 11.
Admergill, Higher, 28.
Agricultural features, 119, 168.
Aire, R., 13.
Alder Hurst Head, 89.
Alluvium, 139, 155, 159. 162.
Alston Hall, 34, 125, 140.
Altham, 74, 84.
Anglezark Lead Mines, 43, 173.
Anglezark Moor, 6, 33, 173.
Angram Green, 17.
Anticlinal axes, 10, 28.
Anticlinal Fault, 9, 88, 100.
Anticlinal of Lothersdale, 48.
Anticlinal of Watersheddles, 50.
Antly Gate, 89.
Arley Brook, 34, 38.
Arley Mine, 54, 73, 76, 82, 92, 96.
Ashnot, 19.
Ashworth, 45.
Asland, R., 5.
Audley Reservoir, 25.
Avenham, 144, 145.
Axletree Edge, 9.
Ayneslack, 86.

B.

Bacup, 6, 45, 56.
Bakestones Clough, 110.
Balderstone Hall, 129, 145.
Balladen Brook, 41.
Ball Grove, 52.
Bamber Bridge, 10, 143.
Bank Hey, 38.
Bannister Hall, 125, 155, 163.
Barden Clough, 81.
Bar House, 107.
Barley, 23, 37.
Barn Hill, 52.
Barnoldswick, 10, 13, 27, 28, 30, 31, 32.
Barrowford, 139.
Bashall, 12.
Bashall Brook, 11, 121, 122, 136.
Bassy Coal, 60.
Baum pots, 62.
Bawmier, 32.
Baxenden, 46.
Baxenden Collieries, 56, 64.
Beacon Hill, 22.
Beater Clough, 62.
Beaumont Clough, 105, 109, 116.
Beech Beck, 69.

Belmout, 6, 33, 34, 37.
Bescar Lane, 127.
Bezza Brook, 123, 125, 147.
Billinge, 42, 44.
Billington, 24, 34, 172.
Bing Mine, 78.
Birkacre, 90, 93.
Birkett Fell, 10.
Birkin Clough, 40.
Birks House, 81.
Blackburn, 5, 42, 44, 46, 56, 63, 64, 65, 129, 134.
Black Castle Clough, 107, 109, 118.
Black Clay Coal, 54, 68.
Black Clough, 39, 40, 85.
Black Dean, 111.
Blackden Bridge, 9.
Blackfield, 35.
Black Hambledon. 6, 9, 40, 88, 101, 128, 137.
Black Hill, 39.
Black Moor, 9, 111.
Blacko, 27, 35, 139.
Black Sears Beck, 50.
Blackstone Edge, 6, 9, 88, 98, 172.
Blackwood Edge, 108, 118.
Blainscough Hall, 95.
Bleara Lowe, 13.
Blindstone Coal, 80.
Blow-up, 66.
Boft Hole, 111.
Bold Venture Quarries, 15, 16.
Bolton Hall, 11, 16, 27, 174, 175.
Bone Coal, 94, 96.
Bonney Barn, 165, 166.
Bonny Inn, 23.
Booth Dean Clough, 107, 108, 118.
Booth Stones, 46.
Boulder Clay near Chorley, 151.
Boulder Clay near Colne, 137.
Boulder Clay near Ellerbeck, 95.
Boulder Clay near Preston, 140.
Boulder Clay of the Ribble Plain, 123, 129, 135.
Boulsworth Hill, 6, 8, 9, 39, 46, 101, 112, 128.
Boulsworth, 137.
Bowland Shales, 21, 47, 49.
Bowley Hill, 84.
Bracewell, 28, 30, 31.
Bradfield, 99.
Bradford Brook, 11.
Bradford, West, 18, 19, 136.
Bradley Fold, 74.
Bradshaw Brook, 42.
Brandwood Moor, 45.
Brast Clough, 25.
Brennaud, 11, 172.
Bridestones, 9, 116.

---

* Not necessarily including the names of places named in the Appendices.

Brinscall, 33, 140, 175.
Broadfield Colliery, 64.
Broad Head Moor, 52.
Brock, R., 11.
Brockholes, 125, 162.
Brogden Hall, 30, 31.
Bromley, 33.
Broughton, 30, 31.
Broughton Pit, 84.
Brown Beck, 32.
Brownlow, 18.
Brun, R., 86.
Brungerley Bridge, 15.
Bullion Coal, 66, 70.
Bullions, 62, 70.
Bullion Clough, 111.
Bunker's Hill, 46.
Bunter Series, 122.
Burnley, 5.
Burnley Basin, Faults of the, 83.
Burnley Basin, Resources of the, 82.
Burnley Coalfield, 53–89.
Burnley Four-foot Coal, 71, 82.
Burn Moor, 27.
Buttock, 23.
Butts Clough, 108.
Butts Green, 108.
Byron Edge, 9.

C.

Cabin End Factory, 64.
Calder River (Lancashire), 5, 6, 10, 11, 12, 34, 36, 135.
Calder River (Yorkshire), 9, 100, 103, 109.
Calder Head, 6, 7, 43, 74.
Cally Coal, 73, 77.
Canker, 21.
Cannel Seam, 80, 94, 96.
Cant Clough, 43, 131.
Carboniferous Limestone, 13.
Carboniferous rocks, Lower, 28.
Carlow Beck, 46.
Carlton Synclinal, 46.
Carton, 13, 46, 47, 137.
Caster Cliff, 67.
Castle Haugh, 17.
Catlow, 55, 68.
Caton, 172.
Cawk veins, 48.
Chalybeate water, 21.
Champion, 11, 19.
Charley Coal, 71.
Charnock Moss, 123.
Charnock Richard, 91, 175.
Chatburn, 14, 133.
Cheesden Brook, 41.
Chelburn Moor, 88.
Cherry Tree, 64, 175.
China Bed, 78.
Chorley, 150.
Chorley Coalfield, 90–98.
Chorley Moor, 91.
Church, 64.
Claude, 27.
Clayton, 74, 76.
Clayton-le-Woods, 153.
Clerk Hill, 25, 34.

Clints Delf, 29, 31.
Clitheroe, 10, 14, 16.
Clitheroe and Blackburn Railway, 38
Cliviger, 40, 71, 74, 78, 84.
Cliviger Colliery, 7, 54.
Cliviger 2-feet Coal, 78.
Clough Foot, 41.
Clough Head Pit, 68.
Clough Mill, 45.
Cludders, 106.
Coalpit Lane, 20, 21.
Coalroad Delf, 34.
Coal-seekers, Warning to, 21.
Coate Flat, 29.
Cock Clough, 35.
Cocker Hill, 35, 37.
Cockridge, 40.
Cold Edge, 8, 9.
Colden Clough, 9, 110, 113.
Colne, 5, 13, 33, 43, 44, 137, 138.
Colne Edge, 39.
Combe Hill, 12, 13, 46, 52, 137.
Contorted Beds in Clitheroe and Blackburn Railway, 39.
Contorted Grits at Salmesbury, 38.
Contortion in Limestones, 18.
Contortions in Yoredale Grits, 24.
Coplow, 15.
Coppul, 90, 92, 94, 175.
Copy Bottom, 39.
Copy Wheel, 135.
Cornfield Colliery, 81.
Cornholme, 45.
Counting Hill, 33.
Cowling, 13.
Cowloughton, 13, 138.
Cowpe Brook, 41.
Cowpe Moss, 42, 45.
Crackers, 72.
Cragg Valley, 117.
Cranberry Moss, 6.
Cranoe, 11, 13.
Cribden Moor, 46.
Crickle, 30, 31.
Crimsden, 99.
Crossens, 158.
Cross Lea, 102.
Cross Stones, 103.
Croston, 157.
Crow Hill, 8, 13, 46, 52, 101, 115, 118.
Cuerden, 123.
Cunliffe Colliery, 64.

D.

Dandy Bed, 73, 77.
Danes House, 87.
Dark Hill Well, 32.
Darwen, 56, 57, 175.
Darwen Moor, 6, 46.
Darwen R., 5, 10, 11, 34, 36, 38, 43, 120, 129, 150, 155, 163.
Dean Brook, 34.
Dean Head, Todmorden, 6, 7.
Dean Valley, Great Harwood, 7.
Dearden's Pasture, 45.
Deepley Hill, 41.
Deerplay Moor, 77, 86.
Deerstone Moor, 44, 101.

INDEX. 221

Denudation, 119.
Denudation by Rivers, 148.
Dimpenley Clough, 7.
Dineley, 54.
Ding, 43.
Dirpley, 6.
Doghole Seam, 71, 82.
Dolerite, 172.
Don R., 40.
Douglas R., 5, 92, 123, 127, 152, 156, 159.
Dovestones, 112.
Downham, 12, 15.
Dowshaw Delf, 46, 48, 49.
Drainage, 8, 11.
Drift Deposits, 128.
Drifts near Colne, 137.
Dry Dam Colliery, 93, 94.
Dulesgate Valley, 41, 45, 56, 86, 175.
Dungeon Top, 109.
Dunscar, 83.
Duxbury Colliery, 90, 93, 94.

E.

Earby, 13, 32, 47.
Easington, 18.
Eastwood, 103.
Eccleston Green, 90, 125, 127, 151.
Edenfield, 45.
Edge, 22.
Edge End Moor, 116.
Edgeworth, 43, 57.
Elevations, 8.
Ellerbeck, 90, 91, 95, 175.
Elslack, 10, 12, 13, 32, 46.
Emmott Hall, 52.
Emmott Moor, 138.
Entwistle, 43, 45, 56.
Erringden, 104, 117.
Estuarine deposits, 163.
Euxton, 122, 135, 143, 151, 152.
Extwistle Moor, 69, 75.

F.

Fall Bank Clough, 45.
Farlton Fell, 15.
Farrington, 123, 141, 160.
Faugh, 115.
Fault in the Darwen, near Arley Brook, 38.
Faults in the Triassic Plain, 169.
Faults of the Burnley Basin, 83.
Feather-edge Coal, 43.
Fence, 29, 75.
Fenniscowles, 5, 64.
Finington Brook, 44.
Firebricks, 56.
Fire-clays, 175.
Fishwick, 148.
Flagstones, 55.
Fleetwood, 128.
Fletcher Bank Quarry, 41.
Flown Scar Hill, 45.
Fool's Syke, 30.
Foot Coal, 55, 66.
Foulridge, 11, 13, 35, 37, 46, 139.
Fox Clough, 69.
Freckleton, 163.

Fulledge, 64, 71, 80, 81, 82, 85.
Fulledge Thin Bed, 80, 83.
Fullwood, 148.

G.

Gablestone Edge, 110.
Gadden Reservoir, 9.
Gagantails, 62.
Galliard, 54.
Gannister Beds, 54, 91.
Gannister Coal, 55, 56, 64, 70, 82.
Garstang, 170, 171.
Gate Brook, 45.
Gauxholme, 9, 40.
Gawthorpe, 71, 79, 81, 87.
Gerna, 12.
Gill, 51, 139.
Gill Church, 11, 13, 29, 32.
Gill Rock, 29.
Gillian, 27.
Gillibrand Hall, 92.
Gingerbread Clough, 7.
Gisburn, 13, 14, 16, 17, 139, 175.
Glacial Deposits around Preston and Chorley, 140.
Glacial Deposits, Sequence of, 154.
Glacial Drifts, 128.
Glaciated Rock Surfaces, 133.
Gledstone, 12, 13, 29.
Goodshaw, 46.
Gorple, 106, 111, 137.
Gorplestones, 9, 89, 101.
Gorpley Wood, 40.
Gorton Brook, 37.
Grane Rake, 45.
Grange Brook, 34.
Great Mine, 78, 80, 83.
Greave Clough, 111, 113.
Greenbill Clough, 85.
Green's Clough, 62.
Grey Fosse Clough, 110.
Greystone Hill, 9, 28, 112, 115.
Grimsargh, 148.
Grindleton, 11, 18, 19, 171, 175.

H.

Habergham, 56, 74, 77, 78, 84.
Habergham Mine, 73.
Hag Gate Colliery, 87.
Half-yard Mine, 57.
Hallan Hill, 51.
Halliwell Field, 92.
Halliwell Fold, 98.
Hall o' th' Hill, 92.
Hambledon, Great, 77.
Hambledon Hill, 6.
Hambledon Scout, 56.
Hammerton Brook, 17.
Han Royd, 115.
Hapton, 74, 84.
Hapton Beck, 6, 56.
Harden Moor, 41.
Hare Law, 49.
Hare Stones, 89, 101.
Harper Clough, 44, 83.

Hartley Naze, 40, 88.
Harwood, 39, 44, 65.
Harwood, Great, 7, 24, 65, 134, 175.
Harwood, Little, 42, 63, 83, 130.
Haslingden, 43, 46.
Haslingden Flags, 42, 44, 65.
Haslingden Moor, 6.
Hathershelf Scout, 118.
Hawkstones Common, 101, 109.
Hawshaw Moor, 13.
Hazle Edge, 40, 43.
Heald Moor, 45.
Heapey, 95.
Hebden, 9, 100, 105, 109, 110, 113, 116, 118.
Heeley Clough, 105.
Hempshaw's, Higher, 33.
Henthorn, 11, 121.
Hesketh Bank, 127, 152.
Heskin, 91, 92.
Heyslacks Clough, 46, 89, 106, 13, 139.
Higham, 75, 87.
High Brown Knowl, 115.
High Close Hill, 31.
Higherford, 44.
Higher Heights, 22.
Higher Hill Delf, 46.
Hill Top, 31, 32, 33.
Hoarstones Brow, 34.
Hodder, R., 11, 12, 16.
Hoddlesden, 57.
Hoghton, 13, 34, 43.
Holcombe, 45.
Holcombe Brook, 41.
Holcombe Moor, 43.
Holden Clough, 11, 16, 19, 22, 174.
Holden Wood, 41.
Hole Brook, 34, 86.
Hole Sike, 111.
Hollin Hall, 24.
Hollinhurst Bridge, 28.
Hollins, 30, 86.
Hollin Top, 35.
Hoofstones Height, 9.
Hook Cliff, 18.
Hoolster Hill, 34.
Horelaw Nook, 6.
Horrocksford Quarries, 14.
Horsebridge Clough, 9, 105, 110.
Horton, 28.
Horwich Moor, 134.
Hough Stone, 102.
Houghton Hay, 87.
Howroyd Clough, 45.
Hudson Moor, 100, 109.
Hunterholme, 79.
Huntroyde, 65, 75, 76.
Hurstwood, 87.
Hyndburn, 83.

I.

Ice-markings on Horwich Moor, 134.
Ickornshaw, 46, 51, 137.
Ightenhill, 81, 87.
Igneous Rocks, 171.
Ingham Clough, 103.
Ingham's Farm, 87.

Ingleborough, 14
Inglewhite, 172.
Ings Beck, 11, 17, 174.
Irwell, 41, 42,

J.

Jack Bridge, 113.
Jack Green, 37.
Johnny Gap, 116.
Joiner Stones, 118.
Jumble Clough, 105.
Jumbles, 58.
Jumps, 116.

K.

Kebcote, 109.
Keelam, 109.
Kelbrook, 13, 32, 46, 49.
Kershaw Coal, 71, 82.
Keuper Marls, 127.
Key-field, 31.
Kinder Scout Grit, 33, 50, 97, 98, 107.
King Coal, 94, 96.
Kirk Clough, 30.
Kirkham, 128, 142, 143.
Kirkless Hall Colliery, 90.
Kirk Sikes, 47.
Kirk's Wife Wood, 30.
Knoll Hill, 13.
Knoll Moor, 45.
Knotts Brook, 34.
Knowlmere Manor, 10, 21, 122
Knowl Top, 18.
Knowl Wood, 40.
Knunk Knowles, 15.
Knuzden, 64.

L.

Lad-law, Boulsworth, 9.
Landslips, 119.
Laneshaw, 12, 51, 52, 137, 139.
Laneside, 175.
Langher, 29.
Langfield Edge, 88, 100, 107.
Langfield Moor, 9.
Lead Mines of the Ribble Valley. 173.
Leyland, 141, 142, 151.
Limestones of Pendle, 16.
Littleborough, 6, 41, 43, 45, 58, 175.
Little Coal, 57.
Little Hoar Edge, 88.
Longridge, 16, 18.
Longton, 165, 169.
Lostock, 156.
Lothersdale, 10, 13, 46, 48.
Loud R., 10.
Low Bottom Coal, 80, 83.
Lower Boulder Clay, 140, 155.
Lower Cyclas Clay, 158.
Lowerford, 56.
Lower Yard Coal, 81, 82.
Lower Yoredale Grit, 20.
Lower Yoredale Shales and Limestones, 30.

INDEX.

Low Moor Mills, 121.
Luddenden, 99, 100, 115, 118.
Luddenden Valley, 8, 9.
Lumb, 117.
Lumb Foot, 45.
Lumbutts, 102, 105, 107.

M.

Malham Cove, 11.
Marsden, 65, 67, 68, 78. 87.
Marsden Four-foot, 73.
Marshaw Bank, 118.
Martin Mere, 157, 161.
Martin Top, 27, 173.
Marton, 29, 31.
Marton Scar, 11, 13, 139.
Mary Anne Seam, 71.
Maudsley Fold, 10.
Mawdesley, 126, 156.
Mearley Hall, 17, 18, 19, 20.
Mellings Clough, 40.
Mellor, 12, 24, 34.
Mere Stones, 110.
Metals, 173.
Middle Drift, 141.
Middle Sand near Ellerbeck, 95.
Middle Sand of the Ribble Plain, 123, 129, 135.
Midgley, 113, 115, 118.
Mill Hill, 64.
Millstone Edge, 33.
Millstone Grit, Divisions of the, 98.
Millstone Grit Series, 33.
Minerals, 172.
Monk Edge, 13.
Moorcock Inn, 19.
Moor Edge, 24.
Moor End, 115.
Moor Isles Clough, 87.
Moreton Park, 36.
Mosleden Height, 8.
Moss Houses Beck, 50.
Mountain Mines, 54, 55, 56, 64, 68, 91.
Musbury Heights, 43.
Mytholmroyd, 9.

N.

Nab, 8.
Nappa, 16.
Nelly Hole, 50.
New a' Nook, 19.
Newchurch-in-Pendle, 6, 35.
Newchurch-in-Rossendale, 42.
Newfield Edge, 20, 22.
Newton, 10, 12, 19.
Nick of Pendle, 23, 34, 175.
Noah Dale, 113.
Northwood, 67.
Noyna, 13.

O.

Oaken Bank, 44.
Oakenshaw, 73, 83.
Oatley Hill, 45.
Offa Hill, 27, 28, 35.

Ogden Clough, 23.
Ogden Valley, 43.
Old Lawrence Rock, 55.
Old Salford Colliery, 82.
Old Yard Coal, 71, 82.
Ormerod Wood, 86.
Orrell Moss, 6.
Oswaldwistle, 56, 63, 64.
Ouseldale, 31.
Over Darwen, 57.
Oxenhope Moors, 8, 115.

P.

Padiham, 5, 79, 81, 84, 87, 130, 134.
Padiham Reservoir, 76.
Parbold, 156.
Park Close, 28.
Park Hall, 92, 93.
Park Head Quarry, 46. 49.
Park House, 14.
Parrock Clough, 104.
Parsley Barn, 25.
Paul Clough, 100, 101.
Paythorne, 16.
Peacock Row, 44.
Peat, 159.
Pebble Beds of the Bunter Series, 123.
Pendle, 17, 20, 23, 25, 26, 34.
Pendle Hill Brook, 18.
"Pendle's brasted hissel," 25.
Pendleside Limestone, 17.
Pendle Range, 12, 168.
Pendle Water, 7, 12, 75.
Pendleton, 18.
Penwortham, 160, 162, 167.
Permian System, 120, 171.
Physical Features, 5, 8, 12.
Piked Edge, 13, 50.
Pike Low, 6, 43, 44.
Pinnaw, 13, 46.
Pisser Clough, 111.
Pleasington, 44, 130.
Plumpton, Great, 144.
Pole Hill, 113.
Portsmouth Valley, 6, 43, 45, 54.
Post-Glacial Drifts, 128, 155, 168.
Pot Brinks Moor, 101, 112.
Presall, 158.
Preston, 123, 140, 142, 145-7, 155, 167.
Proctor Height, 49.
Priestley Ing, 107.
Pyebrook Hall, 93.
Pudsey Clough, 40.

Q.

Quarlton, 57, 58.

R.

Radburn, 37.
Rain Hall, 11, 29, 30, 32.
Ramsbottom, 41.
Ramsden Clough, 41, 86.
Ramsgreave, 34.
Ratchers, 37.
Raven Coal, 96.
Ravensholme, 18.

Raven's Rock, 52.
Raygill, 46, 48.
Read Park, 42, 44.
Reaps, 113.
Reddyshore Scout, 41.
Redscar, 140, 143.
Red Spa Moor, 44.
Red Water Brook, 45.
Red Water Clough, 43.
Reedshaw Moss, 11, 46, 50, 52.
Resburn Fold, 24.
Resources of the Burnley Basin, 82.
Revidge, 44.
Ribble R., 5, 11, 12, 34, 120, 121, 123, 129, 136, 143, 144, 148, 150, 155, 159, 163.
Ribchester Station, 24.
Ribourne, 8, 9.
Ridding Hey, 15.
Riddle Clough, 21.
Riddle Scout, 74.
Riddle Scout Rock, 54, 75.
Ridgaling, 39, 44.
Rig of England, 39.
Rimington, 11, 27.
Ripponden, 100, 108.
Rishworth Hall, 108.
River Terraces, 139.
Rivington, 43.
Rivington Hall, 33.
Rivington Hill, 34.
Rivington Pike, 43.
Roach Bridge, 10, 36, 120, 155, 170.
Robin Wood, 40.
Rochdale, 9, 43.
Roches Moutonnées, 133.
Roddel Chapel, 175.
Roddlesworth, 7, 33, 37, 43, 44.
Roger Moor, 46, 49.
Rossendale, 41, 139.
Roughlee, 12, 36, 37.
Rough Rock, 43, 52, 91.
Rowley Colliery, 86.
Royle, 79, 87.
Royshaw Hill, 44, 83.
Rufford, 126, 143, 156, 157, 165.
Ryal, 46.
Ryburn, 108, 109.
Rye-loaf, 11.

S.

Sabden, 23, 25, 35, 36.
Sabden Brook, 39.
Sabden Valley, 6, 39.
Sabden Valley Shales, 36, 50.
St. Hubert Vein, 174.
Salis Wheel, 34.
Salmesbury, 24, 38, 173.
Salterforth, 31, 32, 46, 138.
Saltonstall Moor, 115.
Salthill, 15.
Samlesbury 36, 125, 143, 147.
Sand-rock Coal, 62.
Saucer Stones, Great, 101, 107, 112.
Sawley, 11, 12, 174.
Scarisbrick, 128, 170.
Scar Limestone, 48.
Scout Moor, 43, 45.

Scout Moor Brook, 41.
Scrobicularia Clay, 158.
Seed Lea, 37.
Shackleton Knowl, 115.
Shales below Rough Rock, 52.
Shales between Yoredale and Kinder Grits, 28.
Shales-with-Limestones, 16.
Shaw Clough, 104.
Shaw Head Beck, 50.
Sheddon Edge, 45.
Shell Coal, 71, 82.
Sheep House Clough, 45.
Shining Moss, 62.
Shirdley Hill Sand, 156.
Shore Farm, 7.
Shore Moor, 6, 45.
Shuttleworth, Higher, 84.
Shuttleworth Moss, 41.
Simonston Hall, 65, 76.
Skeleron, 15, 17, 27, 173.
Skipton, 9, 139.
Skirden Beck, 11.
Slaidburn, 10, 14, 19.
Slaty Coal, 78.
Slitheroe, 108.
Smith Coal, 94, 96.
Smith Hill, 49.
Sough Bridge, 32.
South Field Bridge, 29.
Sowerby, 109.
Sowerby Bridge, 5, 8, 9, 100.
Spen, Lower, 39.
Spence Moor, 25, 26.
Spidden River, 7.
Springs, 32.
Stairs Hill, 8.
Stanally Clough, 100, 102.
Standen Brook, 11, 16.
Stank Top, 27, 28.
Stanworth Edge, 44.
Stiperden House, 89.
Stiperden Moor, 40, 45.
St. Hubert Mine, 174.
St. John's Vale, 9, 100, 107, 118.
Stock Bridge, 31.
Stock Beck, 11, 28.
Stockdale, 11.
Stokes-in-the-Moss, 9.
Stone Edge, 39.
Stone Head Beck, 50.
Stoneyhurst, 12, 136.
Stoodley Clough, 103, 104, 117.
Stoodley Pike, 100, 107.
Stronstrey Bank, 33.
Stubbins, 41.
Swanside Beck, 11.
Sweet Brow, 13, 48.
Swift Place, 108, 109.
Swinden Clough, 40, 132.
Summerseat, 42.
Swinden Water, 43, 67, 69, 86.
Sunderland Hall, 135.
Sun Hill, 8.
Surgill, 48, 50.
Syd Brook, 127.
Sykes, 11.
Synclinal of Carlton, 46.
Synclinal of Reedshaw Moss, 50.

T.

Talbot Bridge, 122.
Tewit Hall, 39.
Thievely, 43, 45, 62, 85.
Thornton, 10, 13, 29, 31, 46, 48.
Thornyholme, 123.
Thursden, 131.
Thursden Brook, 39, 42, 45, 131.
Tidal Deposits, 163.
Till, 140.
Tockholes, 33, 37, 43, 46.
Todber, 175.
Todmorden, 7, 9, 40, 100, 101, 102.
Tom Groove, 106.
Tonacliffe, 45.
Tor Hill, 43.
Tosside Chapel, 11, 172.
Towneley, 82, 175.
Trap Dykes, 171.
Trawden, 12, 46, 68.
Triassic Rocks, 122, 171.
Trippet of Ogden, 41.
Tun Brook, 152.
Turncroft, 57.
Turton, 57, 58.
Turton Heights, 46.
Twiston, 15, 16.

U.

Up-Holland Flags, 91.
Upper Boulder Clay, 135, 141, 151.
Upper Mottled Sandstone (Bunter), 126.
Upper Yoredale Grit, 22.
Upper Yoredale Shales, 50.

V.

Vicarage Mine, 82.
Victoria lead-mine, 19, 174.

W.

Waddington, 12, 18, 19.
Waddington Brook, 11.
Waddow, 12, 14, 121, 136, 175.
Walsden Moor, 9, 88.
Walshaw, 105, 116.
Walshaw Dean Head, 101, 110.
Walton-le-dale, 124, 140, 150, 156.
Wardle, 58.
Warland, 41.
Warley, 115.
Warley Wise, 50.
Warning to Coal-seekers, 21.
Water foot, 41.
Watersheddles, 46, 50.
Watersheds, 6, 11.
Way Stone Edge, 8, 36.

Weather Hill, 45.
Weethead Height, 23, 26.
Weets, 11, 12, 13, 20, 22, 27.
Welsh Whittle, 94.
West Bradford, 137.
Westby Bridge, 144.
Wether Edge, 45.
Wet Moss, 45.
Whalley, 12, 24, 25, 34, 130, 133, 135.
Wheatley Lane, 65.
Whinney Edge, 64.
Whitaker, 60.
Whitaker Pasture, 42.
White Hough Water, 27.
White Moor, 11, 28, 137.
Whitewell Mine, 172.
Whitewell, 15.
Whitley Royd, 103, 109.
Whittle Field Colliery, 84.
Whittle-le-Woods, 13, 43.
Whitworth, 7, 43, 45.
Wickenberry Clough, 102.
Widdop, 106, 111.
Widdop Moor, 9, 11, 40, 89, 137.
Willy Moor, 44, 101.
Winewall, 44, 52, 68.
Winsley, 102, 109.
Winter Hill, 33, 34, 36.
Wiswell, 18, 34.
Withens Clough, 105, 107, 117.
Withens Height, 8.
Withgill, 12, 14.
Withins Height, 110.
Withnell, 33, 44.
Witton, 56.
Wolfstones, Little, 13.
Woodfold Park, 38.
Woodhead Hill Rock, 55, 56, 64, 66.
Woodhouse Lane, 45.
Worsaw, 12, 14, 15, 16, 17.
Worsthorn, 65, 87, 131.
Worsthorn Moor, 56, 69, 89.
Worsthorn Rock, 56.
Worston, 11, 17.
Worth Valley, 137.
Wybersey, 12, 16.
Wycoller, 12.
Wyre R., 5.

Y.

Yard Seams, 57, 81, 82, 94, 96.
Yarlside, 31.
Yarrow R., 90, 91, 95, 120, 122.
Yellison House, 46.
Yeoman Hill, 8.
Yew Tree Inn, 38.
Yoredale Grits, 20, 22, 49.
Yoredale Grit and Shales, 102.
Yoredale Shales, 101.

LONDON:
Printed by GEORGE E. EYRE and WILLIAM SPOTTISWOODE,
Printers to the Queen's most Excellent Majesty.
For Her Majesty's Stationery Office.
[P. 2524.—250.—2/75.]